Lecture Notes in Computer Science

Commenced Publication in 1973
Founding and Former Series Editors:
Gerhard Goos, Juris Hartmanis, and Jan van Leeuwen

James F. Peters   Andrzej Skowron (Eds.)

Transactions on Rough Sets XVII

 Springer

Editors-in-Chief

James F. Peters
University of Manitoba
Winnipeg, MB, Canada
E-mail: james.peters3@umanitoba.ca

Andrzej Skowron
University of Warsaw
Warsaw, Poland
E-mail: skowron@mimuw.edu.pl

ISSN 0302-9743 (LNCS)
ISSN 1861-2059 (TRS)
ISBN 978-3-642-54755-3
DOI 10.1007/978-3-642-54756-0
Springer Heidelberg New York Dordrecht London

e-ISSN 1611-3349 (LNCS)
e-ISSN 1861-2067 (TRS)
e-ISBN 978-3-642-54756-0

Typesetting: Camera-ready by author, data conversion by Scientific Publishing Services, Chennai, India

Printed on acid-free paper

Springer is part of Springer Science+Business Media (www.springer.com)

Preface

Volume XVII of the *Transactions on Rough Sets* (TRS) is a continuation of a number of research streams that have grown out of the seminal work of Zdzisław Pawlak[1] during the first decade of the twenty-first century. The research streams represented in the papers cover both the theory and applications of rough, fuzzy, and near sets as well as their combinations.

Davide Ciucci and Didier Dubois present a comprehensive survey on the connections between three-valued logics and rough sets from the point of view of incomplete information management. Ivo Düntsch and Günther Gediga propose procedures to compute confidence intervals for standard errors of indices such as γ and α to measure quality of approximation in rough set data analysis. Christopher Henry and Garrett Smith present an application to demonstrate descriptive-based approaches to nearness and proximity within the context of digital image analysis. Victor Marek and Andrzej Skowron explore properties of rough sets related to one of the classic structures of combinatorics and computer science, namely, matroid. Mariusz Podsiadło and Henryk Rybiński provide a detailed review of the currently available literature covering applications of rough sets in economy and finance. The classic rough set model and its important extensions are applied to areas of risk management, financial market prediction, valuation, and portfolio management. Sai Prasad and Raghavendra Rao present reduct computation algortihm(s) using a fuzzy rough set approach and the effectiveness of their algorithm(s) is empirically demonstrated by comparative analysis with existing reduct approaches. This volume also includes a long paper by Andrzej Janusz based on his PhD thesis on algorithms for similarity relation learning from high-dimensional data.

The editors would like to express gratitude to the authors of all submitted papers. Special thanks are due to the following reviewers: Jerzy Grzymała-Busse, Chris Cornellis, Ivo Düntsch, Jouni Järvinen, Henryk Rybinski, Sheela Ramanna, Dominik Ślęzak, Marcin Wolski, JingTao Yao, and Yiyu Yao.

The editors and authors of this volume extend their gratitude to Alfred Hofmann and the LNCS staff at Springer for their support in making this volume of the TRS possible.

The Editors-in-Chief were supported by the Polish National Science Centre grants DEC-2011/01/B/ ST6/03867, DEC-2011/01/D/ST6/06981, and DEC-2012/05/B/ST6/03215 as well as by the Polish National Centre for Research and

---

[1] See, *e.g.*, Pawlak, Z., A Treatise on Rough Sets, *Transactions on Rough Sets* IV, (2006), 1-17. See, also, Pawlak, Z., Skowron, A.: Rudiments of rough sets, *Information Sciences* 177 (2007) 3-27; Pawlak, Z., Skowron, A.: Rough sets: Some extensions, *Information Sciences* 177 (2007) 28-40; Pawlak, Z., Skowron, A.: Rough sets and Boolean reasoning, *Information Sciences* 177 (2007) 41-73.

Development (NCBiR) under grant SYNAT No. SP/I/1/77065/10 in the framework of the strategic scientific research and experimental development program: "Interdisciplinary System for Interactive Scientific and Scientific-Technical Information" and by grant No. O ROB/0010/03/001 in the framework of the Defence and Security Programmes and Projects "Modern Engineering Tools for Decision Support for Commanders of the State Fire Service of Poland during Fire and Rescue Operations in the Buildings" as well as by the Natural Sciences and Engineering Research Council of Canada (NSERC) discovery grant 185986.

January 2014   James F. Peters
                                                               Andrzej Skowron

LNCS Transactions on Rough Sets

The *Transactions on Rough Sets* series has as its principal aim the fostering of professional exchanges between scientists and practitioners who are interested in the foundations and applications of rough sets. Topics include foundations and applications of rough sets as well as foundations and applications of hybrid methods combining rough sets with other approaches important for the development of intelligent systems. The journal includes high-quality research articles accepted for publication on the basis of thorough peer reviews. Dissertations and monographs up to 250 pages that include new research results can also be considered as regular papers. Extended and revised versions of selected papers from conferences can also be included in regular or special issues of the journal.

Editors-in-Chief: James F. Peters, Andrzej Skowron
Managing Editor: Sheela Ramanna
Technical Editor: Marcin Szczuka

Editorial Board

Table of Contents

Three-Valued Logics, Uncertainty Management and Rough Sets

Davide Ciucci[1] and Didier Dubois[2]

[1] DISCo - Università di Milano – Bicocca,
Viale Sarca 336 – U14, 20126 Milano Italia
[2] IRIT, Université Paul Sabatier,
118 route de Narbonne, 31062 Toulouse cedex 4 France

Abstract. This paper is a survey of the connections between three-valued logics and rough sets from the point of view of incomplete information management. Based on the fact that many three-valued logics can be put under a unique algebraic umbrella, we show how to translate three-valued conjunctions and implications into operations on ill-known sets such as rough sets. We then show that while such translations may provide mathematically elegant algebraic settings for rough sets, the interpretability of these connectives in terms of an original set approximated via an equivalence relation is very limited, thus casting doubts on the practical relevance of truth-functional logical renderings of rough sets.

1 Introduction

Rough sets have often been studied under a three-valued logic framework and different authors have tried to connect rough sets to different logics: Łukasiewicz [9, 11], Nelson [58, 59], Gödel, Gaines-Rescher three-valued logics [49, 41]. Despite the formal correctness of these approaches, little attention has been devoted to the interpretation of these logics in the rough set context. Moreover, a comprehensive study on the three-valued connectives that can be defined on rough sets is needed and, as we will see, it can be accomplished starting from known results in three-valued logics.

Three-valued logics are apparently simple; they are straightforward generalizations of Boolean logic based on the most simple bipolar scale $\{0, \frac{1}{2}, 1\}$ where 1 (resp. 0) has a positive (resp. negative) flavor, and $\frac{1}{2}$ is neutral. Further, they are widely used in several applied contexts such as logic programming [43], electronic circuits [67], databases [27], and, of course, rough sets. However, there have been several different meanings attached to the third value, some having an epistemic nature. There is not a clear result on the definition of its connectives in connection with this meaning. Here is a list of these interpretations of the third truth-value, different from true and false : *Possible* (due to Łukasiewicz [17]), *Unknown* (Kleene [52]), *Undefined* (also Kleene), *Half-true* (in fuzzy logic [48]), *Borderline* (in logics of vagueness, like in Shapiro [66]), *Inconsistent* (that is both true and false, as in paraconsistent logics or the logic of paradox by Priest

J.F. Peters and A. Skowron (Eds.): Transactions on Rough Sets XVII, LNCS 8375, pp. 1–32, 2014.
© Springer-Verlag Berlin Heidelberg 2014

[63]), or yet *Irrelevant* as in relevance logics [2] or the logic of conditionals [38]. Sometimes, two of these notions are simultaneously used as *Inconsistent* and *Unknown* in Belnap four-valued logic [14].

Three-valued logics go along with three-valued sets having central elements and peripheral ones [46]. However the meaning of such central and peripheral elements depends on the meaning of the third truth-value. It depends on whether it has an epistemic flavor or not; a peripheral element can be understood in one of the following ways:

1. either as an untypical element of a non-classical set,
2. or as an element that cannot be definitely classified as belonging or not to a crisp set due to incomplete information,
3. or as an element that cannot be definitely classified as belonging or not to a crisp set due to conflicting information,
4. or as an element for which membership or non membership makes no sense, due to irrelevance or the dubious existence of such an element.

Case 2 is the one we are concerned with in this paper. Then the three truth-values refer to the epistemic status of otherwise Boolean propositions (provably true, provably false or unknown [39]). This is typically the case of ill-known or interval sets [72], where the central elements are elements that certainly belong to some ill-known set, the third truth-value is assimilated to $\{0, 1\}$ and understood as the hesitancy between membership and non-membership. They are special cases of interval-valued fuzzy sets [77] or twofold fuzzy sets [37]. One of the causes of a set being ill-known can be the lack of precision on the value of some of the attributes that describe it (for instance, a set of single persons is ill-known if the marital status of some of the persons is ill-known).

A rough set, viewed as a pair of nested approximations is a typical example of ill-known set, where the lack of knowledge comes from an equivalence relation between possibly indistinguishable elements, this indistinguishability being due to the use of a language that is not expressive enough (incomplete set of attributes or attributes that are too coarsely defined). This situation contrasts with the case of sets that are ill-known due to the lack of knowledge of attribute values; see Couso and Dubois [28] when the two causes of partial ignorance appear simultaneously.

In recent papers [23–26], we have studied various three-valued logics of partial knowledge, where the third truth-value means *unknown*. It has been shown that a large class of three-valued logics (including Łukasiewicz L_3) is compatible with this understanding of the third truth-value, but their translations into a very elementary modal logic indicate that such three-valued logics cannot account for partial ignorance jointly affecting several Boolean variables: only states of partial ignorance that can be described independently for each variable can be accounted for in a three-valued logic. This is the price paid for truth-functionality.

In this paper, we examine the situation of three-valued logics of rough sets. While the aforementioned limitation is still valid (since rough sets do not behave truth-functionally in general), there is an additional constraint in this case. Namely, the approximation pairs are generated by an equivalence relation, which creates additional interpretive difficulties for truth-functional definitions of conjunction and disjunction [22]. In this paper, we consider the situation of more general three-valued connectives in connection with rough sets.

In the following, we review some results on three-valued logics, in particular we give a list of reasonable connectives on three values that can apply to ill-known sets. Then, these connectives are translated into the language of nested pairs or orthopairs of sets, showing that, from a formal point of view, this translation is correct. On the other hand, some considerations from the interpretation standpoint are put forward casting some doubts on this truth-functional approach to ill-known sets and rough sets. Especially, in the case of rough sets, it seems impossible to interpret combinations of rough sets in terms of pure combinations of the underlying ill-known sets (approximated via an equivalence relation). Finally, we discuss some modal and three-valued logics of rough sets in connection with a recent translation of three-valued logics into a fragment of the KD logic.

2   Aggregation Functions on Three Valued Logics

We denote by $\mathbf{3}$ the set $\{0, \frac{1}{2}, 1\}$ with the usual order: $0 < \frac{1}{2} < 1$. Due to the total order assumption, we can define the idempotent and commutative Kleene conjunction and disjunction, that is, the minimum, denoted by $\sqcap$ and the maximum denoted by $\sqcup$: $x \sqcap y = y \sqcap x = x$ if and only if $x \leq y$ if and only if $x \sqcup y = y \sqcup x = y$. Moreover, Gödel implication is definable by residuation:

$$x \sqcap y \leq z \text{ if and only if } x \leq y \rightarrow_G z.$$

It is such that $y \rightarrow_G z = 1$ if $y \leq z$ and z otherwise. Finally, the intuitionistic negation is obtainable by Gödel implication as $\sim x = x \rightarrow_G 0$. We now report some results [25] about three-valued logics: a list of possible connectives, the logical systems they generate and the links among them.

2.1   Connectives

A maximal family of sensible conjunctions and implications on $\mathbf{3}$ is now recalled, based on some intuitive properties, in the scope of modeling incomplete information. Then, negation and disjunction can be derived respectively as $a \rightarrow 0$ and by De Morgan properties.

Definition 2.1. *A conjunction on $\mathbf{3}$ is a binary mapping $*: \mathbf{3} \times \mathbf{3} \mapsto \mathbf{3}$ that is monotonically increasing in the wide sense, and extends the connective AND in Boolean logic:*

(C1) If $x \leq y$ then $x * z \leq y * z$;
(C2) If $x \leq y$ then $z * x \leq z * y$;
(C3) $0 * 0 = 0 * 1 = 1 * 0 = 0$ and $1 * 1 = 1$.

Bearing in mind our focus on the epistemic understanding of the third truth-value as *unknown*, condition C3 is clearly natural if we notice that, in consequence to this interpretive assumption, 1 must mean "certainly true" and 0 "certainly false", which justifies this requirement of coincidence with Boolean conjunction for truth-values different from $1/2$.

Due to (C3), the monotonicity properties (C1-C2) imply $\frac{1}{2} * 0 = 0 * \frac{1}{2} = 0$. It goes along with the fact that a conjunction is false whenever one of the conjuncts is false, regardless of whether the truth-value of the other conjunct is known or not. If we consider all the possible cases, there are 14 conjunctions satisfying Definition 2.1. Among them, only six are commutative and only five associative. These five conjunctions are already known in the literature and precisely, they have been studied in the following logics: Sette [65], Sobociński [68], Łukasiewicz [17], Kleene [52], Bochvar [15]. The complete list is given in Table 1.

Table 1. All conjunctions on **3** according to Definition 2.1

*	0	$\frac{1}{2}$	1
0	0	0	0
$\frac{1}{2}$	0		
1	0		1

n.	$\frac{1}{2} * \frac{1}{2}$	$1 * \frac{1}{2}$	$\frac{1}{2} * 1$	
1	1	1	1	Sette
2	$\frac{1}{2}$	1	1	quasi conjunction/Sobociński
3	$\frac{1}{2}$	1	$\frac{1}{2}$	
4	$\frac{1}{2}$	$\frac{1}{2}$	1	
5	$\frac{1}{2}$	$\frac{1}{2}$	$\frac{1}{2}$	min/interval conjunction/Kleene
6	0	0	1	
7	0	0	$\frac{1}{2}$	
8	0	0	0	Bochvar external
9	0	$\frac{1}{2}$	0	
10	0	$\frac{1}{2}$	1	
11	0	$\frac{1}{2}$	$\frac{1}{2}$	Łukasiewicz
12	0	1	0	
13	0	1	$\frac{1}{2}$	
14	0	1	1	

Besides Definition 2.1 other possible definitions of conjunction can be found in the literature:

- conjunction of conditional events due to Walker [71]. The required properties are the coincidence with Boolean conjunction on Boolean values $\{0, 1\}$, idempotence and commutativity. Only nine conjunctions satisfy these axioms, among them Sobociński's (it is also Adams quasi-conjunction of conditionals [1]) and the two Kleene ones. The other six are all non-monotonic and only one is associative. Moreover, three of them are such that $\frac{1}{2} * 0 = 1$. All these facts cast some doubts on the interpretability of these six conjunctions on **3** outside the setting of conditional events.

– t-norms, uninorms [54, 29]. A uninorm is a binary operator which is associative, commutative, non-decreasing in each component and with a neutral
element $e : \forall x, e * x = x$. A t-norm is a uninorm such that $e = 1$. Among conjunctions on **3** we have only two t-norms: Gödel and Łukasiewicz and only
one more uninorm: Sobociński. They already appear in the above Table 1.
– t-operators [54]: an associative, commutative binary operators such that $0 *$
$0 = 0, 1 * 1 = 1$ and satisfying 1-smoothness: $x_i * x_{j-1} \leq x_i * x_j$ and if $x_i * x_j =$
x_k then $\{x_{i-1} * x_j, x_i * x_{j-1}\} \subseteq \{x_k, x_{k-1}\}$. Besides Kleene and Łukasiewicz
conjunctions and disjunctions, on three values we get one more operator: the
median $med(x, y, \frac{1}{2})$, which, however, does not generalize Boolean logic.

In the case of implication, we can give a general definition, which extends
Boolean logic and supposes monotonicity (decreasing in the first argument, increasing in the second).

Definition 2.2. *An implication on **3** is a binary mapping $\rightarrow: \mathbf{3} \times \mathbf{3} \mapsto \mathbf{3}$ such
that:*

(I1) If $x \leq y$ then $y \rightarrow z \leq x \rightarrow z$;
(I2) If $x \leq y$ then $z \rightarrow x \leq z \rightarrow y$;
(I3) $0 \rightarrow 0 = 1 \rightarrow 1 = 1$ and $1 \rightarrow 0 = 0$.

From the above definition we derive $x \rightarrow 1 = 1$, $0 \rightarrow 1 = 1$ and $\frac{1}{2} \rightarrow \frac{1}{2} \geq$
$\{1 \rightarrow \frac{1}{2}, \frac{1}{2} \rightarrow 0\}$. There are 14 implications satisfying this definition, listed in
Table 2.

Table 2. All implications according to Definition 2.2

n.	$\frac{1}{2} \rightarrow \frac{1}{2}$	$1 \rightarrow \frac{1}{2}$	$\frac{1}{2} \rightarrow 0$	
1	0	0	0	
2	$\frac{1}{2}$	0	0	Sobociński
3	$\frac{1}{2}$	0	$\frac{1}{2}$	
4	$\frac{1}{2}$	$\frac{1}{2}$	0	Jaśkowski
5	$\frac{1}{2}$	$\frac{1}{2}$	$\frac{1}{2}$	(strong) Kleene
6	1	1	0	Sette
7	1	1	$\frac{1}{2}$	
8	1	1	1	
9	1	$\frac{1}{2}$	1	Nelson
10	1	$\frac{1}{2}$	0	Gödel
11	1	$\frac{1}{2}$	$\frac{1}{2}$	Łukasiewicz
12	1	0	1	Bochvar external
13	1	0	$\frac{1}{2}$	
14	1	0	0	Gaines–Rescher

$\rightarrow$	0	$\frac{1}{2}$	1
0	1	1	1
$\frac{1}{2}$			1
1	0		1

Nine of them are known and have been studied. Besides those implications
named after the five logics mentioned above, there are also those named after
Jaśkowski [50], Gödel [47], Nelson [55], Gaines-Rescher [44].

Finally, there are only three possible negations that extend the Boolean negation, namely, if $0' = 1$ and $1' = 0$:

1. $\sim\frac{1}{2} = 0$. It corresponds to an intuitionistic negation, since it satisfies the law of contradiction, and not the excluded middle. It is of the form $a \to_i 0$ for implications 1, 2, 4, 6, 10, 14.
2. $\neg\frac{1}{2} = \frac{1}{2}$. It is an involutive negation. It is of the form $a \to_i 0$ for implications 3, 5, 7, 11, 13.
3. $-\frac{1}{2} = 1$. It is called a paraconsistent negation, since it satisfies the law of excluded middle, and not the one of contradiction. It is of the form $a \to_i 0$ for implications 8, 9, 12.

2.2 Logical Systems

As mentioned, some of these connectives have already been studied and they are at the basis of known logical formalisms. Here is a (possibly not exhaustive) list:

- *Łukasiewicz logic* $(\to_{11}, *_{11}, +_{11}, \neg)$, where the disjunction $+_{11}$ definable by de Morgan properties as $a +_{11} b := \neg(\neg a *_{11} \neg b)$ is the truncated sum. We also recall that the interpretation given by Łukasiewicz for the third value is *possible* whereas, nowadays, Łukasiewicz logic is mainly used in many valued logics where the third value has a gradual truth meaning.
- *Sobociński logic* $(\to_2, *_2, +_2, \neg)$ where $+_2$ can be defined as $a +_2 b := \neg a \to_2 b$ and designated values are $1, \frac{1}{2}$. In this case, the third value means *irrelevant* and it has been used in the context of relevance logics [2] and conditional events [38]. We recall that conjunction $*_2$ is a discrete uninorm with $\frac{1}{2}$ as neutral element and implication $\to_2$ its residuum [7].
- *Gödel (intuitionistic) logic* $(\to_{10}, *_5(\min), \max, \sim)$ on three values, also known as logic of here-and-there in logic programming [62].
- *Jaśkowski logic* $(\to_4, *_5(\min), \max, \neg)$ has been studied by several authors in the field of paraconsistent logic [32, 2, 4]. The designated values are $\frac{1}{2}$ and 1 and the interpretation of the third value means *inconsistent, paradoxical*, that is, both true and false.
- *Bochvar logic* $(\to_{12}, *_8, +_8, \flat)$ where $x +_8 y$ is 1 if at least one of a and b is equal to 1 and 0 in all other cases. Third value $\frac{1}{2}$ stands for *meaningless*.
- *Sette paraconsistent logic* $(\to_6, *_1, -)$ where $x +_1 y$ takes the value 0 if $x = y = 0$ and 1 otherwise and designated values are $\frac{1}{2}$ and 1. We note that Sette conjunction (n.1) and implication (n.6) correspond to the collapse of the truth-values 1 and $\frac{1}{2}$. The author does not give a clear semantic to the third value and he introduces the logic as a "not absolutely inconsistent" formal system.
- *Nelson logic* $(\to_9, *_5, \max, \neg, -)$ where $-$ is a paraconsistent negation and $\to_9$ Nelson implication. It is the logic of *constructible falsity*, and in this sense it is dual to intuitionistic logic. On five values it is also known as equilibrium logic in the context of logic programming [62].

The reader is referred to the work of Avron [5] for a reconstruction of some of the above three-valued logics (said to be "natural") where the third truth-value is understood as *unknown* or *contradictory*, based on the inferential standpoint.

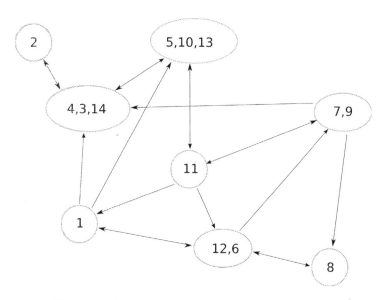

Fig. 1. Outline of all the relations among connectives

2.3 Connections among Logics

Some relations among the above systems are known. For instance, Sette logic has been obtained in [32] from Łukasiewicz logic in order to demonstrate a relationship between many-valued and paraconsistent logics; likewise, it can be proved that Jaśkowski and Sobociński logics are equivalent [5]. However, other connections can be put forward by a systematic study. As a result we can prove that all these systems and more generally, all the 14 conjunctions and implications are inter-definable. More precisely, we consider the following transformations of a binary operator $\odot$ on **3** [35, 36]:

$$a[\mathcal{A}(\odot)]b = b \odot a \qquad\qquad \text{(exchange)} \quad \text{(1a)}$$

$$a[\mathcal{V}(\odot)]b = \neg b \odot \neg a \qquad\qquad \text{(contraposition)} \quad \text{(1b)}$$

$$a[\mathcal{S}(\odot)]b = \neg(a \odot \neg b) \qquad\qquad \text{(material implication)} \quad \text{(1c)}$$

$$a[\mathcal{I}(\odot)]b = \begin{cases} 0 & \nexists s,\ a \odot s \leq b; \\ \sup\{s : a \odot s \leq b\}, & \text{otherwise.} \end{cases} \qquad \text{(residuation)} \quad \text{(1d)}$$

We also define relations among implications through some formulae which are tautologies in Boolean logic:

$$a \rightarrow_{new} b = (a \rightarrow b) \wedge (\neg b \rightarrow \neg a); \tag{2a}$$

$$a \rightarrow_{new} b = b \vee (a \rightarrow b); \tag{2b}$$

$$a \rightarrow_{new} b = a \rightarrow (a \rightarrow b); \tag{2c}$$

$$a \rightarrow_{new} b = (a \rightarrow b) \vee (\neg b \rightarrow \neg a); \tag{2d}$$

$$a \rightarrow_{new} b = \neg a \vee (a \rightarrow b) \tag{2e}$$

In the graph of Figure 1, a representation of all these relationships is given: every circle represents a group of conjunctions/implications related via transformations defined in equations (1), whereas groups are linked by transformations defined in equations (2). These transformations are instrumental to get the following results [23, 25]:

Proposition 2.1. *Let* **3** *be the three-element set with the usual order* $0 < \frac{1}{2} < 1$ *or equivalently,* $\mathbf{3} = (\overline{3}, \wedge, \vee)$, *with* $\overline{3}$ *the set of three elements without the order structure. All the 14 conjunctions and implications can be defined in any of the following systems:*

- $(\mathbf{3}, \neg, \rightarrow_{10}) = (\overline{3}, \wedge, \neg, \rightarrow_{10})$ *(Gödel implication plus the involutive negation);*
- $(\mathbf{3}, \rightarrow_i) = (\overline{3}, \wedge, \vee, \rightarrow_i)$ *where* $i \in I = \{3, 5, 7, 11, 13\}$, *allowing residuation.*

Further, we can also consider a set with three elements without a predefined order (and so without min, max and residuation) and obtain the following proposition.

Proposition 2.2. *We denote by* $\overline{3}$ *the set of three elements without any structure. All the 14 conjunctions and implications can be defined in any of the following systems:*

- $(\overline{3}, \rightarrow_{11}, 0)$ *where* $\rightarrow_{11}$ *is Lukasiewicz implication;*
- $(\overline{3}, \rightarrow_9, \neg)$ *where* $\rightarrow_9$ *is Nelson implication;*
- $(\overline{3}, \rightarrow_5, \sim, 0)$ *where* $\rightarrow_5$ *is Kleene implication and* $\sim$ *the intuitionistic negation.*
- $(\overline{3}, \rightarrow_5, -, 0)$ *where* $\rightarrow_5$ *is Kleene implication and* $-$ *the paraconsistent negation.*

In the two arrays of Table 3 we report how to obtain all the conjunctions and implications starting from the Lukasiewicz implication $\rightarrow_L = \rightarrow_{11}$ and 0; of course $\neg a = a \rightarrow_L 0$. We denote by $\odot$ Lukasiewicz conjunction; moreover, ∇a is an abbreviation for $\neg a \rightarrow_L a$, $\Delta(a)$ stands for $a \odot a = \neg(a \rightarrow_L \neg a)$, and finally $J(a)$ is short for $\neg a *_1 \neg(\neg a *_1 a) = \nabla(\neg a \wedge \neg \nabla(\neg a \wedge a))$. Clearly $\sim a = \neg \nabla a$ and $-a = \neg \Delta a$.

So, the differences among three-valued logics are just apparent. All of them can be interpreted as a fragment of the same logic, such as Lukasiewicz logic, or sometimes a variant thereof with the same expressive power (like Nelson's logic). According to the purpose and to the desired interpretation, we can then choose the proper fragment and connectives.

Table 3. All connectives expressed using Łukasiewicz logic operators

n	$a *_n b$	
1	$\nabla(a \wedge b)$	
2	$\nabla(a \wedge b) \wedge (a \vee b)$	Sobociński
3	$\nabla(a \wedge b) \wedge a$	
4	$\nabla(a \wedge b) \wedge b$	
5	$a \wedge b$	(strong) Kleene
6	$\neg J(a) \odot J(\neg b)$	
7	$\neg b \odot (\neg b \odot a)$	
8	$[\neg b \odot (\neg b \odot a)] \wedge [\neg a \odot (\neg a \odot \neg b)]$	Bochvar external
9	$a \odot (a \odot \neg b)$	
10	$b \wedge [(\neg J(a) \odot J(\neg b)) \vee (J(\neg a) \odot \neg J(b))]$	
11	$a \odot b = \neg(\neg a \to_L b)$	
12	$J(\neg a) \odot \neg J(b)$	
13	$a \wedge [(\neg J(a) \odot J(\neg b)) \vee (J(\neg a) \odot \neg J(b))]$	
14	$(\neg J(a) \odot J(\neg b)) \vee (J(\neg a) \odot \neg J(b))$	

n	$a \to_n b$	
1	$\Delta(\neg a) \vee \Delta(b)$	
2	$(b \vee (a \to_1 b)) \wedge (\neg a \vee (\neg b \to_1 \neg a))$	Sobociński
3	$\neg a \vee [(b \vee (a \to_1 b)) \wedge (\neg a \vee (\neg b \to_1 \neg a))]$	
4	$b \vee (\Delta(\neg a) \vee \Delta(b))$	Jaśkowski
5	$\neg a \vee (\Delta(\neg a) \vee \Delta(b))$	(strong) Kleene
6	$J(b) \to_L J(a)$	Sette
7	$\neg b \to_L (\neg b \to_L \neg a)$	
8	$a \to_L (a \to_L b)) \vee (\neg b \to_L (\neg b \to_L \neg a))$	
9	$a \to_L (a \to_L b)$	Nelson
10	$\neg \Delta((\alpha \to_L \beta) \to_L \beta))$	Gödel
12	$J(\neg a) \to_L J(\neg b)$	Bochvar external
13	$\neg a \vee [(J(\neg a) \to_L J(\neg b)) \wedge (J(b) \to_L J(a))]$	
14	$(J(\neg a) \to_L J(\neg b)) \wedge (J(b) \to_L J(a))$	Gaines–Rescher

2.4 Connectives on Nested Pairs and Orthopairs of Sets

Let $f : X \mapsto \mathbf{3}$ be a three-valued function that may be viewed as a special kind of fuzzy set. Then, from each f, we can induce three (Boolean) subsets forming a partition of the universe X:

$$A_1 := \{x : f(x) = 1\} \qquad \text{The truth domain;}$$
$$A_0 := \{x : f(x) = 0\} \qquad \text{The falsity domain;}$$
$$A_{\frac{1}{2}} := \{x : f(x) = \tfrac{1}{2}\} \qquad \text{The neutral domain.}$$

Formally, we can see any three-valued set f as a pair (A_1, A_0) of classical sets satisfying the property $A_1 \cap A_0 = \emptyset$, i.e., A_1 and A_0 are disjoint sets and (A_1, A_0) is called an *orthopair* [21]. Conversely, given a pair of disjoint sets, we

Table 4. Conjunctions on orthopairs

n	$(A_1, A_0) * (B_1, B_0)$
1	$(A_0^c \cap B_0^c, A_0 \cup B_0)$
2	$((A_1 \cup B_1) \cap A_0^c \cap B_0^c, A_0 \cup B_0)$
3	$(A_1 \cap B_0^c, A_0 \cup B_0)$
4	$(A_0^c \cap B_1, A_0 \cup B_0)$
5	$(A_1 \cap B_1, A_0 \cup B_0)$
6	$(A_0^c \cap B_1, A_0 \cup B_1^c)$
7	$(A_1 \cap B_1, A_0 \cup B_1^c)$
8	$(A_1 \cap B_1, A_1^c \cup B_1^c)$
9	$(A_1 \cap B_1, A_1^c \cup B_0)$
10	$(A_0^c \cap B_1, (A_1^c \cap B_1^c) \cup A_0 \cup B_0)$
11	$(A_1 \cap B_1, (A_1^c \cap B_1^c) \cup A_0 \cup B_0)$
12	$(A_1 \cap B_0^c, A_1^c \cup B_0)$
13	$(A_1 \cap B_0^c, (A_1^c \cap B_1^c) \cup A_0 \cup B_0)$
14	$((A_1 \cup B_1) \cap A_0^c \cap B_0^c, (A_1^c \cap B_1^c) \cup A_0 \cup B_0)$

can define a three-valued sets in an obvious way: $f(x) = 1$ if $x \in A_1$; $f(x) = 0$ if $x \in A_0$ and $f(x) = \frac{1}{2}$ otherwise. So, we have a bijection between the collection of three-valued sets $\mathcal{F}_{\frac{1}{2}}(X) := \{f | f : X \mapsto \mathbf{3}\}$ and the collection of orthopairs of X, $\mathcal{O}(X) := \{(A_1, A_0) | A_1, A_0 \in X; A_1 \cap A_0 = \emptyset\}$.

We note that from (A_1, A_0), another subset $A^* := A_0^c$ of the universe can be defined as the negation of the falsity domain. In other words, renaming A_1 as A_*, an alternative representation of three-valued sets is obtained by means of pairs of nested subsets (A_*, A^*) of X, where $A_* \subseteq A^*$, which can be viewed as upper and lower approximations of some unknown set. We denote by $\mathcal{N}(X)$ the collection of nested pairs of subsets of X.

These constructions are known in the fuzzy set field: orthopairs can be viewed as special cases of so-called "intuitionistic fuzzy sets"[1] of Atanassov [3], and the nested version can be generalized to interval-valued fuzzy sets. They can be equipped with isomorphic structures [31].

Due to the bijection outlined above, we are able to translate all the operations from $\mathcal{F}(X)$ to $\mathcal{O}(X)$, and in particular all the 14 implications and conjunctions defined above. They are listed in Tables 4 and 6. Note that the operations are often easier to understand using nested pairs as seen on Tables 5 and 7.

The three negations, respectively the involutive, intuitionistic and paraconsistent ones, take the following forms:

$$\neg(A_1, A_0) = (A_0, A_1);$$
$$\sim(A_1, A_0) = (A_0, A_0^c);$$
$$-(A_1, A_0) = (A_1^c, A_1).$$

---

[1] Where the word *intuitionistic* does not have the usual meaning [34].

Table 5. Conjunctions on nested pairs

n	$(A_*, A^*) * (B_*, B^*)$
1	$(A^* \cap B^*, A^* \cap B^*)$
2	$((A_* \cup B_*) \cap A^* \cap B^*, A^* \cap B^*)$
3	$(A_* \cap B^*, A^* \cap B^*)$
4	$(A^* \cap B_*, A^* \cap B^*)$
5	$(A_* \cap B_*, A^* \cap B^*)$
6	$(A^* \cap B_*, A^* \cap B_*)$
7	$(A_* \cap B_*, A^* \cap B_*)$
8	$(A_* \cap B_*, A_* \cap B_*)$
9	$(A_* \cap B_*, A_* \cap B^*)$
10	$(A^* \cap B_*, (A_* \cup B_*) \cap A^* \cap B^*)$
11	$(A_* \cap B_*, (A_* \cup B_*) \cap A^* \cap B^*)$
12	$(A_* \cap B^*, A_* \cap B^*)$
13	$(A_* \cap B^*, (A_* \cup B_*) \cap A^* \cap B^*)$
14	$((A_* \cup B_*) \cap A^* \cap B^*, (A_* \cup B_*) \cap A^* \cap B^*)$

So, in the case of orthopairs, the definition of connectives is just a matter of translation from Tables 1 and 2. However, if these orthopairs are viewed as ill-known sets some difficulties with truth-functionality occur. In the case of rough sets, we encounter even more difficulties, as we are going to explain.

3   Three-Valued Connectives on Ill-Known Sets

Ill-known sets are sets whose boundaries are ill-known, namely it is not known whether some elements belong to them or not. The neutral region is then an uncertainty region. A typical situation where ill-known sets are obtained is as follows [37, 28]. Consider X as a set of objects and f a feature (or attribute) mapping : $X \to V$ where V is the domain of the corresponding attribute. So, $\forall x \in X$, $f(x)$ is the attribute value of object x.

Suppose we want to describe the set of objects that satisfy a property represented by a subset $C \subset V$ of values. For instance X is a set of persons, f is the height, and C means *taller than 1.70 m*. The set of persons that satisfy the criterion C is defined by $f^{-1}(C) \subset X$.

Suppose for some reason $f(x)$ is not always known precisely. Let a one-to-many mapping $F : X \to \wp(V)$ represent an imprecise observation of the attribute f. Namely, for each object $x \in X$, all that is known about the attribute value $f(x)$ is that it belongs to the non-empty set $F(x) \subseteq V$. For instance, the heights of some persons x are ill-known, and are described by the sets $F(x)$ of (mutually exclusive) heights. Because of the incompleteness of the information, the subset $A = f^{-1}(C) \subseteq X$ of objects that satisfy the criterion C is an "ill-known set" [37]. Let us first recall the following definition:

Definition 3.1. *([30]) Let X and V be two arbitrary sets and let $F : X \to \wp(V)$ be a multi-valued mapping with non-empty images. Let $C \subseteq V$ be an arbitrary subset of V. The* upper inverse *of C is defined as $F^*(C) = \{x \in X : F(x) \cap C \neq \emptyset\}$. The* lower inverse *of C is defined as $F_*(C) = \{x \in X : F(x) \subseteq C\}$.*

Table 6. Implications on orthopairs, where $A \to B = A^c \cup B$

n	$(A_1, A_0) \Rightarrow (B_1, B_0)$	
1	$(A_0^c \to B_1, (A_0^c \to B_1)^c)$	Sette
2	$(A_0^c \to B_1, [(A_1 \to B_1) \cap (A_0^c \to B_0^c)]^c)$	Sobociński
3	$(A_0^c \to B_1, (A_1 \to B_1)^c)$	
4	$(A_0^c \to B_1, (A_0^c \to B_0^c)^c)$	Jaśkowski
5	$(A_0^c \to B_1, (A_1 \to B_0^c)^c)$	Kleene
6	$(A_0^c \to B_0^c, (A_0^c \to B_0^c)^c)$	
7	$(A_0^c \to B_0^c, (A_1 \to B_0^c)^c)$	
8	$(A_1 \to B_0^c, (A_1 \to B_0^c)^c)$	Bochvar
9	$(A_1 \to B_1, (A_1 \to B_0^c)^c)$	Nelson
10	$((A_1 \to B_1) \cap (A_0^c \to B_0^c), (A_0^c \to B_0^c)^c)$	Gödel
11	$((A_1 \to B_1) \cap (A_0^c \to B_0^c), (A_1 \to B_0^c)^c)$	Łukasiewicz
12	$(A_1 \to B_1, (A_1 \to B_1)^c)$	
13	$((A_1 \to B_1) \cap (A_0^c \to B_0^c), (A_1 \to B_1)^c)$	
14	$((A_1 \to B_1) \cap (A_0^c \to B_0^c), [(A_1 \to B_1) \cap (A_0^c \to B_0^c)]^c)$	Gaines-Rescher

According to this definition, $A = f^{-1}(C)$ can be approximated from above and from below, respectively, by upper and lower inverses of C via F:

- $A^* = F^*(C)$ is the set of objects that possibly belong to $A = f^{-1}(C)$.
- $A_* = F_*(C)$ is the set of objects that surely belong to $A = f^{-1}(C)$.

The interval $[A_*, A^*] = \{B, A_* \subseteq B \subseteq A^*\}$ in the Boolean algebra, called an interval set by Yao [73], contains the ill-known set A. Alternatively, we can consider orthopairs (A_1, A_0) such that $[A_*, A^*] = \{B : A_1 \subseteq B, A_0 \cap B = \emptyset\}$.

If pairs of sets represent constraints on ill-known sets, we would like to compute the knowledge we may have on the result of combining two ill-known sets A and B by means of a three-valued connective merging their approximations (A_1, A_0) and (B_1, B_0). What is aimed at is, for any Boolean connective c, to find the orthopair $(c(A, B)_1, c(A, B)_0)$ representing our knowledge about $c(A, B)$ in the form $c_3((A_1, A_0), (B_1, B_0))$ where c_3 is a three-valued extension of c.

Consider $\frac{1}{2}$ as the set $\{0, 1\}$ (understood as an interval such that $0 < 1$), the other "intervals" being the singletons $\{0\}$ and $\{1\}$. We can define connectives on ill-known sets by extending the Boolean connectives to such three-valued sets understood as interval-valued sets. Indeed this comes down to the following computations [33, 53]:

- For conjunction : $\{0\} \wedge \{0, 1\} = \{0 \wedge 0, 0 \wedge 1\} = \{0\}$;
  $\{1\} \wedge \{0, 1\} = \{1 \wedge 0, 1 \wedge 1\} = \{0, 1\}$, etc.
- For disjunction : $\{0\} \vee \{0, 1\} = \{0 \vee 0, 0 \vee 1\} = \{0, 1\}$;
  $\{1\} \vee \{0, 1\} = \{1 \vee 0, 1 \vee 1\} = \{1\}$, etc.
- For negation: $\neg\{0, 1\} = \{\neg 0, \neg 1\} = \{0, 1\}$.

The set **3** of non-empty intervals on $\{0, 1\}$, equipped with the interval extension of classical connectives is isomorphic to a three-valued Kleene algebra.

Table 7. Implications on nested pairs, where $A \to B = A^c \cup B$

n	$(A_*, A^*) \Rightarrow (B_*, B^*)$	
1	$(A^* \to B_*, A^* \to B_*)$	Sette
2	$(A^* \to B_*, (A_* \to B_*) \cap (A^* \to B^*))$	Sobociński
3	$(A^* \to B_*, A_* \to B_*)$	
4	$(A^* \to B_*, A^* \to B^*)$	Jaśkowski
5	$(A^* \to B_*, A_* \to B^*)$	Kleene
6	$(A^* \to B^*, A^* \to B^*)$	
7	$(A^* \to B^*, A_* \to B^*)$	
8	$(A_* \to B^*, A_* \to B^*)$	Bochvar
9	$(A_* \to B_*, A_* \to B^*)$	Nelson
10	$((A_* \to B_*) \cap (A^* \to B^*), A^* \to B^*)$	Gödel
11	$((A_* \to B_*) \cap (A^* \to B^*), A_* \to B^*)$	Łukasiewicz
12	$(A_* \to B_*, A_* \to B_*)$	
13	$((A_* \to B_*) \cap (A^* \to B^*), A_* \to B_*)$	
14	$((A_* \to B_*) \cap (A^* \to B^*), (A_* \to B_*) \cap (A^* \to B^*))$	Gaines-Rescher

However, using such connectives of Kleene logic to compute a combination of ill-known sets only captures an approximation of the actual result. For instance, even if A is ill-known, $A \cap A^c = \emptyset$ $((\emptyset, X)$ in terms of orthopairs). However, if (A_1, A_0) are constraints on some unknown set A, the orthopair approximation of A^c is (A_0, A_1), but, applying the Kleene conjunction $(A_1, A_0) \cap_5 (A_0, A_1) = (A_1 \cap A_0, A_1 \cup A_0)$ is an imperfect approximation of the expected result $(\emptyset, X)$ since the former is equal to $(\emptyset, A_1 \cup A_0)$.

So we should get $(A_1, A_0) \cap (A_0, A_1) = (\emptyset, X)$ using an appropriate conjunction. This result can be obtained using conjunctions $\cap_i, i > 5$ by checking Table 4 (or Table 5 in terms of nested pairs). But then note that while one expects $(A_1, A_0) \cap (A_1, A_0) = (A_1, A_0)$ this is what is obtained on Table 4 (or Table 5 in terms of nested pairs) only for $\cap_i, i = 2, 3, 4, 5$. So none of the 14 reasonable conjunctions can provide the expected results.

In conclusion, the use of three-valued connectives to reason about ill-known sets looks hopeless: it is not the same to reason truth-functionally on objects made of pairs of sets, and to exploit pairs of sets viewed as constraints on a ill-known set $A \in \{B : A_1 \subseteq B, A_0 \cap B = \emptyset\}$: the former is a coarse approximation of the latter. Note that the same kind of critique applies to interval-valued fuzzy sets where it is often proposed interval extensions of basic connectives [33, 40] to handle uncertainty about gradual membership.

4   Three-Valued Connectives on Rough Sets

A rough set clearly defines a three-valued set, since it can be viewed as an upper and a lower approximation of a set. It was tempting to search for an algebra of rough sets from the three-valued logic literature. The problem of defining a three-valued logic (and especially an implication) for rough sets has been addressed by several authors. Łukasiewicz and Gödel implications have been introduced in rough sets by Pagliani [59] and for abstract versions of rough approximations in [18]. Pagliani also studied rough sets from the standpoint of Nelson algebras [58] and used Nelson implication. Łukasiewicz logic was also considered as the proper setting for rough sets by Banerjee [9] and Iturrioz [49]. On the other hand, the Gaines-Rescher implication is the one adopted in [11] and Kleene implication in [22]. I. Düntsch in [41] introduced a propositional logic for rough set whose algebraic counterpart are double Stone algebras. The objects of this logic are nested pairs of the form (A_*, A^*) and the implication considered is the Gödel one. For a general overview of algebraic structures related to rough sets we refer to [10].

A different and new approach is presented in [6], where the non-deterministic behaviour of rough sets is brought directly into a logical calculus. Indeed, the semantics of the implication is given by the non-deterministic matrix of Table 8.

Table 8. Non-deterministic implication

$\rightarrow$	0	$\frac{1}{2}$	1
0	1	1	1
$\frac{1}{2}$	$\frac{1}{2}$	$\{\frac{1}{2}, 1\}$	1
1	0	$\frac{1}{2}$	1

Clearly, as the authors point out, the two "determinazations" of this situation correspond to Kleene and Łukasiewicz implication. Of course, the problem of non-determinism still remains, it is just shifted on a different level. And while this approach may look "less truth-functional" than the usual ones, its completeness with respect to the calculus of rough sets is unclear.

In this section we study the compatibility between the calculus of rough sets and three-valued connectives. We show that formally, it is possible to express three-valued logic connectives in terms of combinations of rough sets. But our results make it clear that the practical significance of these mathematical results is questionable.

4.1 Some Basics of Rough Sets

In constrast with the scenario for ill-known sets, the starting point of rough sets is usually a set of data about some objects gathered in a so-called Information Table (see for instance [61]).

Definition 4.1. *An* Information Table *is a structure* $\mathcal{K}(X) = \langle X, \mathcal{A}, val, f \rangle$ *where:*

- *the universe X is a non empty set of objects;*
- *$\mathcal{A}$ is a non empty set of attributes;*
- *val is the set of all possible values that can be observed for all attributes;*

- f *(called the* information map*) is a mapping* $X \times A \to val$ *which associates to any pair object* $x \in X$ *and attribute* $a \in A$, *the value* $f(x, a) \in val$ *assumed by* a *for the object* x.

On an Information Table, we define an *Indiscernibility relation* among objects as

$$x R y \quad \text{iff} \quad \forall a \in A \; f(x, a) = f(y, a)$$

The indiscernibility relation is an equivalence relation (reflexive, symmetric, transitive) that partitions the universe into equivalence classes:

$$[x]_R = \{y : x R y\}$$

In the following, we abstract from the notion of Information Table and suppose that an (equivalence) relation is available on a set of objects.

Definition 4.2. *An approximation space is a pair* (X, R) *with* X *a set of objects and* R *an equivalence relation on* X.

On any approximation space, it is possible to define the lower and upper approximation of a given set.

Definition 4.3. *Let* (X, R) *be an approximation space. The* lower *approximation of* $A \subseteq X$ *is*

$$l_R(A) := \{x \in X | [x]_R \subseteq A\}$$

and the upper *approximation of* A *is*

$$u_R(A) := \{x \in X | [x]_R \cap A \neq \emptyset\} \supseteq l_R(A)$$

A rough set is the lower-upper pair $r(A) := (l_R(A), u_R(A))$ *or equivalently the lower-exterior pair* $r_e(A) := (l_R(A), e_R(A)) := (l_R(A), u_R^c(A))$.

A set A is said to be *exact* iff $l_R(A) = A$ or equivalently $A = u_R(A)$. We denote by $RS(X)$ the collection of all lower-upper approximations on X and by $RS_e(X)$ the set of lower-exterior approximations. The lower and upper approximations satisfy some interesting and useful properties. We list here some of them which will be useful later on.

Lemma 4.1. *Let* (X, R) *be an approximation space, and* $A, B \subseteq X$. *Then, the following properties hold.*

1. $l_R(A \cap B) = l_R(A) \cap l_R(B); \; l_R(A \cup B) \supseteq l_R(A) \cup l_R(B)$
2. $u_R(A \cup B) = u_R(A) \cup u_R(B); \; u_R(A \cap B) \subseteq u_R(A) \cap u_R(B)$
3. *If one of* A, B *is exact then* $l_R(A) \cup l_R(B) = l_R(A \cup B)$ *and* $u_R(A \cap B) = u_R(A) \cap u_R(B)$;
4. $l_R(A) \subseteq A \subseteq u_R(A)$;
5. $l_R(l_R(A)) = l_R(A), \; u_R(u_R(A)) = u_R(A)$;
6. $l_R A = u_R^c(A^c)$.

Property 3 does not hold for two non-exact sets, that is $l_R(A) \cup l_R(B) \neq l_R(A \cup B)$ and $u_R(A \cap B) \neq u_R(A) \cap u_R(B)$ in this case. Otherwise stated, l, u are not truth-functional operators [74]; we can say we miss truth-functionality at the "internal level".

On the other hand, what we can try to do is to define truth-functional operators on rough sets viewed as upper-lower pairs $(A_*, A^*) \in RS(X)$, irrespective of the original underlying set. Then we say we have truth-functionality at the "external level". As we will see, this is feasible. Indeed, the lower-upper pair is clearly a nested pair and thus we can carry to this subcase the considerations on operations of the previous section. However, rough sets form a proper subset of nested pairs in the sense that every rough set induces a nested pair of sets in X, generated by a subset H of X through operators l_R, u_R, as $(l_R(H), u_R(H)) \in RS(X)$ but not vice versa [16]. Noticeably, no singleton $\{x\}$ can appear as an equivalence class in the boundary of a rough set, since either $x \in H$ and $\{x\}_R \subset l_R(H)$ or $x \notin H$ and $\{x\}_R \subset u_R(H)^c$.

So a truth-functional operation on orthopairs cannot be simply applied to rough sets. It must be shown that the operation is meaningful, that is:

- closed on the collection of all rough sets $RS(X)$ (or equivalently $RS_e(X)$)
- related to a well-defined combination of the underlying (Boolean) approximated sets.

As we will see, in this process, some interpretability problems of the connectives arise.

4.2   Rough Sets and External Truth-Functionality

Since $RS(X) \subset \mathcal{N}(X)$, the question is whether, once we restrict to $RS(X)$, the implications definable on $\mathcal{N}(X)$ are closed on $RS(X)$. In other words:

If $\odot$ is a three-valued binary operation on pairs $(A_*, A^*), (B_*, B^*) \in RS(X)$, and $A_* = l_R(A), A^* = u_R(A), B_* = l_R(B), B^* = u_R(B)$ for some $A, B \subset X$, does there exist an operation $\cdot$ on 2^X such that
$$(l_R(A \cdot B), u_R(A \cdot B)) = (A_*, A^*) \odot (B_*, B^*)?$$

The answer is not straightforward, since first of all not all nested pairs (A, B) can be generated by a subset H of the universe as $(l_R(H), u_R(H))$, as pointed out before. Moreover it must be clear that the relation R, used to build the partition and then to compute the approximation, is fixed in the above statement.

Let us start from already known results for basic operations [22]. First of all the negation of a set. This case is simple, indeed we have in terms of orthopairs:

$$r(A^c) = (l_R(A^c), u_R(A^c)) = (u_R^c(A), l_R^c(A)) = r^c(A)$$

Thus, the approximation of A^c can be obtained by the approximation of A in a truth-functional way.

In case of intersection $r(A \cap B)$ and union $r(A \cup B)$, corresponding to the min conjunction 5 in Table 4 and the dual disjunction, consider Kleene conjunction and disjunction. Namely, we ask if there exist two sets $C, D \subseteq X$ such that

$$r(C) = r(A) \sqcap r(B) := (l_R(A) \cap l_R(B), u_R(A) \cap u_R(B)) \tag{3a}$$
$$r(D) = r(A) \sqcup r(B) := (l_R(A) \cup l_R(B), u_R(A) \cup u_R(B)) \tag{3b}$$

At least three solutions were proposed in the literature. Bonikowski in [16] showed that the set C can be built according to the following procedure:

1. If $u_R(A) \cap u_R(B) = \emptyset$ then $C = \emptyset$, else $u_R(A) \cap u_R(B)$ is of the form $[x_1] \cup \ldots \cup [x_k]$, where $[x_i]$ are equivalence classes of R.
2. Choose $y_i \in [x_i]$ for all i such that $[x_i] \not\subseteq l_R(A) \cap l_R(B)$ (in the boundary) and build $Y = \{y_i : y_i \in [x_i] \not\subseteq l_R(A) \cap l_R(B)\}$
3. Finally, $C = [l_R(A) \cap l_R(B)] \cup Y$ (disjoint union).

Note that $l_R(C) = l_R(A \cap B)$, since no equivalence class $[x_i]$ in the boundary can be a singleton $\{y_i\}$, any $y_i \in Y$ is an element of a larger equivalence class, and so, $l_R(Y) = \emptyset$. The set D for disjunction in (3) is computed with the same procedure applied to A^c and B^c.

In [45] we can find another definition of internal intersection and union

$$A \cap_1 B = A \cap [l_R(B) \cup (B \cap u_R(A)^c) \cup (u_R(B) \cap l_R(A)^c \cap A) \cup (l_R(A) \cap B)]$$
$$A \cup_1 B = A \cup [l_R(B) \cup (B \cap u_R(A)^c) \cup (u_R(B) \cap l_R(A)^c \cap A) \cup (l_R(A) \cap B)]$$

and again, $r(A \cap_1 B) = r(A) \sqcap r(B), r(A \cup_1 B) = r(A) \sqcup r(B)$. Finally, in [11], the following alternative solution has been proposed.

$$A \cap_2 B = (A \cap B) \cup ((A \cap u_R(B)) \cap (u_R(A \cap B)^c))$$
$$A \cup_2 B = (A \cup B) \cap ((A \cup l_R(B)) \cup (l_R(A \cup B)^c))$$

Note that $A \cap_1 B$ and $A \cap_2 B$ can be written as $[l_R(A) \cap l_R(B)] \cup Y'$, where Y' is the union of proper subsets Y_i of equivalence classes $[x_i]$ not in the intersection of the lower images. So, they are very close to one of the possible solutions of Bonikowski's procedure. Moreover, any solution has this form.

Proposition 4.1. *Any set C whose upper and lower approximations are respectively the intersections of the upper and of the lower approximation of A and B is of the form $[l_R(A) \cap l_R(B)] \cup Y'$, where $Y' = \cup_{i=1}^n Y_i$ and $\emptyset \neq Y_i \subset [x_i], \forall i = 1, \ldots, n$.*

Proof. Indeed, if Y' does not contain at least one element of each equivalence class outside the intersection of the lower approximations and inside the intersection of their upper approximations, then the upper approximation of C is not the intersection of the upper approximations of A and B. If Y' contains one equivalence class outside the intersection of lower approximations and inside their union, then its lower approximation is larger than the intersection of lower approximations of A and B.

Now, we want to extend these definability results to all other three-valued operations introduced in the previous Section 2. Let us start from the negations whose proof is straightforward.

Proposition 4.2

$$\neg(l_R(A), u_R(A)) = (u_R^c(A), l_R^c(A)) = r(A^c);$$
$$\sim(l_R(A), u_R(A)) = (u_R^c(A), u_R^c(A)) = r(l_R(A^c));$$
$$-(l_R(A), u_R(A)) = (u_R(A^c), u_R(A^c)) = r(u_R(A^c)).$$

Now, as far as implications are concerned, some of them have already been studied in literature and it has been shown that they are closed on rough sets. These results are summarized in the following proposition.

Proposition 4.3

$$r(A) \Rightarrow_5 r(B) = r((A \to B) \cap ((A \to l_R(B)) \cup (l_R(A \to B)^c)));$$
$$r(A) \Rightarrow_9 r(B) = r(l_R(A) \to B);$$
$$r(A) \Rightarrow_{10} r(B) = r((u_R(A) \to B) \cup [l_R(A) \to u_R^c(B)]^c);$$
$$r(A) \Rightarrow_{11} r(B) = r((l_R(A) \to B) \cap (A \to u_R(B)));$$
$$r(A) \Rightarrow_{14} r(B) = r(((u_R(A) \to u_R(B)) \cap (l_R(A) \to l_R(B)))).$$

Proof. The cases 5,10,11, respectively Kleene, Gödel and Łukasiewicz implications, are proved in [22]. The Nelson (case 9) implication immediately follows by its definition (and see also [59]). Finally, the Gaines-Rescher implication 14 has been studied in [11], where it is defined as

$$r(A) \Rightarrow r(B) = (\neg \Box r(A) \cup \Box r(B)) \cap (\neg \Diamond r(A) \cup \Diamond r(B)), \qquad (4)$$

with $\Box r(H) = (l_R(H), l_R(H))$ and $\Diamond r(H) = (u_R(H), u_R(H))$. So, first of all, let us note that equation 4 is equivalent to the one in Table 7, as can be easily proven. Then, from the definition in [11], we have $r(A) \Rightarrow_{14} r(B) = [r(l_R^c(A)) \sqcup r(l_R(B))] \sqcap [r(u_R^c(A)) \sqcup r(u_R(B))]$ from which we arrive at the thesis.

In order to study the other implications and conjunctions, the following result concerning the application of the transformations (1) can be given.

Proposition 4.4. *Let $\odot$ be a closed operation on $\mathcal{R}(X)$. Then, also $a[\mathcal{A}(\odot)]b$, $a[\mathcal{V}(\odot)]b$, $a[\mathcal{S}(\odot)]b$ are closed on $\mathcal{R}(X)$.*

Proof. The case of $\mathcal{A}(\odot)$ is trivial since it is the same operation as $\odot$ with different arguments. Operations $\mathcal{V}(\odot)$ and $\mathcal{S}(\odot)$ are a composition of $\odot$ and involutive negation $\neg$ which is closed by proposition 4.2. So, we will have that $r(X)[\mathcal{V}(\odot)]r(Y) = r(Y^c) \odot r(X^c)$ and $r(X)[\mathcal{S}(\odot)]r(Y) = \neg[r(X) \odot r(Y^c)]$.

By the above propositions, we immediately get that also other implications and conjunctions are well defined, since as shown in [23, 25] they can be obtained by equation system (1) from the above implications in Proposition 4.3.

Corollary 4.1. *Conjunctions $7, 9, 10, 11, 13, 14$ are closed on $\mathcal{R}(X)$ and the following hold:*

$$r(A) *_7 r(B) = r(A \cap l_R(B));$$
$$r(A) *_9 r(B) = r(l_R(A) \cap B);$$
$$r(A) *_{10} r(B) = r([l_R(A) \cup l_R(B)] \cap u_R(A) \cap B);$$
$$r(A) *_{11} r(B) = r([l_R(A) \cup l_R(B)] \cap A \cap B);$$
$$r(A) *_{13} r(B) = r([l_R(A) \cup l_R(B)] \cap A \cap u_R(B));$$
$$r(A) *_{14} r(B) = r((l_R(A) \cup l_R(B)) \cap u_R(A) \cap u_R(B)).$$

Further, implications $7, 13$ are closed on $\mathcal{R}(X)$ and we have:

$$r(A) \Rightarrow_7 r(B) = r(A \to u_R(B));$$
$$r(A) \Rightarrow_{13} r(B) = r(A \to l_R(B) \cup [l_R^c(A) \to u_R^c(B)]^c).$$

We now prove that all the remaining implications and conjunctions are closed on $\mathcal{RS}(X)$.

Proposition 4.5

$$r(A) \Rightarrow_1 r(B) = r(u_R(A) \to l_R(B));$$
$$r(A) \Rightarrow_2 r(B) = r([A \to l_R(B)] \cap [u_R(A) \to B]);$$
$$r(A) \Rightarrow_3 r(B) = r(A \to l_R(B));$$
$$r(A) \Rightarrow_6 r(B) = r(u_R(A) \to u_R(B));$$
$$r(A) \Rightarrow_8 r(B) = r(l_R(A) \to u_R(B)).$$

Proof. Only $\Rightarrow_2$ deserves some explanation, the others being trivial. By Table 6, we get $(l_R(A), u_R(A)) \Rightarrow_2 (l_R(B), u_R(B)) = (u_R^c(A) \cup l_R(B), (u_R(B) \cup u_R^c(A)) \cap (l_R^c(A) \cup l_R(B)))$, which can be re-written as $((u_R^c(A) \cup l_R(B)) \cap (u_R^c(A) \cup l_R(B)), (u_R(B) \cup u_R^c(A)) \cap (l_R^c(A) \cup l_R(B)))$. Applying equations (3), we obtain $[r(A^c) \sqcup r(l_R(B))] \sqcap [r(l_R(A^c)) \sqcup r(B)]$ and by Lemma 4.1 we have the thesis.

Based on the implications in Proposition 4.5, it is possible to construct other conjunctions and implications (see [23, 25]). So, due to Proposition 4.4 the following corollary holds.

Corollary 4.2. *Conjunctions $1, 2, 3, 4, 6, 8, 12$ are closed on $\mathcal{R}(X)$ and the following hold:*

$$r(A) *_1 r(B) = r(u_R(A) \cap u_R(B));$$
$$r(A) *_2 r(B) = r([A \cap u_R(B)] \cup [u_R(A) \cap B]);$$
$$r(A) *_3 r(B) = r(A \cap u_R(B));$$
$$r(A) *_4 r(B) = r(u_R(A) \cap B);$$
$$r(A) *_6 r(B) = r(u_R(A) \cap l_R(B));$$
$$r(A) *_8 r(B) = r(l_R(A) \cap l_R(B));$$
$$r(A) *_{12} r(B) = r(l_R(A) \cap u_R(B)).$$

Implications $4, 12$ *are closed on* $\mathcal{R}(X)$ *and the following hold:*

$$r(A) \Rightarrow_4 r(B) = r(u_R(A) \to B);$$
$$r(A) \Rightarrow_{12} r(B) = r(l_R(A) \to l_R(B)).$$

The above results show that the original question of finding subsets of X that underlie all 14 three-valued conjunctions and implications applied to upper and lower approximations of sets in the sense of rough sets can be answered in the affirmative. However the reader may observe that the definition of such subsets, the approximations of which are constructed by such connectives, always involve lower and/or upper approximations of the two underlying sets to be combined.

4.3 The Interpretability of External Truth-Functional Operations on Rough Sets

In [22], we started an investigation on the significance of existing truth-functional three-valued logics of ill-known sets described by pairs of disjoint (or pairs of nested) subsets. This work strongly suggested that while, from a mathematical standpoint, such three-valued logics are consistent with a rough set view, their interpretation with respect to reasoning about the original data tables is questionable. The operators analyzed in that work were Kleene conjunction and disjunction (min/max) on three values and three different implications: Lukasiewicz, Gödel and Kleene. However, the concerns already raised for these known connectives seem to carry over to all the 28 three-valued connectives recalled in this paper, as the results obtained here in the previous section indicate.

Let us consider two sets of items A, B defined in extension, the approximations $r(A)$ and $r(B)$ of which we want to aggregate with one of the three-valued connectives laid bare in this paper, say $\odot$. Concerning the existence of a set C such that $r(C) = r(A) \odot r(B)$, we have seen that such an underlying set C always exists. However, C does not depend exclusively on A and B but strongly depends on the partition chosen (that is on the equivalence relation R of the approximation space and finally on the set of attributes of an Information Table) because it depends on the lower and/or upper approximations of A and B as well. Moreover, even inside the same partition, several choices of C are possible.

This difficulty is due, in some sense, to the presence of two languages: the fine-grained one needed to distinguish elements of X and the (more restricted) one based on the attributes of the information table, that only allows to describe approximations of any subset of such elements. Combining approximations of ill-known sets A and B truth-functionally yields well-behaved pairs of nested sets, but the corresponding internal combination of A and B that makes the external truth-functional combination meaningful is problematic.

Indeed, in the setting of rough sets, A and B are known in extension (they are in some sense the actual entities referred to) whereas, using the coarser attribute language instrumental to describe them, their intensions are available only through their approximations. So, the intension depends on the coarse language: the more (less) numerous the attributes, the finer (coarser) the description. Results in the previous section show that the set $C = A \cdot B$ displayed in the previous section for the 14 conjunctions and the 14 implications such that $r(A) \odot r(B) = r(A \cdot B)$ laid bare in this paper always depends on the partition induced by R and so on the attributes defining the coarser language. Changing the coarser language (i.e., attributes in the Information Table) will alter the set $A \cdot B$ but not A, B. So, while we can interpret the external truth-functionality of operations on approximation pairs as providing approximation pairs of definable combinations of subsets of X, these subsets are definable only if the coarser language is fixed (in fact they need both languages, since $A \cdot B$ is potentially a Boolean set-theoretic combination of $A, B, l_R(A), l_R(B), u_R(A), u_R(B)$). As a consequence, we lose the interpretability of the results since these inner combinations are not intrinsic to A and B, and depend on the indistinguishability relation.

5   Rough Sets: From Modal Logic to Three-Valued Logics

Apart from many-valued logics, a natural logical rendering of rough sets is through modal logics. This possibility has been addressed by several authors taking into account different variants of rough sets [57, 42, 69, 70, 11, 75, 76, 8, 51]. This section provides some hints toward relating the three-valued and the modal logic views of rough sets, in connection with recent works translating three-valued logics into fragments of the modal logic KD.

5.1   The Standard Modal Approach to Rough Sets

We now recall a modal logic for handling approximations of sets generated by an equivalence relation [56].

Its language $\mathcal{L}_M$ is the usual one of propositional logic plus necessity $\Box$ and possibility $\Diamond$. That is, we have a set of propositional variables $\mathcal{V} = \{a, b, c, \ldots\}$ and the connectives $\wedge, ', \Box$. As usual, disjunction $\alpha \vee \beta$ stands for $(\alpha' \wedge \beta')'$, implication $\alpha \to \beta$ stands for $\alpha' \vee \beta$, tautology $\top$ for $\alpha \vee \alpha'$ and $\Diamond \alpha = (\Box \alpha')'$. Well formed formulae are built in the standard way.

The axioms are those of propositional logic plus the axioms to characterize the modal connectives.

1. $\phi \to (\psi \to \phi)$
2. $(\psi \to (\phi \to \mu)) \to ((\psi \to \phi) \to (\psi \to \mu))$
3. $(\phi' \to \psi') \to (\psi \to \phi)$
(K) $\Box(\alpha \to \beta) \to (\Box\alpha \to \Box\beta)$
(T) $\Box\alpha \to \alpha$
(5) $\Diamond\alpha \to \Box\Diamond\alpha$

Finally, rules are modus ponens and necessitation: If $\vdash \alpha$ then $\vdash \Box\alpha$. The above system is called S5, and its semantics is in terms of equivalence relations [20]. It is thus the natural logical setting for rough sets [56].

The semantics is given through a model $M = (X, R, v)$, where (X, R) is an approximation space and v is a mapping from formulae to 2^X. In standard modal logic terminology, X is the set of possible worlds, R the accessibility relation and $v(\alpha)$ represents the set of possible worlds where α holds. The interpretation v is recursively defined on propositional connectives as usual as:

$$v(\alpha') = v(\alpha)^c$$
$$v(\alpha_1 \wedge \alpha_2) = v(\alpha_1) \cap v(\alpha_2)$$
$$v(\alpha_1 \vee \alpha_2) = v(\alpha_1) \cup v(\alpha_2)$$

and modal operators are mapped to lower and upper approximations:

$$v(\Box\alpha) = L_R(v(\alpha)) = \{x \in X : [x]_R \subseteq v(\alpha)\} = \{x \in X : \forall w, x\mathcal{R}w, w \in v(\alpha)\}$$
$$v(\Diamond\alpha) = U_R(v(\alpha)) = \{x \in X : [x]_R \cap v(\alpha) \neq \emptyset\} = \{x \in X : \exists w, x\mathcal{R}w, w \in v(\alpha)\}$$

Note that, in the S5 approach, one can represent sets ("objective" formulae α) and their lower ($\Box\alpha$) and upper ($\Diamond\alpha$) approximations.

This approach can easily be extended to rough set models based on a relation that is not necessarily an equivalence one [75, 76]. Indeed, it is well known in modal logic [20] that, once fixed the basic axioms 1-3 and (K), then to any additional modal axiom according to Table 9 corresponds a specific property of the accessibility relation.

Table 9. Correspondence between modal axioms and relation properties

Name	Axiom	Property
T	$\Box\alpha \to \alpha$	Reflexive
4	$\Box\alpha \to \Box\Box(\alpha)$	Transitive
5	$\Diamond\alpha \to \Box(\Diamond(\alpha))$	Euclidean
D	$\Box\alpha \to \Diamond\alpha$	Serial
B	$\alpha \to \Box\Diamond\alpha$	Symmetric

Another extension of the basic approach is the logic DAL [42], meant to deal with approximation spaces with more than one equivalence relation (X, R_i). Each relation represents a different attribute, for instance "having the same number of circles", "having the same number of crosses".

5.2   The Three-Valued Modal Approach

A different approach is given by the so-called *Pre-Rough Logic (PRL)* and its corresponding algebra called *pre-rough algebra* [11], which is based on a 3-valued logic. Atoms of the logic are three-valued entities, which represent nested approximation pairs. They can be obtained from the S5 logic of the previous section by considering a weaker notion of logical equivalence in S5. Namely, Banerjee and Chakraborty speak of *rough equivalence* of two propositional formulae α and β whenever $\Box\alpha$ is semantically equivalent to $\Box\beta$ and $\Diamond\alpha$ is semantically equivalent to $\Diamond\beta$. They consider the result of quotienting the language with the rough equivalence relation, and each equivalence class corresponds to an approxima-tion pair, which becomes a formula of the pre-rough logic PRL. Note that by doing so, the underlying set, the approximations of which are given by modal formulae, is lost: we can no longer distinguish between propositional formulae that are roughly equivalent.

For the sake of clarity, we denote by μ, ν, ρ the formulae of PRL. Primitive connectives of the logic are negation, intersection and necessity, respectively denoted by $\neg, \wedge, \blacksquare$, from which we derive the disjunction $\vee$ through de Morgan properties, the dual modality $\blacklozenge\mu = \neg\blacksquare\neg\mu$ and the implication as $\mu \rightarrow \nu = (\neg\blacksquare\mu \vee \blacksquare\nu) \wedge (\neg\blacklozenge\mu \vee \blacklozenge\nu)$. The axioms of the logic are:

RL1 $\mu \rightarrow \mu$
RL2 $\neg\neg\mu \leftrightarrow \mu$
RL3 $\mu \wedge \nu \rightarrow \mu$
RL4 $\mu \wedge \nu \rightarrow \nu \wedge \mu$
RL5 $\mu \wedge (\nu \vee \rho) \leftrightarrow (\mu \wedge \nu) \vee (\mu \wedge \rho)$
RL6 $\blacksquare\mu \rightarrow \mu$
RL7 $\blacksquare(\mu \wedge \nu) \leftrightarrow \blacksquare\mu \wedge \blacksquare\nu$
RL8 $\blacksquare\mu \rightarrow \blacksquare\blacksquare\mu$
RL9 $\blacklozenge\blacksquare\mu \rightarrow \blacksquare\mu$
RL10 $\blacksquare(\mu \vee \nu) \leftrightarrow \blacksquare\mu \vee \blacksquare\nu$

A sequent calculus for this logic is provided by Sen and Chakraborty [64].

The semantics is three-valued, and some connectives are based on Kleene logic, using ternary valuations t such that [11]

$$t(\neg\mu) = \neg t(\mu) \text{ (Kleene negation)} \tag{5}$$
$$t(\mu \wedge \nu) = t(\mu) \sqcap t(\nu) \text{ (Kleene conjunction)} \tag{6}$$
$$t(\blacksquare\mu) = \neg - t(\mu) \text{ (using the paraconsistent negation } -) \tag{7}$$
$$t(\mu \rightarrow \nu) = (t(\neg\blacksquare\mu) \cup t(\blacksquare\nu)) \cap (t(\neg\blacklozenge\mu) \cup t(\blacklozenge\nu)) \tag{8}$$

In connection with the S5-based rough set logic of the previous section, a non-modal formula μ in PRL corresponds to an approximation pair $(A(\mu)_*, A(\mu)^*)$ over possible worlds in X (Boolean interpretations), both $A(\mu)_*$ and $A(\mu)^*$ being exact sets of such valuations (formulae α such that $v(\Box\alpha) = v(\Diamond\alpha)$, whenever $v(\alpha) = A(\mu)_*$ or $A(\mu)^*$). The operator $\blacksquare$ corresponds to extracting the core $A(\mu)_*$ of the three-valued set over X induced by μ. In terms of fuzzy sets, $\blacksquare\mu$ corresponds to the core of μ and $\blacklozenge\mu$ correspond to its support. It can thus be easily seen that

- $t(\blacksquare\mu)$ is two-valued and $\blacksquare\mu$ corresponds to the (exact) approximation pair $(A(\mu)_*, A(\mu)_*)$;
- $t(\blacklozenge\mu)$ is two-valued and $\blacklozenge\mu$ corresponds to the (exact) approximation pair $(A(\mu)^*, A(\mu)^*)$
- $t(\mu \rightarrow \nu)$ is two-valued: it is Gaines-Rescher implication expressing the double inclusion of upper and lower approximations $A(\mu)_* \subseteq B(\nu)_*$ and $A(\mu)^* \subseteq B(\nu)^*$.

In view of the above semantical considerations, RL6-RL10 are expected. RL6 and RL9 correspond to S5 axioms T, and 5 respectively, while RL7 and RL8 are valid in S5. However, due to RL10, ■ and ◆ are deviant modalities, that are not trivial because the logic is 3-valued [19].

The connection between S5 and PRL is maintained by noticing that an approximation pair $(A(\mu)_*, A(\mu)^*)$ underlies a crisp set A of X that is approximated by this pair. In order to maintain this view throughout all the formulae in PRL, Banerjee and Chakraborty make it clear what is the set approximated by for instance $(C(\mu \wedge \nu)_*, C(\mu \wedge \nu)^*)$ when A and B are the sets approximated by $(A(\mu)_*, A(\mu)^*)$ and $(B(\nu)_*, B(\nu)^*)$. They do it by introducing the intersection $A \cap_2 B$ already discussed in Subsection 4.2, and that does not depend solely on A and B.

The fact that in the PRL syntax, we no longer explicitly refer to the approximated set and maintain truth-functionality for evaluating formulae expressing approximation pairs is thus paid by the fact that it is no longer possible to intrinsically define the approximated set referred to by a compound PRL formula in terms of the approximated sets of its elementary sub-formulae. On the other hand, while the S5 setting avoids this pitfall, one may find it unrealistic to represent at the same time the approximated pairs with the approximated set in the language, as the point made by rough set theory is that sets are only described in intension through the available attributes, while their precise extension is out of reach. In this sense, while S5 seems to precisely capture the formal setting of rough sets, the PRL logic looks more faithful to the way rough sets can be used in practice, that is, it refers to the situation where we know the approximations, but neither the underlying set nor the equivalence relation. Unfortunately, the PRL rendering of rough set theory, and logical combinations of upper and lower approximations, in terms of a three-valued logic looks like an approximation as well.

Interestingly, connections between PRL and major three-valued logics have been laid bare:

- In [11] it is shown that the algebra of PRL (a bounded lattice structure $(L, 0, 1)$ equipped with Kleene connectives $(\sqcup, \neg, \sqcap)$ the modality ■ and Gaines-Rescher implication $\rightarrow_{14}$, is equivalent to semi-simple Nelson algebras, that is, a bounded lattice structure $(L, 0, 1)$ equipped with Kleene connectives, the paraconsistent negation $-$ and Nelson implication $\rightarrow_9$. Indeed they notice that $-\mu = \neg \blacksquare \mu$ and $\mu \rightarrow_9 \nu = \neg \blacksquare \mu \sqcup \nu$. Conversely, $\blacksquare \mu = \neg - \mu$ and $\mu \rightarrow_{14} \nu = \neg - (\mu \rightarrow_9 \nu)$.

- It has been proved in [9], that PRL is equivalent to three-valued Łukasiewicz logic. Especially, Banerjee points out that the Łukasiewicz implication can be written as $(\blacklozenge \neg \mu \sqcup \nu) \sqcap (\neg \mu \sqcup \blacklozenge \nu)$. Conversely, PRL connectives $\wedge, \vee, \Diamond$ are defined as usual $\mu \sqcup \nu = ((\mu \rightarrow_{11} \nu) \rightarrow_{11} \nu)$, $\mu \sqcap \nu = \neg(\neg \mu \sqcup \neg \nu)$ and $\blacklozenge \mu = \neg(\mu \rightarrow_{11} \neg \mu)$.

But these findings are not surprising at all given the results in [25] whereby from Kleene connectives plus paraconsistent negation, one can reconstruct all

28 monotonic three-valued conjunctions and implications that extend Boolean ones (as per the last item of Proposition 2.2).

5.3   From Three-Valued Rough Set Logic to Modal Logic

We have seen two different modal approaches concerning rough sets: a Boolean one based on S5 and a many-valued one, the PRL logic, the latter being based on a clustering of roughly equivalent formulae of the former. Conversely, Banerjee [9] proves that PRL is embeddable in S5 in the sense that PRL formulae can be expressed as S5 formulae via a translation operation $(\cdot)^\tau$ such that $\vdash_{PRL} \mu$ iff $\vdash_{S5} \mu^\tau$.

Namely, the PRL negation $\neg$ becomes the classical negation $'$, the PRL necessity operator $\blacksquare$ becomes the classical one $\Box$, and

$$(\mu \sqcap \nu)^\tau = (\mu^\tau \wedge \nu^\tau) \vee (\mu^\tau \wedge \Diamond\nu^\tau \wedge (\Diamond(\mu^\tau \wedge \nu^\tau))). \tag{9}$$

The latter encodes the set supposedly upper and lower approximated by the intersection of upper and lower approximations of two sets, already met in previous sections as $\cap_2$. This translation can only yield a fragment of S5.

There is another way of capturing the semantics of three-valued logics in a modal setting, whenever the third truth-value stands for unknown [24, 26]. It is enough to use a fragment of KD (or of S5) called MEL [12]. In particular, we can translate three-valued Łukasiewicz logic L3 into MEL, while preserving L3 theorems. Since PRL is equivalent to L3, it is interesting to translate PRL into MEL as well.

The language MEL [12, 13] is a very limited fragment of the modal logic S5. It uses a sublanguage $\mathcal{L}_\Box$ of S5 defined by encapsulating propositional formulae from a modality-free propositional language $\mathcal{L}$ (using the same notations as in Subsection 5.1):

$$\mathcal{L}_\Box = \Box\alpha : \alpha \in \mathcal{L}|\neg\phi|\phi \wedge \psi|\phi \vee \psi|\phi \to \psi.$$

Note that $\mathcal{L}_\Box \cap \mathcal{L} = \emptyset$ and $\mathcal{L}_\Box \subset \mathcal{L}_M$ (the set of all modal formulae, including nested ones). MEL is equipped with the following axioms:

1. $\phi \to (\psi \to \phi)$.
2. $(\psi \to (\phi \to \mu)) \to ((\psi \to \phi) \to (\psi \to \mu))$.
3. $(\phi' \to \psi') \to (\psi \to \phi)$
(K) $\Box(p \to q) \to (\Box p \to \Box q)$.
(D) $\Box\alpha \to \Diamond\alpha$.
(N) if $\vdash_{PL} \alpha$ then $\Box\alpha$.

and the inference rule is modus ponens. As usual, the *possible* modality $\Diamond$ is defined as $\Diamond\alpha \equiv (\Box\alpha')'$. The first three axioms are those of PL and the other those of modal logic KD. Axiom (N) is inspired from the necessitation rule that cannot be written in MEL. The following axioms (M) and (C) are implied by the above system:

$$(M)\,\Box(\alpha \wedge \beta) \to (\Box\alpha \wedge \Box\beta);$$

$$(C)\,(\Box\alpha \wedge \Box\beta) \rightarrow \Box(\alpha \wedge \beta).$$

MEL is the subjective fragment of KD (or S5) without modality nesting.

The MEL semantics is very simple [12]. Let Ω be the set of $\mathcal{L}$ interpretations: $\{\omega : \mathcal{V} \rightarrow \{0,1\}\}$. The set of models of α is $[\alpha] = \{\omega : \omega \models \alpha\}$. A (meta)-interpretation of $\mathcal{L}_\Box$ is a non-empty set $E \subseteq \Omega$ of interpretations of $\mathcal{L}$ interpreted as an epistemic state. We define satisfiability as follows:

- $E \models \Box\alpha$ if $E \subseteq [\alpha]$ (α is certainly true in the epistemic state E)
- $E \models \phi \wedge \psi$ if $E \models \phi$ and $E \models \psi$;
- $E \models \phi'$ if $E \models \phi$ is false.

MEL is sound and complete with respect to this semantics [13].

We remark that in this framework, uncertainty modeling is Boolean but possibilistic. The satisfiability $E \models \Box\alpha$ can be written as $N([\alpha]) = 1$ in the sense of a necessity measure computed with the possibility distribution given by the characteristic function of E. Axioms (M) and (C) lay bare the connection with possibility theory [39], as they state the equivalence between $(\Box\alpha \wedge \Box\beta)$ and $\Box(\alpha \wedge \beta)$.

We can justify the choice of this minimal modal formalism. It is the most simple logic to reason on incomplete propositional information. We only need to express that a proposition in PL is certainly true, certainly false or unknown as well as all the logical combinations of these assertions.

In [24, 26] we have proposed to translate three-valued logics of incomplete information into MEL, provided that the third truth-value refers to the idea of unknown Boolean truth-value. Let a be a Boolean variable and $t(a)$ indicate the knowledge we have about a, that is:

- **1** certainly true, the Boolean value of a is 1;
- **0** certainly false, the Boolean value of a is 0;
- $\frac{1}{2}$ unknown, the Boolean value of a is 0 or 1.

For the sake of clarity, we have used different symbols $\mathbf{0}, \mathbf{1}, \frac{1}{2}$ for epistemic truth-values with respect to ontic ones $0, 1$. Under this understanding of the three epistemic truth-values, we can naturally translate three-valued truth-assigments to atomic propositions as follows:

$$\mathcal{T}(t(a) = \mathbf{1}) = \Box a$$
$$\mathcal{T}(t(a) = \mathbf{0}) = \Box a'$$
$$\mathcal{T}(t(a) = \tfrac{1}{2}) = \Diamond a \wedge \Diamond a'$$
$$\mathcal{T}(t(a) \geq \tfrac{1}{2}) = \Diamond a$$
$$\mathcal{T}(t(a) \leq \tfrac{1}{2}) = \Diamond a'$$

Now, we want to map three-valued formulae to MEL formulae. Compound formulae are managed recursively. In the case of Łukasiewicz logic, we have:

$$\mathcal{T}(t(\alpha \sqcap \beta) \geq i) = \mathcal{T}(t(\alpha) \geq i) \wedge \mathcal{T}(t(\beta) \geq i), i \geq \tfrac{1}{2}$$

$$\mathcal{T}(t(\alpha \sqcup \beta) \geq i) = \mathcal{T}(t(\alpha) \geq i) \vee \mathcal{T}(t(\beta) \geq i), i \geq \tfrac{1}{2}$$

$$\mathcal{T}(t(\neg\alpha) = \mathbf{1}) = \mathcal{T}(t(\alpha) = \mathbf{0}) = (\mathcal{T}(t(\alpha) \geq \tfrac{1}{2}))'$$

$$\mathcal{T}(t(\neg\alpha) \geq \tfrac{1}{2}) = \mathcal{T}(t(\alpha) \leq \tfrac{1}{2}) = (\mathcal{T}(t(\alpha) = \mathbf{1}))'$$

$$\mathcal{T}(t(\alpha \rightarrow_L \beta) = \mathbf{1}) = [\mathcal{T}(t(\alpha) = \mathbf{1}) \rightarrow \mathcal{T}(t(\beta) = \mathbf{1})]$$
$$\wedge [\mathcal{T}(t(\alpha) \geq \tfrac{1}{2}) \rightarrow \mathcal{T}(t(\beta) \geq \tfrac{1}{2})]$$

$$\mathcal{T}(t(\alpha \rightarrow_L \beta) \geq \tfrac{1}{2}) = \mathcal{T}(t(\alpha) = \mathbf{1}) \rightarrow \mathcal{T}(t(\beta) \geq \tfrac{1}{2})$$

Note that even if we use the same symbols for connectives in L3 and MEL, we are moving from three-valued variables (formulae) to Boolean ones. In particular, in the case of atoms, we have

$$\mathcal{T}(t(a \rightarrow_L b) \geq \tfrac{1}{2}) = (\Box a)' \vee \Diamond b = \Box a \rightarrow \Diamond b$$

and

$$\mathcal{T}(t(a \rightarrow_L b) = \mathbf{1}) = ((\Box a)' \vee \Box b) \wedge ((\Diamond a)' \vee \Diamond b) = \Box a' \vee \Box b \vee ((\Box a)' \wedge \Diamond b))$$
$$= (\Box a \rightarrow \Box b) \wedge (\Diamond a \rightarrow \Diamond b).$$

It can be easily shown that by this translation, only a fragment of MEL can be captured by Łukasiewicz logic. Namely, $\mathcal{L}_\Box^{\mathcal{L}} = \Box a | \Box a' | \phi' | \phi \vee \psi | \phi \wedge \psi$. That is, we can only have modalities in front of literals.

Finally, we note that this translation makes sense; that is tautologies are preserved by the translation and we can reason in three-valued logic inside MEL. More formally, the following two theorems hold [24, 26]:

Theorem 5.1. *If α is an axiom of Łukasiewicz (but also, Gödel, Nelson) logic, then $\mathcal{T}(t(\alpha))$ is a tautology in MEL.*

Theorem 5.2. *Let α be a formula in Łukasiewicz (but also, Nelson) logic L_3 and B_L a knowledge base in this logic. Then, $B_L \vdash \alpha$ in L_3 iff $\mathcal{T}(B_L) \vdash \mathcal{T}(t(\alpha) = \mathbf{1})$ in MEL.*

Due to the equivalence between L3 and PRL, we can also translate PRL into MEL. Let us start from the connectives: $\vee, \wedge, \neg$ are the same as in Łukasiewicz logic, the only difference is the necessity which is not a primitive operator in L3 but can be derived as follows:

$$\mathcal{T}(t(\blacksquare\mu)) = \mathbf{1}) = \mathcal{T}(t(\blacksquare\mu) \geq \tfrac{1}{2})) = \mathcal{T}(t(\neg(\mu \rightarrow_L \neg\mu)) = \mathbf{1}) = \mathcal{T}(t(\mu) = \mathbf{1})$$

which on atoms corresponds to $\mathcal{T}(t(\blacksquare a) = \mathbf{1})) = \mathcal{T}(t(\blacksquare a)) \geq \tfrac{1}{2})) = \Box a$.

Consequently, PRL implication (i.e., Gaines-Rescher $\rightarrow_{14}$) becomes in MEL:

$$\mathcal{T}(t(\mu \rightarrow_{14} \nu) = \mathbf{1}) = \mathcal{T}(t(\mu \rightarrow_{14} \nu) \geq \tfrac{1}{2}) = (\Box\mu \rightarrow \Box\nu) \wedge (\Diamond\mu \rightarrow \Diamond\nu)$$

which, as expected corresponds to the translation of any residuated implication into MEL.

Due to the above results, in particular Theorem 5.1, it is also possible to see that most PRL axioms translate into MEL tautologies.

Corollary 5.1. *If ϕ is an axiom of PRL logic (but for RL8 and RL9), then $\mathcal{T}(t(\phi) = 1)$ is a tautology in MEL.*

Axioms RL8 and RL9, which involve nested modalities, are not directly expressible in MEL logic. However, in PRL, $\blacksquare\blacksquare\mu$ and $\blacklozenge\blacksquare\mu$ are equivalent to $\blacksquare\mu$ so that they do not need to be translated.

It is interesting to comment on the difference between the two translations of PRL into S5 and into MEL:

- The translation of PRL into MEL yields the L3-fragment of MEL, and we know from [24, 26] that the two logics are equivalent. In particular, the MEL translation does not involve at all the logical rendering of the approximated set underlying the pair $(\blacksquare\alpha, \blacklozenge\alpha)$ in PRL. The lack of expressiveness of PRL with respect to S5 for rough sets is highlighted by the small fragment of S5 attained by the exact translation of PRL into MEL.
- The translation of PRL into S5 highlights a possible logical expression of the set approximated by the combination of approximation pairs, but this expression involves modalities, and does not refer to a set expressible in the pure propositional language $\mathcal{L}$ contained in the one of S5 (see equation (9)). So, even if this translation reaches a language richer than MEL, it carries over to the logical level the semantic difficulty of applying a truth-functional view on approximation pairs, as pointed out in the previous section.

6   Conclusion

The main lesson of this paper is that there is a gap between rough sets and three-valued calculi of approximation pairs regardless of the chosen rich enough algebraic setting, since there are several equivalent ones in three-valued logics. Whether we use Łukasiewicz 3-valued MV algebra, Nelson semi-simple algebras or the pre-rough setting, we can only imperfectly capture the modal logic of rough sets, that is, S5, even if the three-valued approaches can be embedded in the modal setting.

Our exploration of the 28 basic implications and conjunctions acting on orthopairs of sets show that there is no way to find binary connectives on orthopairs that would correspond exactly to the approximation pair enclosing some appropriate Boolean combination of the two sets approximated by each orthopair. This Boolean combination either does not exist, or must involve the chosen equivalence relation in some way, and then it is not even unique. This result is a generalisation of Bonikowski [16] old finding, systematized to all reasonable three-valued conjunctions, disjunctions and implications. It also echoes early remarks on the impossibility of representing rough sets by three-valued fuzzy sets put forward by Pawlak himself [60] and more recently by Yao [74].

Among the perspectives of this paper, one may point out the potential of the MEL language for representing and reasoning about information tables with missing values. Information tables are often encoded in a logical format using languages such as Datalog. In the case of missing values, the L3 fragment of MEL with modalities in front of literals could typically represent information about objects in intension, as an alternative to the use of Kleene logic proposed very early by Codd [27]. It would be of interest to reconsider proposals for defining rough sets under incomplete information in the light of this modal translation of three-valued logics.

References

1. Adams, E.: The Logic of Conditionals. D. Reidel, Dordrecht (1975)
2. Asenjo, F.G., Tamburino, J.: Logic of antinomies. Notre Dame Journal of Formal Logic 16, 17–44 (1975)
3. Atanassov, K.: Intuitionistic Fuzzy Sets. Physica-Verlag, Heidelberg (1999)
4. Avron, A.: On an implication connective of RM. Notre Dame Journal of Formal Logic 27, 201–209 (1986)
5. Avron, A.: Natural 3-valued logics - characterization and proof theory. J. Symb. Log. 56(1), 276–294 (1991)
6. Avron, A., Konikowska, B.: Rough sets and 3–valued logics. Studia Logica 90, 69–92 (2008)
7. Baets, B.D., Fodor, J.C.: Residual operators of uninorms. Soft Computing 3, 89–100 (1999)
8. Balbiani, P., Vakarelov, D.: A modal logic for indiscernibility and complementarity in information systems. Fundam. Inform. 50(3-4), 243–263 (2002)
9. Banerjee, M.: Rough sets and 3-valued Lukasiewicz logic. Fundamenta Informaticae 31(3/4), 213–220 (1997)
10. Banerjee, M., Chakraborty, K.: Algebras from rough sets. In: Pal, S., Skowron, A., Polkowski, L. (eds.) Rough-Neural Computing, pp. 157–188. Springer (2004)
11. Banerjee, M., Chakraborty, M.: Rough sets through algebraic logic. Fundamenta Informaticae 28, 211–221 (1996)
12. Banerjee, M., Dubois, D.: A simple modal logic for reasoning about revealed beliefs. In: Sossai, C., Chemello, G. (eds.) ECSQARU 2009. LNCS, vol. 5590, pp. 805–816. Springer, Heidelberg (2009)
13. Banerjee, M., Dubois, D.: A simple logic for reasoning about incomplete knowledge. International Journal of Approximate Reasoning 55, 639–653 (2014)
14. Belnap, N.D.: A useful four-valued logic. In: Dunn, J.M., Epstein, G. (eds.) Modern Uses of Multiple-Valued Logic, pp. 8–37. D. Reidel (1977)
15. Bochvar, D.A.: On a three-valued logical calculus and its application to the analysis of the paradoxes of the classical extended functional calculus. History and Philosophy of Logic 2, 87–112 (1981)
16. Bonikowski, Z.: A certain conception of the calculus of rough sets. Notre Dame Journal of Formal Logic 33(3), 412–421 (1992)
17. Borowski, L. (ed.): Selected works of J. Lukasiewicz. North-Holland, Amsterdam (1970)
18. Cattaneo, G., Ciucci, D.: Algebraic structures for rough sets. In: Peters, J.F., Skowron, A., Dubois, D., Grzymała-Busse, J.W., Inuiguchi, M., Polkowski, L. (eds.) Transactions on Rough Sets II. LNCS, vol. 3135, pp. 208–252. Springer, Heidelberg (2004)

19. Cattaneo, G., Ciucci, D., Dubois, D.: Algebraic models of deviant modal operators based on de Morgan and Kleene lattices. Inf. Sci. 181(19), 4075–4100 (2011)

20. Chellas, B.F.: Modal Logic, An Introduction. Cambridge University Press, Cambridge (1988)

21. Ciucci, D.: Orthopairs: A Simple and Widely Used Way to Model Uncertainty. Fundam. Inform. 108(3-4), 287–304 (2011)

22. Ciucci, D., Dubois, D.: Truth-Functionality, Rough Sets and Three-Valued Logics. In: Proceedings ISMVL, pp. 98–103 (2010)

23. Ciucci, D., Dubois, D.: Relationships between connectives in three-valued logics. In: Greco, S., Bouchon-Meunier, B., Coletti, G., Fedrizzi, M., Matarazzo, B., Yager, R.R. (eds.) IPMU 2012, Part I. CCIS, vol. 297, pp. 633–642. Springer, Heidelberg (2012)

24. Ciucci, D., Dubois, D.: Three-valued logics for incomplete information and epistemic logic. In: del Cerro, L.F., Herzig, A., Mengin, J. (eds.) JELIA 2012. LNCS, vol. 7519, pp. 147–159. Springer, Heidelberg (2012)

25. Ciucci, D., Dubois, D.: A map of dependencies among three-valued logics. Information Sciences 250, 162–177 (2013), Corrigendum: Information Sciences 256, 234-235 (2014)

26. Ciucci, D., Dubois, D.: A modal theorem-preserving translation of a class of three-valued logics of incomplete information. Journal of Applied Non-Classical Logics 23, 321–352 (2013)

27. Codd, E.F.: Extending the database relational model to capture more meaning. ACM Trans. Database Syst. 4(4), 397–434 (1979)

28. Couso, I., Dubois, D.: Rough sets, coverings and incomplete information. Fundamenta Informaticae 108(3-4), 223–247 (2011)

29. De Baets, B., Fodor, J.C., Ruiz-Aguilera, D., Torrens, J.: Idempotent uninorms on finite ordinal scales. International Journal of Uncertainty, Fuzziness and Knowledge-Based Systems, 1–14 (2009)

30. Dempster, A.P.: Upper and lower probabilities induced by a multivalued mapping. The Annals of Statistics 28, 325–339 (1967)

31. Deschrijver, G., Kerre, E.E.: On the relationship between some extensions of fuzzy set theory. Fuzzy Sets and Systems 133(2), 227–235 (2003)

32. D'Ottaviano, I.M.L., da Costa, N.C.A.: Sur un problème de Jaśkowski. Comptes Rendus de l'Académie des Sciences 270, 1349–1353 (1970)

33. Dubois, D.: Degrees of truth, ill-known sets and contradiction. In: Bouchon-Meunier, B., Magdalena, L., Ojeda-Aciego, M., Verdegay, J.-L., Yager, R.R. (eds.) Foundations of Reasoning under Uncertainty. STUDFUZZ, vol. 249, pp. 65–83. Springer, Heidelberg (2010)

34. Dubois, D., Gottwald, S., Hájek, P., Kacprzyk, J., Prade, H.: Terminological difficulties in fuzzy set theory - the case of intuitionistic fuzzy sets. Fuzzy Sets and Systems 156(3), 485–491 (2005)

35. Dubois, D., Prade, H.: Fuzzy-set-theoretic differences and inclusions and their use in the analysis of fuzzy equations. Control and Cybernetics 13(3), 129–146 (1984)

36. Dubois, D., Prade, H.: A theorem on implication functions defined from triangular norms. Stochastica VIII, 267–279 (1984)

37. Dubois, D., Prade, H.: Twofold fuzzy sets and rough sets –some issues in knowledge representation. Fuzzy Sets and Systems 23, 3–18 (1987)

38. Dubois, D., Prade, H.: Conditional Objects as Nonmonotonic Consequence Relationships. IEEE Transactions on Systems, Man, and Cybernetics 24(12), 1724–1740 (1994)

39. Dubois, D., Prade, H.: Possibility theory, probability theory and multiple-valued logics: A clarification. Ann. Math. and AI 32, 35–66 (2001)
40. Dubois, D., Prade, H.: Gradualness, uncertainty and bipolarity: Making sense of fuzzy sets. Fuzzy Sets and Systems 192, 3–24 (2012)
41. Düntsch, I.: A logic for rough sets. Theoretical Computer Science 179, 427–436 (1997)
42. Fariñas del Cerro, L., Orlowska, E.: Dal - a logic for data analysis. Theor. Comput. Sci. 36, 251–264 (1985)
43. Fitting, M.: A Kripke-Kleene semantics for logic programs. J. Log. Program. 2(4), 295–312 (1985)
44. Gaines, B.R.: Foundations of fuzzy reasoning. Int. J. of Man-Machine Studies 6, 623–668 (1976)
45. Gehrke, M., Walker, E.: On the structure of Rough Sets. Bulletin Polish Academy of Science (Mathematics) 40, 235–245 (1992)
46. Gentilhomme, M.Y.: Les ensembles flous en linguistique. Cahiers de Linguistique Théorique et Appliquée, Bucarest 47, 47–65 (1968)
47. Gödel, K.: Zum intuitionistischen Aussagenkalkül. Anzeiger Akademie der Wissenschaften Wien 69, 65–66 (1932)
48. Hájek, P.: Metamathematics of Fuzzy Logic. Kluwer, Dordrecht (1998)
49. Iturrioz, L.: Rough sets and three valued structures. In: Orlowska, E. (ed.) Logic at work, pp. 596–603. Springer (1999)
50. Jaśkowski, S.: Propositional calculus for contradictory deductive systems. Studia Logica 24, 143–160 (1969)
51. Khan, M.A., Banerjee, M.: A logic for complete information systems. In: Sossai, C., Chemello, G. (eds.) ECSQARU 2009. LNCS, vol. 5590, pp. 829–840. Springer, Heidelberg (2009)
52. Kleene, S.C.: Introduction to Metamathematics. North–Holland Pub. Co., Amsterdam (1952)
53. Lawry, J., González Rodríguez, I.: A bipolar model of assertability and belief. Int. J. Approx. Reasoning 52(1), 76–91 (2011)
54. Mas, M., Mayor, G., Torrens, J.: t-Operators and Uninorms on a Finite Totally Ordered Set. International Journal of Intelligent Systems 14, 909–922 (1999)
55. Nelson, D.: Constructible Falsity. J. of Symbolic Logic 14, 16–26 (1949)
56. Orlowska, E.: A logic of indiscernibility relations. In: Skowron, A. (ed.) SCT 1984. LNCS, vol. 208, pp. 177–186. Springer, Heidelberg (1985)
57. Orlowska, E., Pawlak, Z.: Representation of nondeterministic information. Theor. Comput. Sci. 29, 27–39 (1984)
58. Pagliani, P.: Rough Sets and Nelson Algebras. Fundamenta Informaticae 27(2,3), 205–219 (1996)
59. Pagliani, P.: Rough set theory and logic–algebraic structures. In: Incomplete Information: Rough Set Analysis, pp. 109–190. Physica–Verlag, Heidelberg (1998)
60. Pawlak, Z.: Rough sets and fuzzy sets. Fuzzy Sets and Systems 17, 99–102 (1985)
61. Pawlak, Z., Skowron, A.: Rudiments of rough sets. Information Sciences 177, 3–27 (2007)
62. Pearce, D.: Equilibrium logic. Annals of Mathematics and Artificial Intelligence 47, 3–41 (2006)
63. Priest, G.: The logic of paradox. The Journal of Philosophical Logic 8, 219–241 (1979)
64. Sen, J., Chakraborty, M.: A study of interconnections between rough and 3-valued Lukasiewicz logics. Fundam. Inform. 51, 311–324 (2002)

65. Sette, A.M.: On propositional calculus P_1. Math. Japon. 16, 173–180 (1973)
66. Shapiro, S.: Vagueness in Context. Oxford University Press (2006)
67. Smith, K.C.: The prospects for multivalued logic: A technology and applications view. IEEE Trans. Computers 30(9), 619–634 (1981)
68. Sobociński, B.: Axiomatization of a partial system of three-value calculus of propositions. J. of Computing Systems 1, 23–55 (1952)
69. Vakarelov, D.: A modal logic for similarity relations in Pawlak knowledge representation systems. Fundam. Inform. 15(1), 61–79 (1991)
70. Vakarelov, D.: Modal logics for knowledge representation systems. Theor. Comput. Sci. 90(2), 433–456 (1991)
71. Walker, E.A.: Stone algebras, conditional events, and three valued logic. IEEE Transactions on Systems, Man, and Cybernetics 24(12), 1699–1707 (1994)
72. Yao, Y.: Interval sets and interval-set algebras. In: Proceedings of the 8th IEEE International Conference on Cognitive Informatics, pp. 307–314 (2009)
73. Yao, Y.Y.: Interval-set algebra for qualitative knowledge representation. In: Abou-Rabia, O., Chang, C.K., Koczkodaj, W.W. (eds.) ICCI, pp. 370–374. IEEE Computer Society (1993)
74. Yao, Y.: Semantics of fuzzy sets in rough set theory. In: Peters, J.F., Skowron, A., Dubois, D., Grzymała-Busse, J.W., Inuiguchi, M., Polkowski, L. (eds.) Transactions on Rough Sets II. LNCS, vol. 3135, pp. 297–318. Springer, Heidelberg (2004)
75. Yao, Y., Lin, T.: Generalization of rough sets using modal logic. Automation and Soft Computing 2(2), 103–120 (1996)
76. Yao, Y., Wang, S., Lin, T.: A review of rough set models. In: Polkowski, L., Skowron, A. (eds.) Rough Sets and Data Mining: Analysis for Imprecise Data, pp. 47–75. Kluwer Academic Publishers (1997)
77. Zadeh, L.A.: The concept of a linguistic variable and its application to approximate reasoning — part I. Information Sciences 8, 199–251 (1975)

Standard Errors of Indices in Rough Set Data Analysis

Günther Gediga[1] and Ivo Düntsch[2,*]

[1] Department of Psychology, Institut IV, Universität Münster
Fliednerstr. 21, Münster, Germany
gediga@uni-muenster.de
[2] Brock University, St. Catharines, Ontario, Canada, L2S 3A1
duentsch@brocku.ca

Abstract. The sample variation of indices for approximation of sets in the context of rough sets data analysis is considered. We consider the γ and α indices and some other ones – lower and upper bound approximation of decision classes. We derive confidence bounds for these indices as well as a two group comparison procedure. Finally we present procedures to compare the approximation quality of two sets within one sample.

1 Introduction

Rough sets were introduced by Pawlak [1] as a means to approximate sets relative to a given granulation of knowledge: If U is a finite set, $X \subseteq U$, and θ an equivalence relation on U, the *lower approximation of X* (with respect to θ), denoted by $Low(X)$, is the union of all classes of θ contained in X, and the *upper approximation of X*, denoted by $Upp(X)$, is the union of all classes whose intersection with X is not empty. Intuitively, $Low(X)$ is the set of elements which are certainly in X and $Upp(X)$ is the set of elements which are possibly in X with the knowledge given by the granularity provided by the classes of θ. $U \setminus Upp(X)$ is the set of elements certainly not in X. The *boundary of X* is the set $Upp(X) \setminus Low(X)$. A *rough set* is a pair $\langle Low(X), Upp(X) \rangle$, where $X \subseteq U$.

The standard data structures of rough set data analysis (RSDA) are information systems, a relative of relational databases: An *information system* is a structure $\mathscr{I} = \langle U, \Omega, \{V_p : p \in \Omega\}, \{f_p : p \in \Omega\} \rangle$ where U is a nonempty finite set of objects, Ω a nonempty finite set of attributes, each V_p the set of values that attribute p can take, and $f_p : U \to V_p$ an information function. A *decision system* is a special kind of information system, where one or more attributes are singled out as decision attributes; in this case, we write $\mathscr{I} = \langle U, \Omega, d, \{V_p : p \in \Omega\}, V_d, \{f_p : p \in \Omega\}, f_d \rangle$, and the attributes in Ω predictor (or independent) attributes, and d the decision attribute. It is one aim of RSDA to find optimal attribute sets which explain – or approximate – the decision attribute, in other words, with which accuracy membership in a prescribed decision class can be predicted.

* Ivo Düntsch gratefully acknowledges support by the Natural Sciences and Engineering Research Council of Canada.

J.F. Peters and A. Skowron (Eds.): Transactions on Rough Sets XVII, LNCS 8375, pp. 33–47, 2014.

Each set P of attributes induces an equivalence relation θ_P on U by

$$x\theta_P y \iff (\forall p \in P)[f_p(x) = f_p(y)].$$

Similarly, the decision attribute induces the equivalence relation θ_d. An equivalence class X of θ_Y is called *deterministic* if there is an equivalence class Y of θ_d such that $X \subseteq Y$.

Two statistics are prevalent in RSDA to measure success (or failure): Suppose that $P \subseteq \Omega$ and $Y \subseteq U$.

$$\gamma(P) = \frac{|\bigcup\{X : X \text{ is a deterministic class of } \theta_P\}|}{|U|} \qquad \text{Approximation quality,}$$

$$\alpha(Y) = \frac{|Low(Y)|}{|Upp(Y)|} \qquad \text{Accuracy.}$$

Indices such as α and γ are used to measure the quality of approximation. One should be aware, however, that these indices are point estimates of parameters based on one specific sample, and a generalization to a second sample with exactly the same structure is usually not valid. In the sequel we assume that the underlying structure remains the same, but the sample generation is based on simple random sampling. In this way we distinguish between the true index in the population (e.g. γ) and the estimated index in a sample (e.g. $\hat{\gamma}$). As we fix the structure of deterministic and indeterministic rules, only the occurrence of the rules is based on random variation. In other words, the relative frequencies of the elements (of e.g. the lower bounds) can be described by a multinomial random process.

The basic assumption is that the estimated α or γ values are based on frequencies of categories (like lower bound or the boundary of a set) which are sampled from a population. The frequencies of these categories are assumed to be multinomial distributed, which simply means that the frequencies of elements which can be assigned to category in a sample depends only on a constant probability.

Using this rather natural – and simplest – assumption, we can use standard methods of statistics to estimate the errors of various indices connected to RSDA. In order to keep to the spirit of the rough set model, such estimation will be done by nonparametric methods, see e.g. [2].

2   Statistical Notation and Preliminaries

Suppose that U is a base set, and $Y_1, \ldots, Y_i, \ldots, Y_k, k \geq 2$, is a partition $\mathscr{P}$ of U. We think of the classes Y_i as decision classes; they will also be called *categories*. We let $|U| = n$ and $|Y_i| = n_i$; furthermore, $nl_i = |Low(Y_i)|$ and $nu_i = |Upp(Y_i)|$.

For each $1 \leq i \leq k$, $\Pi_i : \mathscr{P} \to \{0, 1\}$ is a random variable which can take the values 0 or 1 such that $\Pi_i(x) = 1 \iff x \in Y_i$ randomly. π_i is the expectation of Π_i and $\hat{\pi}_i$ an estimator of π_i, for example, the well known (and natural) Maximum Likelihood estimator $\hat{\pi}_i = \frac{n_i}{n}$, which we shall use throughout the paper.

Suppose that $f(\Pi_1, \ldots \Pi_k)$ is a real–valued function ("statistic") of multinomially distributed random variables; therefore, we assume that $\Pi_i = 1$ implies $\Pi_j = 0$ for all $j \neq i$. The expectation of f is denoted by $E(f) = f(\pi_1, \ldots \pi_k)$, and $\hat{f}(\pi_1, \ldots \pi_k) = f(\hat{\pi}_1, \ldots \hat{\pi}_k)$ is an estimator of $f(\pi_1, \ldots \pi_k)$. The variance of $\hat{f}$ is denoted by $Var(\hat{f})$, and $SE(\hat{f}) = \sqrt{Var(\hat{f})}$ is the standard estimate error (deviation) of the expectation of f. Note that $Var(\hat{f})$ and $SE(\hat{f})$ are functions of $\pi_1, \ldots, \pi_k$. $\widehat{Var}(\hat{f})$ is an estimator of the variance of $\hat{f}(\pi_1, \ldots \pi_k)$, and $\widehat{SE}(\hat{f})$ is an estimator of the standard error of $\hat{f}(\pi_1, \ldots \pi_k)$. Both $\widehat{Var}(\hat{f})$ and $\widehat{SE}(\hat{f})$ are functions of $\hat{\pi}_1, \ldots, \hat{\pi}_k$, respectively, $n_1, \ldots, n_k$.

In the following we will use simple random sampling to simulate the composition of different samples. Since we may consider the frequencies of more than two classes, our method of choice is multinomial sampling. We shall use the Delta method ([3], [4] p. 587-591) to find the variances of expectations of nonlinear functions of random variables: Assuming that $f(\pi_1, \ldots, \pi_k)$ is partially differentiable in all variables, we can approximate the variance of the estimator $\hat{f} = f(\hat{\pi}_1, \ldots, \hat{\pi}_k)$ by the first order Taylor series expansion around the observed empirical means for the estimation of $\overline{\pi}_i$ as

$$f(\hat{\pi}_1, \ldots, \hat{\pi}_k) \approx f(\overline{\hat{\pi}}_1, \ldots, \overline{\hat{\pi}}_k) + \sum_{i=1}^{k} \frac{\partial f}{\partial x_i}(\overline{\hat{\pi}}_1, \ldots, \overline{\hat{\pi}}_k) \cdot (\hat{\pi}_i - \overline{\hat{\pi}}_i),$$

$$Var(f(\hat{\pi}_1, \ldots, \hat{\pi}_k)) \approx E\left[(f(\hat{\pi}_1, \ldots, \hat{\pi}_k) - f(\overline{\hat{\pi}}_1, \ldots, \overline{\hat{\pi}}_k))^2\right]$$

$$\approx E\left[(\sum_{i=1}^{k} \frac{\partial f}{\partial \pi_i}(\overline{\hat{\pi}}_1, \ldots, \overline{\hat{\pi}}_k) \cdot (\hat{\pi}_i - \overline{\hat{\pi}}_i))^2\right].$$

Note that $\frac{\partial f}{\partial \pi_i}(\overline{\hat{\pi}}_1, \ldots, \overline{\hat{\pi}}_k)$ is a constant as no random variable is included. From the basic facts of probability theory that

$$Var[\sum_i Y_i] = \sum_i Var[Y_i] + 2\sum_{i>j} COV[Y_i, Y_j],$$

$$Var[cY] = c^2 Y,$$

$$COV[cY_i, dY_j] = c \cdot d\, COV[Y_i, Y_j],$$

we arrive at

$$Var(f(\hat{\pi}_1, \ldots, \hat{\pi}_k)) \approx \sum_{i=1}^{k} Var(\hat{\pi}_i) \cdot \left(\frac{\partial f}{\partial \pi_i}\right)^2 + 2\sum_{i>j} COV(\hat{\pi}_i, \hat{\pi}_j) \cdot \frac{\partial f}{\partial \pi_i} \cdot \frac{\partial f}{\partial \pi_j}$$

$$= \sum_{i=1}^{k} SE(\hat{\pi}_i)^2 \cdot \left(\frac{\partial f}{\partial \pi_i}\right)^2 + 2\sum_{i>j} COV(\hat{\pi}_i, \hat{\pi}_j) \cdot \frac{\partial f}{\partial \pi_i} \cdot \frac{\partial f}{\partial \pi_j}.$$

If the sample size is n and the π_i are the probabilities in the multinomial model, we obtain

$$Var(\hat{f}) \approx \sum_{i=1}^{k} \frac{\pi_i(1 - \pi_i)}{n} \left(\frac{\partial f}{\partial \pi_i}\right)^2 - 2\sum_{i>j} \frac{\pi_i \pi_j}{n} \cdot \frac{\partial f}{\partial \pi_i} \cdot \frac{\partial f}{\partial \pi_j}.$$

To estimate $\widehat{Var}(\hat{f})$ we substitute the estimates $\overline{\hat{\pi}}_i$ for π_i. The estimator of the standard error now is obtained – as usual – by $\widehat{SE}(\hat{f}) = \sqrt{\widehat{Var}(\hat{f})}$.

In this paper we are concerned with the estimator of the standard error $\widehat{SE}(\hat{f})$. Since in multinomial sampling we may assume that $\hat{f}$ is approximately normally distributed for moderately large n, we obtain the 95% confidence interval as $[\hat{f} - 1.96 \cdot \widehat{SE}(\hat{f}), \hat{f} + 1.96 \cdot \widehat{SE}(\hat{f})]$.

3   Simple Precision and the Approximation Quality

Throughout suppose that $Y_1, \ldots, Y_i, \ldots, Y_k$ are decision classes and that lower and upper approximations are taken with respect to θ_P for a fixed attribute set P. We usually write γ instead of $\gamma(P)$.

3.1   Approximation Quality and Accuracy

Given a decision class Y_i, the relative precision of deterministic membership in Y_i is defined by

$$p_i := \frac{|Low(Y_i)|}{|Y_i|} = 1 - \frac{|Y_i| - |Low(Y_i)|}{|Y_i|},$$

i.e. the percentage of correctly predicted elements of Y_i, whereas $1 - p_i$ is the percentage of elements of Y_i which cannot be predicted by the deterministic rules. The classical rough approximation quality is now the weighted sum

$$\gamma = \sum_{i=1}^{k} \frac{|Y_i|}{|U|} \cdot p_i,$$

see [5]. Similarly, one can define a precision index for the upper approximation by

$$p^i = \frac{|Y_i|}{|Upp(Y_i)|} = 1 - \frac{|Upp(Y_i)| - |Y_i|}{|Upp(Y_i)|},$$

which is the percentage of elements possibly in Y_i as captured by indeterministic rules. Furthermore, $1 - p^i$ is the percentage of those elements certainly not in Y_i. Since $0 \leq |Y_i| \leq |Upp(Y_i)| \leq n$, we see that always $p^i \neq 0$ unlike p_i.

As mentioned in Section 1, the accuracy is usually defined by the index

$$\alpha_i = \frac{|Low(Y_i)|}{|Upp(Y_i)|}.$$

We note that α_i is the product of p_i and p^i; therefore, the geometric mean $\sqrt{\alpha_i}$ may be a more adequate index for accuracy than α_i.

Example 1. We will use the data of this example throughout the paper. Suppose we have three decision classes with the frequencies shown in Table 1. Using these data we obtain the values of the indices shown in Table 2.   □

Table 1. Results of a rough set analysis

	Y_1	Y_2	Y_3	Sum
$n_i = \|Y_i\|$	30	50	120	200
$nl_i = \|Low(Y_i)\|$	15	40	90	145
$nu_i = \|Upp(Y_i)\|$	45	60	150	irr.

Table 2. Indices for the results of a rough set analysis

$p_1 = 0.5 \quad p_2 = 0.8 \quad p_3 = 0.75$
$p^1 = 0.667 \quad p^2 = 0.833 \quad p^3 = 0.8$
$\alpha_1 = 0.333 \quad \alpha_2 = 0.667 \quad \alpha_3 = 0.6$
$\gamma = 0.725$

3.2 Standard Error of One Index

Assuming a multinomial model for the cardinalities of the classes, the standard error of (the estimation of) γ is easily seen to be the root of

$$\widehat{Var}(\hat{\gamma}) = \frac{\hat{\gamma}(1 - \hat{\gamma})}{n}$$

Example 2. Continuing Example 1,

$$\widehat{SE}(\hat{\gamma}) = \sqrt{\frac{0.725(1 - 0.725)}{200}} = 0.0326$$

with the 95% confidence interval $[0.633, 0.787]$.

   With the given data and the point estimate $\hat{\gamma} = 0.725$ it is not plausible to assume that γ will fall below 0.633 or rise above 0.787 in a comparable sample. The statement "In this sample, 90% of cases can be predicted by deterministic rules" can be rejected with a two–sided significance of 5%, since the confidence interval does not contain 0.9. Similarly, the hypothesis "In this sample, 60% of cases can be predicted by deterministic rules" can be rejected. In the first case the percentage is too high, while in the second it is too low. □

We use the Delta method for determine the standard errors of the other indices. Starting with p_i we obtain

$$\hat{p}_i = \frac{|low(Y_i)|}{|low(Y_i)| + |Y_i \setminus low(Y_i)|} = \frac{nl_i}{nl_i + (n_i - nl_i)} \tag{3.1}$$

and the parametrization

$$p_i = \frac{a}{a+b} = f(a,b)$$

with the expectations

$$\hat{a} = \frac{|Low(Y_i)|}{n} = \frac{nl_i}{n} \qquad \hat{b} = \frac{|Y_i \setminus Low(Y_i)|}{n} = \frac{(n_i - nl_i)}{n}.$$

Computing the derivatives

$$\frac{\partial p_i}{\partial a} = \frac{a+b-a}{(a+b)^2} = \frac{b}{(a+b)^2} \qquad \frac{\partial p_i}{\partial b} = -\frac{a}{(a+b)^2}$$

we obtain

$$Var(\widehat{p}_i(a,b)) = \frac{a \cdot (1-a)}{n} \cdot \frac{b^2}{(a+b)^4} + \frac{b \cdot (1-b)}{n} \cdot \frac{a^2}{(a+b)^4} + 2 \cdot \frac{a \cdot b}{n} \cdot \frac{a \cdot b}{(a+b)^4}$$

$$= \frac{1}{n(a+b)^4} \cdot \left(a \cdot (1-a) \cdot b^2 + b \cdot (1-b) \cdot a^2 + 2 \cdot a^2 \cdot b^2\right)$$

$$= \frac{a \cdot b^2 + b \cdot a^2}{n \cdot (a+b)^4}$$

$$= \frac{a \cdot b}{n \cdot (a+b)^3}.$$

Substituting the frequencies to find the estimation of $\widehat{Var}(\widehat{p}_i)$, we obtain $\hat{a} + \hat{b} = \frac{n_i}{n}$ and therefore,

$$\widehat{Var}(\widehat{p}_i) \approx \frac{\frac{nl_i}{n} \cdot \frac{(n_i - nl_i)}{n}}{n \cdot \left(\frac{n_i}{n}\right)^3}$$

$$= \frac{nl_i(n_i - nl_i)}{n_i^3}.$$

Example 3. Continuing Example 1,

$$\widehat{SE}(p_1) = \sqrt{\frac{15 \cdot 15}{30^3}} = 0.091 \qquad 95\% \text{ CI} = [0.321, 0.679],$$

$$\widehat{SE}(p_2) = \sqrt{\frac{40 \cdot 10}{50^3}} = 0.057 \qquad 95\% \text{ CI} = [0.689, 0.911],$$

$$\widehat{SE}(p_3) = \sqrt{\frac{90 \cdot 30}{120^3}} = 0.040 \qquad 95\% \text{ CI} = [0.673, 0.827].$$

$\square$

If c is a constant, it is well known that $Var(\widehat{p}_i - c) = Var(\widehat{p}_i)$ holds. Given the hypothesis $p_i = c$, we observe that

$$\frac{\widehat{p}_i - c}{SE(\widehat{p}_i)} \sim N(0,1)$$

is approximately standard normally distributed, and

$$\frac{(\widehat{p}_i - c)^2}{Var(\widehat{p}_i)} \sim \chi_1^2$$

is approximately χ^2 distributed with one degree of freedom. Given k categories with estimates $\widehat{p}_1, ..., \widehat{p}_k$ and assuming that all $p_i = c$, we observe

$$\sum_{i=1}^{k} \frac{(\widehat{p}_i - c)^2}{Var(\widehat{p}_i)} \sim \chi_k^2.$$

Using $\hat{c} = \hat{\gamma}$ instead of a constant c, we observe that $\hat{\gamma}$ is a linear function of $\hat{p}_1, ..., \hat{p}_k$, and thus the number of degrees of freedom of the χ^2 is reduced by one. Therefore,

$$\sum_{i=1}^{k} \frac{(\hat{p}_i - \hat{\gamma})^2}{Var(\hat{p}_i)} \sim \chi^2_{k-1}.$$

Example 4. Continuing Example 1, we shall test whether the p_i are approximately the same – in other words, how well γ measures the approximation quality for a particular decision class.

$$\chi^2 = \left(\frac{0.5 - 0.725}{0.091}\right)^2 + \left(\frac{0.8 - 0.725}{0.057}\right)^2 + \left(\frac{0.75 - 0.725}{0.040}\right)^2$$
$$= (-2.473)^2 + 1.316^2 + 0.625^2 = 8.23$$

Since the critical value $\chi^2_2(.95) = 5.99$, we see that the approximation of the three sets differ – and as the approximation of the first set differs by -2.473 standard deviations from the mean value, we further conclude that the first class is not as well approximated as the other classes.

Given this result it is problematic to state that γ is <u>the</u> approximation quality for the classes of the decision attributes, as – obviously – the approximation differs among the classes.   □

Next, we turn to α_i, and note that the structure of formula 3.1 is repeated in the estimation of α_i:

$$\hat{\alpha}_i = \frac{|Low(Y_i)|}{|Upp(Y_i)|} = \frac{|Low(Y_i)|}{|Low(Y_i)| + |Bnd(Y_i)|} = \frac{nl_i}{nl_i + nb_i}.$$

Therefore there is no need for further application of the Delta method – we simply replace the frequencies for the estimation of $\widehat{Var}(\hat{\alpha}_i)$ and obtain

$$\widehat{Var}(\hat{\alpha}_i) = \frac{nl_i \cdot nb_i}{nu_i^3}.$$

Example 5. Continuing Example 1,

$$\widehat{SE}(\alpha_1) = \sqrt{\frac{15 \cdot 30}{45^3}} = 0.070 \qquad 95\% \text{ CI} = [0.196, 0.471],$$

$$\widehat{SE}(\alpha_2) = \sqrt{\frac{40 \cdot 20}{60^3}} = 0.061 \qquad 95\% \text{ CI} = [0.547, 0.786],$$

$$\widehat{SE}(\alpha_3) = \sqrt{\frac{90 \cdot 60}{150^3}} = 0.040 \qquad 95\% \text{ CI} = [0.522, 0.678].$$

□

Finally, the standard error of p^i has to be estimated. Once again, we see that

$$\hat{p}^i = \frac{|Y_i|}{|Upp(Y_i)|} = \frac{|Y_i|}{|Y_i| + |Upp Y_i \setminus Y_i|}$$

has the same structure as formula 3.1, and we obtain

$$\widehat{Var}(\hat{p^i}) = \frac{n_i(nu_i - n_i)}{(n_i + (nu_i - n_i))^3}$$
$$= \frac{n_i(nu_i - n_i)}{nu_i^3}.$$

Example 6. Continuing Example 1,

$$\widehat{SE}(\hat{p^1}) = \sqrt{\frac{30(45-30)}{45^3}} = 0.07 \qquad 95\% \ CI = [0.529, 0.804],$$

$$\widehat{SE}(\hat{p^2}) = \sqrt{\frac{50(60-50)}{60^3}} = 0.048 \qquad 95\% \ CI = [0.739, 0.927],$$

$$\widehat{SE}(\hat{p^3}) = \sqrt{\frac{120(150-120)}{150^3}} = 0.033 \quad 95\% \ CI = [0.736, 0.864]. \qquad \square$$

4 Two Sample Comparisons

Having solved the problem of variation in one sample, it is straightforward to solve the comparison of indices which arise from two different samples S_1 and S_2. For an index f, we form the difference $f(S_1, S_2) := f(S_1) - f(S_2)$. Assuming independent sampling, the standard error of the difference estimator is

$$SE(\hat{f}(S_1, S_2)) = \sqrt{Var(\hat{f}(S_1)) + Var(\hat{f}(S_2))}.$$

This result holds regardless which index is used. Note that $SE(\hat{f}(S_1, S_2))$ can be used to test the difference to be zero (or any other value in $[-1, 1]$), which allows us in particular to test whether $f(S_1) > f(S_2)$.

Example 7. A second data set using the same decision variable and the same rules shows the following results:

	Y_1	Y_2	Y_3	sum
$n_i = \|Y_i\|$	40	40	100	180
$nl_i = \|Low(Y_i)\|$	10	35	60	105
$nu_i = \|Upp(Y_i)\|$	60	60	135	irr.

The results are shown in Table 3 on the facing page. $\qquad \square$

5 Comparing Categories

The questions arise whether a category Y_i is approximated better than a class Y_j, or whether a category Y_i is better approximated by one sample than by another when the rule system is the same. Furthermore, one needs to ask whether Y_i is better approximated by rule system A than by rule system B using the same sample.

Table 3. Indices for two samples

	Sample 1	Sample 2	Difference
p_1	0.5	0.25	0.25
SE	0.070	0.068	0.114
95%-CI	$[.362,.638]$	$[.116,.384]$	$[.026,.474]$
p_2	0.8	0.875	-0.075
SE	0.048	0.052	0.077
95%-CI	$[.706,.894]$	$[.773,.977]$	$[-.226,.076]$
p_3	0.75	0.6	0.15
SE	0.033	0.049	0.063
95%-CI	$[.686,.814]$	$[.504,.696]$	$[.027,.273]$
γ	0.725	0.583	0.142
SE	0.032	0.037	0.048
95%-CI	$[.663,.787]$	$[.511,.655]$	$[.048,.237]$
α_1	.333	.167	.167
SE	0.070	0.048	0.085
95%-CI	$[.196,.471]$	$[.072,.261]$	$[-.000,.334]$
α_2	.667	.583	.083
SE	0.061	0.064	0.088
95%-CI	$[.547,.786]$	$[.459,.708]$	$[-.089,.256]$
α_3	.6	.444	.156
SE	0.04	0.043	0.059
95%-CI	$[.522,.678]$	$[.361,.528]$	$[.041,.270]$

A note on a technical detail in this section: As we deal with dependent random variates based on frequencies, it turns out that determining the variance of a fraction of frequencies is most of the time more convenient than working with the differences. Furthermore, taking the logarithm reduces variance heterogeneity between both variates and normalizes the resulting random variable. This is a valuable property because the computation of the confidence intervals needs a good approximation by the standard normal distribution. Note that this approach is similar to the treatment of odds and odds ratios ([4], p. 70). As above, we shall use the Delta method.

5.1 Comparing p_i

To compare p_i and p_j we use the ratio $\frac{p_i}{p_j}$, and obtain

$$
\frac{p_i}{p_j} = \frac{|Low(Y_i)|/|Y_i|}{|Low(Y_j)|/|Y_j|}
$$
$$
= \frac{|Low(Y_i)|/(|Low(Y_i)| + |Y_i \setminus Low(Y_i)|)}{|Low(Y_j)|/(|Low(Y_j)| + |Y_j \setminus Low(Y_j)|)}.
$$

This leads to the parametrization

$$
\frac{p_i}{p_j} = \frac{a/(a+b)}{c/(c+d)},
$$

and, linearizing by taking the logarithm,

$$f(a,b,c,d) = \ln\left(\frac{p_i}{p_j}\right) = \ln(a) - \ln(a+b) - \ln(c) + \ln(c+d)$$

Using the estimators

$$\hat{a} = \frac{|Low(Y_i)|}{n} \qquad\qquad \hat{b} = \frac{|Y_i| - |Low(Y_i)|}{n}$$

$$\hat{c} = \frac{|Low(Y_j)|}{n} \qquad\qquad \hat{d} = \frac{|Y_j| - |Low(Y_j)|}{n}$$

and the partial derivatives

$$\frac{\partial f}{\partial a} = \frac{1}{a} - \frac{1}{a+b} \qquad\qquad \frac{\partial f}{\partial c} = -\frac{1}{c} + \frac{1}{c+d}$$

$$\frac{\partial f}{\partial b} = -\frac{1}{a+b} \qquad\qquad \frac{\partial f}{\partial d} = \frac{1}{c+d}$$

we can compute the variance

$$
\begin{aligned}
Var(\hat{f}) &= \frac{a\cdot(1-a)}{n}\left(\frac{1}{a} - \frac{1}{a+b}\right)^2 + \frac{b\cdot(1-b)}{n}\left(-\frac{1}{a+b}\right)^2 \\
&+ \frac{c\cdot(1-c)}{n}\left(-\frac{1}{c} + \frac{1}{c+d}\right)^2 + \frac{d\cdot(1-d)}{n}\left(\frac{1}{c+d}\right)^2 \\
&- 2\frac{a\cdot c}{n}\left(\frac{1}{a} - \frac{1}{a+b}\right)\left(-\frac{1}{c} + \frac{1}{c+d}\right) - 2\frac{a\cdot b}{n}\left(\frac{1}{a} - \frac{1}{a+b}\right)\left(-\frac{1}{a+b}\right) \\
&- 2\frac{a\cdot d}{n}\left(\frac{1}{a} - \frac{1}{a+b}\right)\left(\frac{1}{c+d}\right) - 2\frac{c\cdot b}{n}\left(-\frac{1}{c} + \frac{1}{c+d}\right)\left(-\frac{1}{a+b}\right) \\
&- 2\frac{c\cdot d}{n}\left(-\frac{1}{c} + \frac{1}{c+d}\right)\left(\frac{1}{c+d}\right) - 2\frac{b\cdot d}{n}\left(-\frac{1}{a+b}\right)\left(\frac{1}{c+d}\right) \\
&= \frac{1}{n}\left(\frac{b}{a\cdot(a+b)} + \frac{d}{c\cdot(c+d)}\right)
\end{aligned}
$$

Substituting the estimators we obtain

$$
\begin{aligned}
\widehat{Var}(\hat{f}) &= \frac{n_i - nl_i}{nl_i n_i} + \frac{n_j - nl_j}{nl_j n_j} \\
&= \frac{1}{nl_i} - \frac{1}{n_i} + \frac{1}{nl_j} - \frac{1}{n_j}.
\end{aligned}
$$

The next task is to check how one category differs from the others. In terms of accuracy the question is whether p_i is different from the (weighted) mean of the accuracy of the other categories, which is γ restricted to $U \setminus Y_i$. We first define the sum of the cardinalities of the lower bounds of the other classes by

$$Nl_i = \sum_{j\neq i} nl_j$$

and then restrict γ to these classes:

$$\hat{\gamma}_i = \frac{Nl_i}{n - n_i}$$

The comparison

$$\frac{\hat{p}_i}{\hat{\gamma}_i} = \frac{nl_i/n_i}{Nl_i/(n - n_i)}$$

can be described by the parametrization

$$f(a,b,c) = \ln(a) - \ln(a+b) - \ln(c) + \ln(1-a-b)$$

with the estimates

$$\hat{a} = nl_i/n$$
$$\hat{b} = (n_i - nl_i)/n$$
$$\hat{c} = 1/n \sum_{j \neq i} nl_j.$$

Once again the Delta method is used. First, we find the derivatives:

$$\frac{\partial f}{\partial a} = \frac{1}{a} - \frac{1}{a+b} - \frac{1}{1-a-b} = \frac{1}{a} - \frac{1}{(a+b)(1-a-b)}$$

$$\frac{\partial f}{\partial b} = -\frac{1}{(a+b)(1-a-b)}$$

$$\frac{\partial f}{\partial c} = -\frac{1}{c},$$

and then obtain

$$Var(\hat{f}) = \frac{a \cdot (1-a)}{n} \cdot \left(\frac{1}{a} - \frac{1}{(a+b) \cdot (1-a-b)} \right)^2$$

$$+ \frac{b \cdot (1-b)}{n} \left(\frac{1}{(a+b) \cdot (1-a-b)} \right)^2 + \frac{1-c}{n \cdot c}$$

$$+ \frac{2 \cdot a \cdot b}{n} \cdot \left(\frac{1}{a} - \frac{1}{(a+b) \cdot (1-a-b)} \right) \cdot \left(\frac{1}{(a+b) \cdot (1-a-b)} \right)$$

$$+ \frac{2 \cdot a}{n} \cdot \left(\frac{1}{a} - \frac{1}{(a+b) \cdot (1-a-b)} \right) - \frac{2 \cdot b}{n} \cdot \left(\frac{1}{(a+b) \cdot (1-a-b)} \right)$$

$$= \frac{1}{n} \cdot \frac{b \cdot c \cdot (1-b) + a \cdot b \cdot (1-b-2 \cdot c) + a^2 \cdot (1-2 \cdot b - c) - a^3}{a \cdot c \cdot (1-a-b) \cdot (a+b)}.$$

Example 8. In our example we compare p_1 with p_2 and p_3:

Comparison	Fraction	$Var(p_1/p_j)$	Standard error	95%CI
$\hat{p}_1/\hat{p}_2$	0.625	0.038	0.196	$[0.426, 0.917]$
$\hat{p}_1/\hat{p}_3$	0.667	0.036	0.190	$[0.459, 0.968]$

We observe that the approximation quality of category Y_1 is in fact lower than those of the other categories.

$\hat{\gamma}_i$	Comparison	Fraction	$Var(p_i/\gamma_i)$	Standard error	95%CI
0.765	$\hat{p}_1/\hat{\gamma}_1$	0.654	0.035	0.187	$[0.453, 0.944]$
0.700	$\hat{p}_2/\hat{\gamma}_2$	1.143	0.008	0.089	$[0.961, 1.360]$
0.688	$\hat{p}_3/\hat{\gamma}_3$	1.091	0.008	0.092	$[0.911, 1.306]$

We conclude that p_1 is smaller than the mean approximation quality of the other categories, as the right border of the 95%CI is smaller than 1.   □

5.2 Comparing α_i

For the upper bounds of two different classes we find the following disjoint representation of the upper bounds; disjoint union is denoted by $\uplus$. Observe that for $i \neq j$ we have $Low(Y_i) \cap Upp(Y_i) = \emptyset$.

$$Upp(Y_i) = Low(Y_i) \uplus (Upp(Y_i) \cap Upp(Y_j)) \uplus [Upp(Y_i) \setminus ((Upp(Y_i) \cap Upp(Y_j)) \cup Low(Y_i))],$$
$$Upp(Y_j) = Low(Y_j) \uplus (Upp(Y_i) \cap Upp(Y_j)) \uplus [Upp(Y_j) \setminus ((Upp(Y_i) \cap Upp(Y_j)) \cup Low(Y_j))].$$

The parametrization of $\frac{\alpha_i}{\alpha_j}$ is therefore

$$f(a,b,c,d,e) = \ln\left(\frac{\frac{a}{a+b+c}}{\frac{d}{d+b+e}}\right)$$

where

$$\hat{a} = \frac{|Low(Y_i)|}{n} \qquad\qquad \hat{b} = \frac{|Upp(Y_i) \cap Upp(Y_j)|}{n}$$

$$\hat{c} = \frac{|Upp(Y_i) \setminus ((Upp(Y_i) \cap Upp(Y_j)) \cup Low(Y_i))|}{n} \qquad \hat{d} = \frac{|Low(Y_j)|}{n}$$

$$\hat{e} = \frac{|Upp(Y_j) \setminus ((Upp(Y_i) \cap Upp(Y_j)) \cup Low(Y_j))|}{n}.$$

The partial derivatives are

$$\frac{\partial f}{\partial a} = \frac{1}{a} - \frac{1}{a+b+c} \qquad\qquad \frac{\partial f}{\partial b} = -\frac{1}{a+b+c} + \frac{1}{d+b+e}$$
$$\frac{\partial f}{\partial c} = -\frac{1}{a+c+d} \qquad\qquad \frac{\partial f}{\partial d} = -\frac{1}{d} + \frac{1}{d+b+e}$$
$$\frac{\partial f}{\partial e} = \frac{1}{d+b+e}.$$

Computing $Var(\hat{f})$ is straightforward, if somewhat tedious; we have used Mathematica [6] for the last line.

$$Var(\hat{f}) = \frac{a(1-a)}{n}\left(\frac{1}{a} - \frac{1}{a+b+c}\right)^2 + \frac{b(1-b)}{n}\left(-\frac{1}{a+b+c} + \frac{1}{d+b+e}\right)^2$$

$$+ \frac{c(1-c)}{n}\left(-\frac{1}{a+b+c}\right)^2 + \frac{d(1-d)}{n}\left(-\frac{1}{d} + \frac{1}{d+b+e}\right)^2$$

$$+ \frac{e(1-e)}{n}\left(\frac{1}{d+b+e}\right)^2$$

$$- 2\frac{a\cdot b}{n}\left(\frac{1}{a} - \frac{1}{a+b+c}\right)\left(-\frac{1}{a+b+c} + \frac{1}{d+b+e}\right)$$

$$- 2\frac{a\cdot c}{n}\left(\frac{1}{a} - \frac{1}{a+b+c}\right)\left(-\frac{1}{a+b+c}\right)$$

$$- 2\frac{a\cdot d}{n}\left(\frac{1}{a} - \frac{1}{a+b+c}\right)\left(-\frac{1}{d} + \frac{1}{d+b+e}\right)$$

$$- 2\frac{a\cdot e}{n}\left(\frac{1}{a} - \frac{1}{a+b+c}\right)\left(\frac{1}{d+b+e}\right)$$

$$- 2\frac{b\cdot c}{n}\left(-\frac{1}{a+b+c} + \frac{1}{d+b+e}\right)\left(-\frac{1}{a+b+c}\right)$$

$$- 2\frac{b\cdot d}{n}\left(-\frac{1}{a+b+c} + \frac{1}{d+b+e}\right)\left(-\frac{1}{d} + \frac{1}{d+b+e}\right)$$

$$- 2\frac{b\cdot e}{n}\left(-\frac{1}{a+b+c} + \frac{1}{d+b+e}\right)\left(\frac{1}{d+b+e}\right)$$

$$- 2\frac{c\cdot d}{n}\left(-\frac{1}{a+b+c}\right)\left(-\frac{1}{d} + \frac{1}{d+b+e}\right)$$

$$- 2\frac{c\cdot e}{n}\left(-\frac{1}{a+b+c}\right)\left(\frac{1}{d+b+e}\right)$$

$$- 2\frac{d\cdot e}{n}\left(-\frac{1}{d} + \frac{1}{d+b+e}\right)\left(\frac{1}{d+b+e}\right)$$

$$= \frac{a^2\cdot(b+e) + (b+c)\cdot d\cdot(b+d+e) + a\cdot(b^2+c\cdot e+b\cdot(c-2\cdot d+e))}{n(a\cdot(a+b+c)\cdot d\cdot(b+d+e))}$$

Example 9. For our example we compare α_1 with α_2 and α_3, assuming an overlap of the upper bounds of 5 elements.

Comparison	Fraction	$Var(\alpha_1/\alpha_j)$	Standard error	95%CI
α_1/α_2	0.500	0.049	0.222	$[0.324, 0.772]$
α_1/α_3	0.556	0.047	0.218	$[0.363, 0.851]$

If we take the upper bounds in account, we find a similar result as before: The approximation of category Y_1 is not as good as the approximation of the other categories.

□

5.3   Comparing p^i-Values

It is straightforward to find a parametrization of

$$\frac{\hat{p}^i}{\hat{p}^j} = \frac{|Y_i|/|Upp(Y_i)|}{|Y_j|/|Upp(Y_j)|}$$

by using

$$\frac{p^i}{p^j} = \frac{(a+b)/(a+b+c+d+e)}{(e+f)/(a+d+e+f+g)}$$

But as p^i and its relatives have not been explored up to now, we skip the developments of the standard error of this fraction.

6   Discussion

The paper discusses how sample variation influences indices of rough set approximation. We use the lower approximation p_i of a class Y_i, which is the one-set-counterpart of γ, and the upper approximation p^i of a class Y_i, which is a new construction as far as we know. The "classical" indices γ and α_i are derivations of p_i and p^i.

Computing the one sample variation of these indices results in a confidence interval for the indices based on the assumption of simple random sampling. A 95% confidence interval will cover the population value of the respective index with probability .95, and will tell us something about the mutual position of the population index. Any value outside the interval is quite unreasonable, but – and this is at least as interesting – any value within the CI is plausible. Inspecting the left end of the CI, we see whether a high value of the point estimate is really "high" in the population. Inspecting the right end of the CI tells us whether a low point estimate is really "low" in the population.

Once we have generated standard errors of indices in one sample, it is straightforward to compare the results of two samples. This helps us to check the reliability of a learned rule system, using e.g. a simple hold-out-sample technique.

Finally, we propose some procedures to compare the approximation of two categories in one sample, which may be useful the check the relative precision of a rule system with respect of categories.

Using real life data, we have to consider possible problems with the approach:

1. The applied techniques are based on asymptotically correct approximations of the sampling error. Hence, if the frequencies in the nominator or denominator of the measures) are low, say, less than 20, the approximation cannot be used. In that case a simulation based on the same assumptions of a multinomial model will be a better choice to measure the sampling variance of the estimates.
2. RSDA is often used to find optimal relationships. As the frequencies do not vary in terms of a multinomial distribution, we cannot use the approach. This is not a problem unique to RSDA, but to all methods of machine learning. In this case simple techniques like using a hold-out sample should be used to measure the sampling effect.

3. All the proposed techniques assume a fixed rule system. Therefore, none of these can be used to compare different rule systems – this is a challenge for future research. The same restriction holds for the derivation of confidence intervals for probabilistic enhancements of RSDA such as the variable precision model [7] or the λ-precision model [8].

An R package to compute confidence intervals for indices in rough set data analysis as presented in this paper is available at www.roughsets.net.

References

1. Pawlak, Z.: Rough Sets. Internat. J. Comput. Inform. Sci. 11, 341–356 (1982)
2. Efron, B.: Nonparametric estimates of standard error: The jackknife, the bootstrap and other methods. Biometrika 68, 589–599 (1981)
3. Oehlert, G.: A note on the Delta method. American Statistician 46, 27–29 (1992)
4. Agresti, A.: Categorical Data Analysis, 3rd edn. Wiley, New York (2012)
5. Gediga, G., Düntsch, I.: Rough approximation quality revisited. Artificial Intelligence 132, 219–234 (2001)
6. Research, I. W.: Mathematica Edition: Version 8.0. Wolfram Research, Inc., Champaign (2010)
7. Ziarko, W.: Variable precision rough set model. Journal of Computer and System Sciences 46, 39–59 (1993)
8. Düntsch, I., Gediga, G.: Weighted λ precision models in rough set data analysis. In: Proceedings of the Federated Conference on Computer Science and Information Systems, Wrocław, Poland, pp. 309–316. IEEE (2012)

Proximity System: A Description-Based System for Quantifying the Nearness or Apartness of Visual Rough Sets*

Christopher J. Henry and Garrett Smith

University of Winnipeg, Department of Computer Science,
Winnipeg, Manitoba R3B 2E9, Canada
ch.henry@uwinnipeg.ca, garrettwhsmith@gmail.com

Abstract. This article introduces the Proximity System, an application developed to demonstrate descriptive-based approaches to nearness and proximity within the context of digital image analysis. Specifically, the system implements the descriptive-based intersection, compliment, and difference operations defined on sets of pixels representing regions of interest. These sets of pixels can be considered visual rough sets, since the results of the descriptive-based operators are always defined with respect to a set of probe functions, which induce a partition of the objects (pixels) being considered. The contribution of this article is an overview of the Proximity System, its use of visual rough sets as description-based operands, its ability to quantify the nearness or apartness of visual rough sets, and a practical application to the problem of human visual search.

Keywords: Digital images, visual rough sets, description-based operators, metric-free distance measure, probe functions, image analysis, nearness, neighbourhoods, proximity, human visual search.

1 Introduction

The problem considered in this article is the implementation of descriptive-based operators [17] for use in digital images. The solution to this problem comes by way of the Proximity System, a cross-platform system for digital image analysis using descriptive-based operators. The inspiration for the approach implemented in the Proximity System is an observation in [58] that the concept of nearness[1] is a generalization of set intersection. The idea follows from the notion of set description [36, §4.3], which is a collection of the unique feature vectors (n-dimensional real-valued feature vectors representing characteristics of the objects) associated with all the objects in the set. Describing sets in this manner, at

---

* This research has been supported by the Natural Sciences and Engineering Research Council of Canada (NSERC) grant 418413.
[1] Introduced within the context of a descriptive extension of Efremovič's proximity space theory [56], which carries forward the basic idea of near and far begun by Riesz's [34].

J.F. Peters and A. Skowron (Eds.): Transactions on Rough Sets XVII, LNCS 8375, pp. 48–73, 2014.

some level, matches the human approach to describing sets of objects. Furthermore, in comparing disjoint sets of objects, we must at some level be performing a comparison of the descriptions we associate with the objects within the sets. Thus, a natural approach for quantifying the degree of similarity (*i.e.* the *nearness* or *apartness*) between two sets would be to look at the intersection of the sets containing their unique feature vectors.

The approach reported here builds on the work of many others. The idea of sets of similar sensations was first introduced by Poincaré in which he reflects on experiments performed by Weber in 1834, and Fechner's insight in 1850 [67, 62, 2, 23]. Poincaré's work was inspired by Fechner, but the key difference is Poincaré's work marked a shift from stimuli and sensations to an abstraction in terms of sets together with an implicit idea of tolerance. Next, the idea of tolerance is formally introduced by Zeeman [71] with respect to the brain and visual perception. Zeeman makes the observation that a single eye cannot identify a 2D Euclidean space because the Euclidean plane has an infinite number of points. Instead, we see things only within a certain tolerance. This idea of tolerance is important in mathematical applications where systems deal with approximate input and results are accepted with a tolerable level of error, an observation made by Sossinsky [67], who also connected Zeeman's work with that of Poincaré's. In addition to these ideas on tolerance, Riesz first published a paper in 1908 on the nearness of two sets [34, 37], initiating the mathematical study of proximity spaces and the eventual discovery of descriptively near sets. Specifically, Near set theory was inspired by a collaboration in 2002 by Pawlak and Peters on a poem entitled "How Near" [46], which lead to the introduction of descriptively near sets [49, 50]. Next, tolerance near sets were also introduced by Peters [52, 53], which combines near set theory with the ideas of tolerance spaces and relations. Finally, a tolerance-based nearness measure was introduced in [15, 16].

The sets considered in the Proximity System are obtained from digital images. In particular, four types of regions of interest are identified as operands for the descriptive operators reported in [17], namely a simple set of pixels, a spatial neighbourhood, a descriptive neighbourhood, and a hybrid approach in which the neighbourhood is formed by spatial and descriptive characteristics of the objects [36, 17, 55]. In each case, the regions of interest constitute a set of pixels. In addition, the results of the descriptive-based operators are always defined with respect to a set of probe functions, which induce a partition of the objects (pixels) being considered. As a result, these sets of pixels can be considered in the light of rough set theory, and can be considered visual rough sets.

Rough sets were introduced by Pawlak during the early 1980s [44, 45] and elaborated in [47, 63, 44, 38, 39]. The rough set-based approach to image analysis dates back to the early 1990s. The application of rough sets in image analysis was launched in a seminal paper published by Mrózek and Plonka [32]. The early work on the use of rough sets in image analysis can be found in [40, 57, 42, 4, 3, 24, 33]. A review of rough sets and near sets in medical imaging can be found in [13]. More recently, Sen and Pal [65] introduced an entropy based, average image ambiguity measure to quantify greyness and spatial ambiguities in images. This measure has been used for standard image processing tasks such as enhancement, segmentation and edge detection. Forms of rough entropy have also been used in a clustering approach to image segmentation [27, 30, 41, 10, 28, 29, 31].

The contribution of this article is a discussion on the Proximity System, its use of visual rough sets as description-based operators, its ability to quantify the nearness or apartness of visual rough sets (with examples), and the application of descriptive-based intersection to the problem of Human Visual Search. This article is organized as follows: Section 2 defines the visual rough sets used in the Proximity System, Section 3 provides background on the description-based operators implemented in the Proximity System, Section 4 details the nearness measure used by the Proximity System to assess the nearness or apartness of visual rough sets, Section 5 discusses the Proximity System in detail, Section 6 contains examples using the Proximity System to obtain the descriptive intersection of visual rough sets, and Section 7 discusses the application of the descriptive intersection to the problem of Human Visual Search.

2   Visual Rough Sets

This section briefly presents visual rough sets using an approach similar to [16]. Consider Fig. 1(a) as a starting point for an example of a visual rough set. Each pixel in the image has an associated tuple consisting of three values that specify its colour using the RGB colour model. These values can be used to partition the image using an equivalence relation. In image processing terms, the partition is a segmentation of the image into regions, where the members of each region contain pixels with equal colour values. For example, the unnaturally coloured pixels in Fig. 1(b) represent the partition of Fig. 1(a), which was obtained using the blue component of the RGB colour model. Similarly, an example of a single equivalence class is given by pixels of the colour ■ in Fig. 1(c).

The partition in Fig. 1(b) is obtained by considering a very simple example of an indiscernibility (equivalence) relation[2] $\sim_\phi$ defined by

$$\sim_\phi = \{(x, y) \in \Im \times \Im : \phi(x) = \phi(y)\}, \text{ where,}$$
$$\Im = \{x : x = \text{ digital image pixel}\},$$
$$\phi : X \to \mathbb{R}, \text{ defined by}$$
$$\phi(x) = y(\text{blue component intensity}), \; y \in [0, 255].$$

The notation $x_{/\sim_\phi}$ denotes a equivalence class containing x and $\Im_{/\sim_\phi}$ denotes the set of all equivalence classes (quotient set) of the partition (see, *e.g.*, Fig. 1(c) & 1(b), respectively). The image partition is a rich source of examples

---

[2] The indiscernibility relation introduced by Z. Pawlak [45] was defined in terms of attributes (partial functions) of objects in an information system. With the advent of probe functions in a perceptual view of feature extraction [50, §3, pp. 414-415], a form of perceptual indiscernibility relation was introduced in the context of a perceptual system [59, §2.2, p. 53], elaborated in [60].

(a)   (b)

(c)

Fig. 1. Sample visual rough set. (a) Natural image [1], (b) partition, and (c) class displayed with the colour ▦ superimposed on the image from (a).

of rough sets. Let B denote a set of probe functions used to extract feature values from an object such as a digital image pixel. Briefly, a rough set is obtained by considering the lower approximation (denoted by B_*X) and upper approximation (denoted by B^*X) of a nonempty set X and the approximation boundary (denoted by Bnd_BX).

Overview of Rough Sets:

$X \subseteq O =$ nonempty set (where O is the universe of objects),

$\quad B =$ set of probe functions,

$\quad \phi \in B,$

$\quad \sim_B = \{(x, y) \in X \times X : \forall \phi \in B.\ \phi(x) = \phi(y)\}$ (indiscernibility relation),

$\quad x_{/\sim_\phi} = \{y \in X : \forall \phi \in B.\ \phi(x) = \phi(y)\}$ (class),

$\quad B_*X = \bigcup_{x_{/\sim_\phi} \subseteq X} x_{/\sim_\phi}$ (lower approximation),

$\quad B^*X = \bigcup_{x_{/\sim_\phi} \cap X \neq \emptyset} x_{/\sim_\phi}$ (upper approximation),

$\quad Bnd_BX = B^*X - B_*X$ (approximation boundary).

With this in mind, one can observe many rough sets[3] in Fig. 1(b). Let O represent a database of images. Next, consider, for example, the set X shown embedded in the digital image $\Im$ showing a lake shore scene in Fig. 2 and the same set X shown in Fig. 3.

Fig. 2. Rough set X in lake image

Fig. 3. Rough set X in lake image partition (upper and lower approximations only)

Assume $\phi(x) = y$(blue component intensity), and consider the equivalence class represted by the purple ■ coloured pixels in Fig. 3. Notice that the members of the ■ equivalence class are partly in and partly outside the set X. In effect,

$$\blacksquare_{/\sim_\phi} \cap X \neq \emptyset.$$

In fact, the situation with the class denoted by $\blacksquare_{/\sim_\phi}$ is true of every class in the image $\Im$ shown in Fig. 3. Hence, X in Fig. 3 is an example of a visual rough set[4].

---

[3] This experiment with partitioning and identifying rough sets in a partition of a digital image $\Im$, is the result of using the Eq option in NEAR system, available at http://wren.ece.umanitoba.ca

[4] For other examples of visual rough sets, see [59, 15, 54, 64, 16].

3   Description-Based Set Operators

Many interesting properties of rough sets can be considered by introducing the description of a set. The following gives definitions of operators considered in the light of object descriptions (as originally reported in [17]). A logical starting point for introducing descriptive-based operators begins with establishing a basis for describing elements of sets. All sets in this work consist of *perceptual objects*, which is anything that has its origin in the physical world with characteristics observable to the senses such that they can be measured and are knowable to the mind. In keeping with the approach to pattern recognition suggested by Pavel [43], the features of a perceptual object are quantified by probe functions. In particular, a *feature* characterizes some aspect of the makeup of a perceptual object [48], and a *probe function* is a real-valued function representing a feature of a perceptual object [49, 51].

Next, a perceptual system is a set of perceptual objects, together with a set of probe functions.

Definition 1 Perceptual System [60]. *A perceptual system* $\langle O, \mathbb{F} \rangle$ *consists of a non-empty finite set* O *of sample perceptual objects and a non-empty finite ordered n-tuple* $\mathbb{F} = (\phi_1, \phi_2, \ldots, \phi_n)$ *of real-valued functions* $\phi \in \mathbb{F}$ *such that* $\phi : O \to \mathbb{R}$.

Combining the concepts of objects and probe functions, the description of a perceptual object within a perceptual system can be defined as follows.

Definition 2 Object Description. *Let* $\langle O, \mathbb{F} \rangle$ *be a perceptual system, and let* $\mathcal{B} \subseteq \mathbb{F}$ *be a set of probe functions. Then, the* $\mathbb{F}$-*description of a perceptual object* $x \in O$ *is a feature vector given by*

$$\Phi_{\mathbb{F}}(x) = (\phi_1(x), \phi_2(x), \ldots, \phi_i(x), \ldots, \phi_l(x)),$$

where l *is the length of the vector* $\Phi_{\mathbb{F}}$, *and each* $\phi_i(x)$ *in* $\Phi_{\mathcal{B}}(x)$ *is a probe function value that is part of the description of the object* $x \in O$. *The* $\mathcal{B}$-*description of* x *is a subsequence of* $\Phi_{\mathbb{F}}(x)$ *having as probe functions all and only elements from* $\mathcal{B}$.

Note, the idea of a feature space is implicitly introduced along with the definition of object description. An object description is the same as a feature vector as described in traditional pattern classification [7]. The description of an object can be considered a point in an l-dimensional Euclidean space $\mathbb{R}^l$ called a feature space. Further, a collection of these points, *i.e.*, a set of objects $A \subseteq O$, is characterized by the unique description of each object in the set.

Definition 3 Set Description [36, §4.3] [17]. *Let* A *be a set. Then the* set description *of* A *is defined as*

$$\mathcal{D}(A) = \{\Phi(a) : a \in A\}.$$

Example 1. Let $\langle O, \mathbb{F} \rangle$ be a perceptual system, where O contains the pixels in Fig. 4, $A \subseteq O$, and $\mathcal{B} \subseteq \mathbb{F}$ contains probe functions based on the RGB colour model. Then, the set description of A is $\mathcal{D}(A) = \{ \blacksquare, \blacksquare, \blacksquare, \blacksquare, \square \}$, where each coloured box represents the 3-dimensional real-valued RGB vector associated the box's colour.

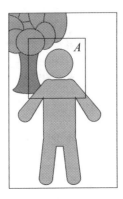

Fig. 4. Example demonstrating Definition 3

Next, J. Peters and S. Naimpally observed that, from a spatial point of view, the idea of nearness is a generalization of set intersection [58]. In other words, when considering the metric proximity, two sets are near each other when their intersection is not the empty set. Furthermore, they applied this idea to the concept of descriptive nearness in [36, §4.3] by focusing on the descriptions of objects within the sets. In this case, two sets are considered near each other if the intersection of their descriptions is not the empty set.

Definition 4 Descriptive Set Intersection [58, 36, 17]. *Let A and B be any two sets. The* descriptive (set) intersection *of A and B is defined as*

$$A \underset{\Phi}{\cap} B = \{ x \in A \cup B : \Phi(x) \in \mathcal{D}(A) \text{ and } \Phi(x) \in \mathcal{D}(B) \}.$$

Observe, the descriptive intersection differs from the standard set intersection $A \cap B = \{ x : x \in A \text{ and } x \in B \}$ in that A and B can be disjoint, *i.e.* $A \cap B = \emptyset$. In fact, the intended applications of the descriptive intersection use disjoint sets. However, in the case where the sets are not disjoint $A \cap B \subseteq A \underset{\Phi}{\cap} B$.

Example 2. Let $\langle O_1, \mathbb{F} \rangle$ and $\langle O_2, \mathbb{F} \rangle$ be perceptual systems corresponding to Fig. 5(a) & 5(c), respectively, where the perceptual objects and probe functions are defined in the same manner as Example 1. Moreover, let the blue rectangles in Fig. 5(b) (resp. Fig. 5(d)) represent two sets, A, B, for which the descriptive intersection is considered. Then, the inverted pixels (*i.e.* $p_i = \{ c_i, r_i, 255 - R_i, 255 - G_i, 255 - B_i)^{\mathrm{T}} \}$) within these sets represent their descriptive intersection, *i.e.* the inverted pixels represent the objects with matching descriptions in both sets.

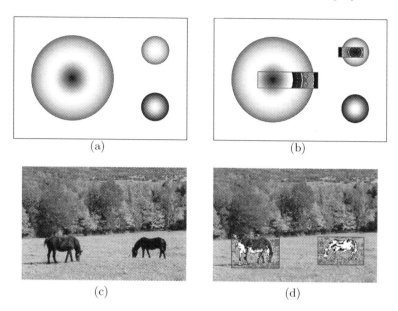

(a)   (b)

(c)   (d)

Fig. 5. Example demonstrating Definition 4

Definition 5 Descriptive Set Difference [17]. *The descriptive (set) differ-ence (or descriptive difference set) between two sets A and B is defined as*

$$A \setminus_\Phi B = \{x \in A : \Phi(x) \notin \mathcal{D}(B)\}.$$

Observe, the descriptive difference differs from the standard set difference $A \backslash B = \{x \in A \mid x \notin B\}$ in that A and B are also intended to be disjoint. Similarly, in the case where the sets are not disjoint $A \setminus_\Phi B \subseteq A \backslash B$.

Example 3. The descriptive difference between the sets introduced in Example 2 are given Fig. 6. In this case, the inverted pixels represent all the objects that do not have matching descriptions in the other set.

Definition 6 Descriptive Set Complement [17]. *The descriptive (set) com-plement of a set A in the universe U is defined as*

$$\complement_\Phi (A) = U \setminus_\Phi A$$

The biggest difference between the descriptive complement and the standard set complement $\complement(A) = U \backslash A$ is that the descriptive complement of the descrip-tive complement of a set is not the original set (see, *e.g.*, [17]).

Example 4. Considering the perceptual systems introduced in Example 2, the descriptive complement of each set represented by a blue rectangles in Fig. 7

(a)   (b)

Fig. 6. Example demonstrating Definition 5

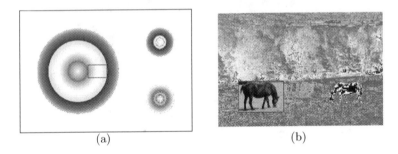

(a)   (b)

Fig. 7. Example demonstrating Definition 6

is given by the inverted pixels. In other words, the inverted pixels represent objects that do not have matching descriptions to those contained inside the blue rectangle.

4   Metric-Free Nearness Measure

Next, a metric-free description-based nearness measure using the descriptive operators introduced in Section 3 is presented, which is related to work on a tolerance-based nearness measure reported in [15, 16]. Furthermore, the approach presented here has direct application to image analysis and is related to the rough set image analysis approaches reported in [40, 57, 42, 4, 3, 24, 33, 13]. Similarly, this measure has been applied to the problem of Content-Based Image Retrieval (CBIR) [66] by Henry *et al.* [18]. The approach to applying this measure to CBIR is analogous to the tolerance nearness measure approached taken in [21, 19, 20]. As in the case of the tolerance nearness measure, both approaches aim to quantify the similarity between sets of objects based on object description. However, the tolerance nearness measure is obtained using tolerance classes (see, *e.g.*, [61]) obtained from the union of the sets under consideration, while the description-based nearness measure is based on the descriptive operators reported in [17]. The idea that motivated this measure comes from the observation in [58] that nearness is considered a generalization of intersection.

Intuitively speaking, we perceive sets of objects to be similar or near in some manner when they share common characteristics. Thus, if considering set descriptions (as given in Definition 3), the descriptive intersection should not be empty if we consider the sets to be similar with respect to one or more features. Keeping these ideas in mind, a metric-free description-based nearness measure, dNM, is defined as follows.

Definition 7 Metric-Free Description-Based Nearness Measure [17]. *Let* $X, Y \subseteq O$ *be sets of perceptual objects within a perceptual system. Then, a* metric-free description-based nearness measure *is defined as*

$$dNM(X,Y) = 1 - \frac{|X \underset{\Phi}{\cap} Y|}{|X \cup Y|},$$

where the operator $|\cdot|$ *represents the cardinality of a set. The nearness measure produces values in the interval* $[0,1]$, *where, for a pair of sets* X, Y, *a value of 0 represents complete resemblance, and a value of 1 indicates no resemblance.*

5   Proximity System

The Proximity System is an application developed to demonstrate descriptive-based topological approaches to nearness and proximity within the context of digital image analysis. The Proximity System grew out of the work of S. Naimpally and J. Peters [37, 46, 49, 50, 34, 35, 53, 52, 58, 36], was also influenced by work reported in [14–16, 21], and has resulted in one publication [17]. The Proximity System was written in Java and is intended to run in two different operating environments, namely on Android smartphones and tablets, as well as desktop platforms running the Java Virtual Machine. Fig. 8 gives screenshots of the interfaces for the two different environments. With respect to the desktop environment, the Proximity System is a cross-platform Java application for Windows, OSX, and Linux systems, which has been tested on Windows 7 and Debian Linux using the Sun Java 6 Runtime. In terms of the implementation of the theoretical approaches presented in the article, both the Android and the desktop based applications use the same back-end libraries (written in Java) to perform the description-based calculations, where the only differences is the user interface.

The Proximity System contains six main tabs, *Regions, Neighbourhoods, Intersection, Complement, Difference* and *Upper Approximation*, allowing users to perform description-based operations on regions of interest within a digital image. The user is also able to choose probe functions (from the left pane) that form the basis of the description-based operations and induce a partition of the pixels using the indiscernibility relation. These regions of interest correspond to user-defined sets of pixels that may be rough, *i.e.* the sets may be inexactly defined by equivalence classes in the partition formed by the selected probe functions (see, *e.g.*, Fig. 3). The first tab, *Regions*, allows users to add square, round, or mouse-click defined sets of pixels. The second tab allows the user to

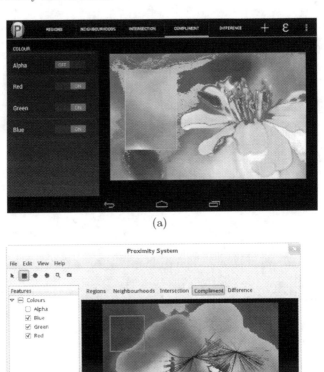

Fig. 8. Screen shots of the Proximity System. (a) Android OS version, and (b) PC version.

view description-based neighbourhood within the selected sets of pixels (see, e.g., [17, 22]). The remaining three tabs allow users to perform the operations corresponding to Defn. 4 – 6, respectively. A detailed manual for the Proximity System is given in [22].

Additional probe functions may be added or removed using the Proximity System GUI to facilitate research in different areas of application. By default, the Proximity System contains six probe functions, namely the output of the four components of the RGBA colour model (i.e., RGB and an opacity channel), an edge detection probe function (implemented using the Weber Law Differential Excitation [18, 5]), and a texture probe function (implemented using homogeneity defined with respect to a grey level co-occurrence matrix [11, 12, 16]). Probe functions are appended to the Proximity System using the *ProbeFunc* Java class, which is used to map a perceptual object to a probe function (feature) value.

The *ProbeFunc* class is a generic abstract class with a map method that maps perceptual objects to their corresponding probe function output. This class also has a minimum and maximum value used to normalize the result of the map method, which is necessary to ensure that the output of probe functions with large magnitude do not dominate the Euclidean distance calculation. The *toString* method is used to provide the name of the feature within both applications. An example ProbeFunc is given in [22].

Finally, memory resources are tightly regulated by the Android OS to ensure the system remains responsive. As a result, the code used to calculate the description-based operations presented here needed to be optimized in order to run to completion without running out of memory. In particular, the Proximity System allows users to find the result of description-based operations using a tolerance relation rather than the indiscernibility relation defined in Section 2. A tolerance relation presents a view of the world without transitivity (see, *e.g.*, [61]). In this case, the Euclidean distance between object descriptions must simply be within some ε in order to be included in the result. Consequently, the approach to performing these descriptive-based operations can be more computationally complex than when using the indiscernibility relation.

Algorithm 1. Descriptive Intersection Algorithm

Input : $A, B, \mathcal{D}(A), \mathcal{D}(B), \varepsilon$
Output: $A \underset{\Phi}{\cap} B$

1   Find $\mathcal{D}(A)$ (The unique colours in A);
2   Find $\mathcal{D}(B)$ (The unique colours in B);
3   $\mathcal{D}(C) \leftarrow \mathcal{D}(B)$;
4   $\mathcal{D}(A \underset{\Phi}{\cap} B) \leftarrow \emptyset$;

5   **for** $\Phi(a) \in \mathcal{D}(A)$ **do**
6   **for** $\Phi(b) \in \mathcal{D}(B)$ **do**
7   **if** $\| \Phi(a) - \Phi(b) \|_2 \leq \varepsilon$ **then**
8         $\mathcal{D}(A \underset{\Phi}{\cap} B) \leftarrow \mathcal{D}(A \underset{\Phi}{\cap} B) \cup \Phi(a)$;
9         $\mathcal{D}(A \underset{\Phi}{\cap} B) \leftarrow \mathcal{D}(A \underset{\Phi}{\cap} B) \cup \Phi(b)$;
10         $\mathcal{D}(C) \leftarrow \mathcal{D}(C) \backslash \Phi(b)$;
11   **if** $|\mathcal{D}(C)| > 0$ **then**
12   **for** $\Phi(c) \in \mathcal{D}(C)$ **do**
13   **if** $\| \Phi(a) - \Phi(c) \|_2 \leq \varepsilon$ **then**
14               $\mathcal{D}(A \underset{\Phi}{\cap} B) \leftarrow \mathcal{D}(A \underset{\Phi}{\cap} B) \cup \Phi(c)$;
15             $\mathcal{D}(C) \leftarrow \mathcal{D}(C) \backslash \Phi(c)$;

16   break;

17   //For each description in the result, add the pixels that have this colour
     $A \underset{\Phi}{\cap} B = \{ x \in A \cup B : \Phi(x) \in \mathcal{D}(A \underset{\Phi}{\cap} B) \}$;

The *PerceptualSystem* class[5] implements the methods used to calculate the output of probe functions. Perceptual objects and probe functions must be added to a *PerceptualSystem* object before probe function values can be calculated. Each perceptual object is given an index number when it is added to the *PerceptualSystem* object. Probe function calculations are then made using object indices. This allows arrays to be used rather than list objects, which removes some overhead (especially where look ups are concerned).

Maps of unique object description are created between object indices and their corresponding object descriptions, which greatly reduced the overhead with comparisons. Specifically, as a result of this modification, comparisons between sets are made solely on set descriptions (*i.e.* lists of unique object descriptions associated with a set), and then the lists of objects associated with the matched descriptions can be combined into the final result. This optimization was particularly important since these comparisons were found to be one of the most time consuming part of the calculation.

An example of the optimized algorithm for finding the descriptive intersection is given in Alg. 1. Here, the descriptive intersection is calculated on two sets A and B by comparing $\mathcal{D}(A)$ to $\mathcal{D}(B)$ using an additional set $\mathcal{D}(C)$, which is a copy of $\mathcal{D}(B)$. Object descriptions will subsequently be removed from $\mathcal{D}(C)$ during calculation of the descriptive intersection, causing it to become a subset of $\mathcal{D}(B)$. During the calculation process, an object description $\Phi(a) \in \mathcal{D}(A)$ is compared to each object description $\Phi(b) \in \mathcal{D}(B)$. If a matching description[6] is found, both descriptions are marked as matched and the description from $\mathcal{D}(B)$ is removed from $\mathcal{D}(C)$, *i.e.* $\mathcal{D}(C) \leftarrow \mathcal{D}(C) \backslash \Phi(b)$. Once a match occurs, $\Phi(a)$ is then compared to the remaining object descriptions in $\mathcal{D}(C)$, starting at the index of b. Any additional matches are also removed from $\mathcal{D}(C)$. Each new iteration starts by comparing an object in $\mathcal{D}(A)$ with $\mathcal{D}(B)$, and only switches to the reduced set $\mathcal{D}(C)$ if a match is found. In this way, the problem of comparing two descriptions that have both already been found to be within the descriptive intersection is avoided and the number of comparisons is reduced.

6   Proximity System: Visual Rough Set Examples

This section presents examples of using the Proximity System to obtain the descriptive intersection to visual rough sets, using the framework for defining visual rough sets given in Section 2. First, a simple example is given in Fig. 9, where $\phi(x) = y$(green component intensity), $y \in [0, 255]$. The sets in Fig. 9(b) are clearly visual rough sets since the image is partitioned based on colours, and neither set can be completely defined by classes from the partition.

---

[5] Note, in the following discussion, we distinguish between perceptual objects (*i.e.*, pixels in the case of the Proximity System), and objects defined in traditional Java programming.

[6] Recall, this includes the case where the difference in object descriptions is within some ε.

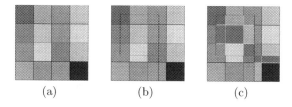

(a) (b) (c)

Fig. 9. Sample visual rough set example. (a) Simple image, (b) user-defined sets, and (c) description-based intersection.

(a) (b)

(c) (d)

Fig. 10. Natural image visual rough set example. (a) User-defined sets of a tree and its reflection, (b) description-based intersection of sets in (a), (c) user-defined sets of two trees, and (d) description-based intersection of sets in (c).

Next, Fig. 9(c) contains the descriptive intersection of Fig. 9(b), which is represented by the pixels with inverted colours. Notice, only pixels from the two sets that share the same green component from the RGB colour model appear in the result. Also, the black boundary lines are a different shade of black than the square in the lower right corner. Hence, they appear in the result, while the black square does not. Similarly, Fig. 10(b) & 10(d) also contain the result of descriptive intersection. In this case, there are two comparisons, namely a tree on the shore is compared with its reflection in the lake, as well as another tree on the shore. Observe, the trees on the shore have more descriptions in common (*i.e.*, there are more pixels that share the same shade of green) than the tree and its reflection, which, in the context of the selected probe functions, makes sense since the water would alter the colour of the reflected light incident on the camera sensors. Next, Table 1 gives the metric-free nearness measure values for the sets depicted in Fig. 9(c), 10(b), & 10(d). Notice, the measure also indicates the trees on the shore are more near each other (in terms of green component of the RGB colour model) than the tree and its reflection.

Table 1. Nearness measure values for the images in Fig. 9–15

Image	dNM	Image	dNM	Image	dNM
Fig. 9(c)	0.47	Fig. 11(d)	0.40	Fig. 12(d)	0.06
Fig. 10(b)	0.40	Fig. 11(h)	0.84	Fig. 12(h)	1.00
Fig. 10(d)	0.02	Fig. 11(l)	0.78	Fig. 12(l)	0.90
		Fig. 11(p)	1.00	Fig. 12(p)	0.88
		Fig. 11(t)	1.00	Fig. 12(t)	0.51
Fig. 13(d)	0.85	Fig. 14(d)	1.00	Fig. 15(d)	1.00
Fig. 13(h)	0.20	Fig. 14(h)	0.78	Fig. 15(h)	1.00
Fig. 13(l)	0.69	Fig. 14(l)	1.00	Fig. 15(l)	0.64
Fig. 13(p)	0.99	Fig. 14(p)	1.00		
Fig. 13(t)	0.22	Fig. 14(t)	1.00		
		Fig. 14(x)	1.00		

Lastly, Fig. 11–15 contain further examples of performing the descriptive intersection on visual rough sets, and Table 1 contains the metric-free nearness measures values for these examples. The probe functions used to create these examples are based on the following image features: image colour using the RGB colour model ($\phi_{Red}, \phi_{Green}, \phi_{Blue}$); image grey levels using perceptual greyscale (ϕ_{Grey}); image edge detection using the Weber Law Descriptor [18, 5] (ϕ_{DiffE}); and image texture using homogeneity defined with respect to a grey level co-occurrence matrix [11, 12, 16] (ϕ_{Homog}). Note, the colour of the equivalence classes in the inner columns are random. Thus, as depicted in Fig. 13(f) & 13(g), two different colours can refer to the same class.

7   Application: Human Visual Search

This section presents a perceptual application of the descriptive operators, where the perceptual basis of these operators is one of four psychological theories of visual attention [70]. Namely, feature integration theory [68], Guided Search Model (GSM) [69], biased competition [6], and integrated competition [8, 9]. This article focuses on the GSM [69, §1]. The aim of the GSM is to explain the processes involved in the visual search problem, which is the task of searching and identifying targets in a visual field full of distractor items. This task, which we perform repeatedly and effortlessly, is not easily achieved by artificial systems (see, *e.g.*, [70]). In order to understand this processes it is important to realize that the human visual system does not contain enough neural hardware to process all the visual information presented to our senses [69]. Instead, the visual scene is preprocessed in parallel, and certain stimuli are identified as important depending on one of two attentive processes, namely bottom-up and top-down activation. Activation is based on the concept of feature maps, which are representations of basic visual features that respond to local differences in the visual field. For example, the features colour and orientation are two distinct maps. Each feature map is further divided into broadly tuned channels. For instance, colour consists of red, yellow, green, and blue channels, and orientation consists of left, steep,

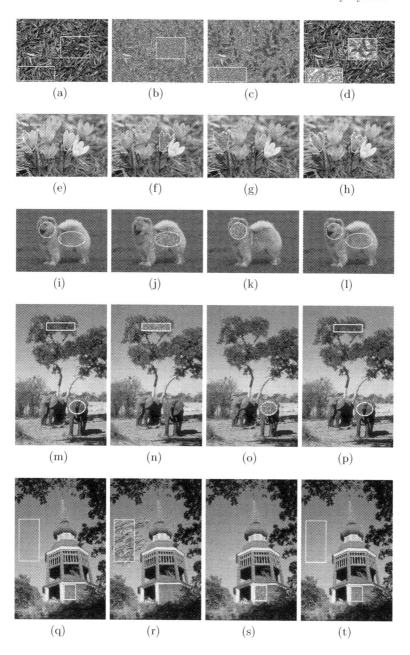

Fig. 11. Visual rough set example. Columns (left to right): User-defined sets, upper approximation of 1st set, upper approximation of 2nd set, and descriptive-based intersection. Rows: (a) $\mathcal{B} = \{\phi_{\text{Red}}\}$, (e)–(m) $\mathcal{B} = \{\phi_{\text{Red}}, \phi_{\text{Green}}, \phi_{\text{Blue}}\}$, and (q) $\mathcal{B} = \{\phi_{\text{Red}}, \phi_{\text{Green}}, \phi_{\text{Blue}}, \phi_{\text{DiffE}}\}$.

Fig. 12. Visual rough set example. Columns (left to right): User-defined sets, upper approximation of 1st set, upper approximation of 2nd set, and descriptive-based intersection. Rows: (a) $\mathcal{B} = \{\phi_{\text{Grey}}\}$, (e) $\mathcal{B} = \{\phi_{\text{Grey}}, \phi_{\text{Red}}\}$, and (i)–(q) $\mathcal{B} = \{\phi_{\text{Grey}}, \phi_{\text{DiffE}}\}$.

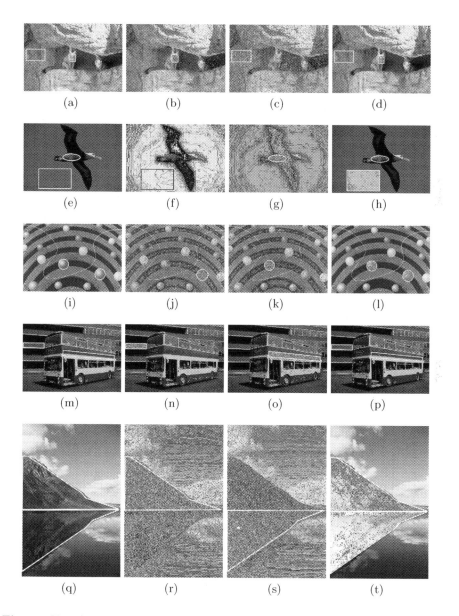

Fig. 13. Visual rough set example. Columns (left to right): User-defined sets, upper approximation of 1st set, upper approximation of 2nd set, and descriptive-based intersection. Rows: (a)–(q) $\mathcal{B} = \{\phi_{\text{DiffE}}\}$.

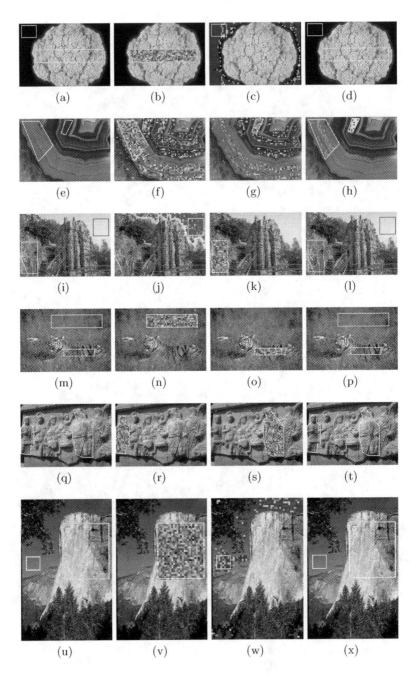

Fig. 14. Visual rough set example. Columns (left to right): User-defined sets, upper approximation of 1st set, upper approximation of 2nd set, and descriptive-based intersection. Rows: (a)–(u) $\mathcal{B} = \{\phi_{\text{Homog}}\}$.

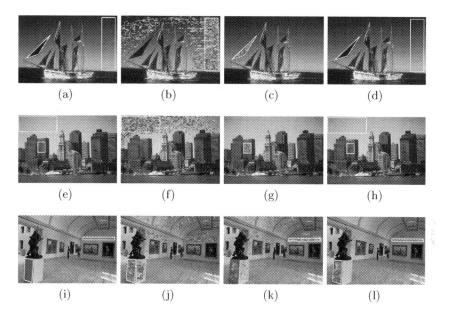

Fig. 15. Visual rough set example. Columns (left to right): User-defined sets, upper approximation of 1st set, upper approximation of 2nd set, and descriptive-based intersection. Rows: (a) $\mathcal{B} = \{\phi_{\text{blue}}, \phi_{\text{Homog}}\}$, (e)$\mathcal{B} = \{\phi_{\text{Grey}}, \phi_{\text{Homog}}\}$, and (i) $\mathcal{B} = \{\phi_{\text{DiffE}}, \phi_{\text{Homog}}\}$.

right, and shallow channels (with respect to vertical) [69]. Finally, the output of feature maps are combined by way of a weighed average into an activation map (see, *e.g.*, Fig. 16[7]). Then, attention is directed to the loci of these maps using either top-down or bottom up activation.

The top-down and bottom-up activation processes are differentiated by the presence or absence of intent. Bottom-up activation is stimulus-driven and is used to identify differences in the current field. Attention is then directed to these differences. For example, bottom-up activation is of use when identifying a target line of orientation x among distractors of orientation y [69]. In contrast, top-down activation is user-driven and used to guide attention to targets in the visual field that have specific feature values. Top-down activation requires specifying the output of one or more broadly tuned channels per feature [69]. An example where top-down activation is of use is in searching for a red vertical target among red horizontal and green vertical items. In this case, bottom-down activation is of no use because half the items are red, half are green, half are vertical, and half horizontal, which means there are no local differences [69]. The top-down approach is used to demonstrate the perceptual qualities of the descriptive-based operators presented here.

---

[7] The original figure contains more input channels and features maps, which were not given to save space.

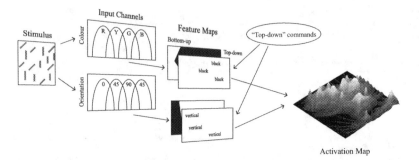

Fig. 16. The GSM model [Wolfe 1994, §1]

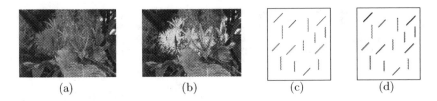

(a)   (b)   (c)   (d)

Fig. 17. Examples demonstrating descriptive intersection in visual search

The first relationship of descriptive-based operators to the human visual search problem can be found in the observation that guided search is the search for a target object, not the search for features within the visual field [69, §2]. For example, the search for the red vertical object, not a search for redness and verticality. Thus, use of *perceptual objects* within a *perceptual system* is in keeping with the GSM, and supports the claim that using collections of unique feature vectors associated with objects in a set, at some level, matches the human approach to describing sets of objects.

Next, the goal of the top-down approach is to pick the channel that best differentiates the target from the distractors [69]. This is not necessarily the channel that gives the largest response (*i.e.* it is possible to suppress the contribution of specific stimuli). Moreover, channels can be combined to specify the target in the visual search problem (as in the red vertical object example). The top-down process is akin to envisioning a set description $\mathcal{D}(A)$ containing descriptions of target objects we wish to find in the visual field, where A contains either concrete objects from memory, or some abstract thought such as a blue object. Then, the descriptive intersection between $\mathcal{D}(A)$ and the set of objects identified by the feature maps is used to find the target. Finally, attention is directed to areas of the activation map where the intersection is not empty (starting with the largest locus). Notice, a comparison between descriptions we associate with the targets and the output of feature maps is vital to performing a visual search. Thus, in comparing disjoint sets of objects, the claim we must, at some level, be performing a comparison of the descriptions we associate with the objects within the sets is also in keeping with the GSM.

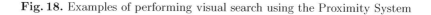

Fig. 18. Examples of performing visual search using the Proximity System

Two examples of using the descriptive-based set intersection to perform visual search are given in Fig. 17. Let the pixels in Fig. 17(a) represent a field of view, and chose red as the colour channel that best differentiates the target from the distractors. In terms of the descriptive framework given in this article, let $\langle O, \mathbb{F} \rangle$ be a perceptual system corresponding to Fig 17(a), where the perceptual objects and probe functions are defined in the same manner as the example in Section 3. Let the top-down channel selection be given by Fig. 18, *i.e.* $\mathcal{D}(A)$ consists of 3-dimensional object descriptions containing values for each shade of red given in Fig. 18. Then, the result of a visual search using descriptive intersection is given in Fig. 17(b) (by the inverted pixels). A similar example is given in Fig. 17(c) & 17(d) where $\mathcal{D}(A)$ consists of 4-dimensional object descriptions with values for the RGB colour green and line orientation of 45°. In this case the selected pixels are coloured black (for increased contrast). Note, the edge orientation was obtained using using Mallat's multiscale edge detection method [26, 25], which is the reason for only partial selection of the 45° line, as well as that the very tips of the vertical green lines.

8 Conclusion

This article presented details on the Proximity System, background on description-based set operations and a metric-free nearness measure, and the use of the Proximity System with respect to visual rough sets. Also, a discussion on the perceptual similarities between the descriptive-based set intersection and the visual processes that occur during human visual search was presented by way of a practical example. The contribution of the article was a discussion of the Proximity System and its use as a description-based system for quantifying the nearness or apartness of visual rough sets. A particularly nice feature of this tool is the ability for users to define there own probe functions. This tool has already proved vital in the study of descriptive-based topological approaches to nearness and proximity within the context of digital image analysis, as can be seen by results reported in [17]. Future work will include the use of the Proximity System and the nearness measure for performing content-based image retrieval on databases containing thousands of images, and further investigation into the observations that in comparing disjoint sets of objects, at some level humans are performing a comparison of the descriptions we associate with the objects within the sets.

References

1. The Corel Stock Photo Library (PC/MAC ver.). Corel Corporation, Ottawa, Ontario, Canada (1994)
2. Benjamin Jr., L.T.: A Brief History of Modern Psychology. Blackwell Publishing, Malden (2007)

3. Borkowski, M.: 2D to 3D Conversion with Direct Geometrical Search and Approximation Spaces. Ph.D. thesis, University of Manitoba (2007)
4. Borkowski, M., Peters, J.F.: Matching 2D image segments with genetic algorithms and approximation spaces. In: Peters, J.F., Skowron, A. (eds.) Transactions on Rough Sets V. LNCS, vol. 4100, pp. 63–101. Springer, Heidelberg (2006)
5. Chen, J., Shan, S., He, C., Zhao, G., Pietikäinen, M., Chen, X., Gao, W.: Wld: A robust local image descriptor. IEEE Transactions on Pattern Analysis and Machine Intelligence 32(9), 1705–1720 (2010)
6. Desimone, R., Duncan, J.: Neural mechanisms of selective visual attention. Annual Review of Neuroscience 18, 193–222 (1995)
7. Duda, R.O., Hart, P.E., Stork, D.G.: Pattern Classification, 2nd edn. Wiley (2001)
8. Duncan, J.: Converging levels of analysis in the cognitive neuroscience of visual attention. Philosophical Transactions: Biological Sciences 353(1373), 1307–1317 (1998)
9. Duncan, J., Humphreys, G., Ward, R.: Competitive brain activity in visual attention. Current Opinion in Neurobiology 7(2), 255–261 (1997)
10. Fang, Y., Zhen, Z., Huang, Z., Zhang, C.: Multi-objective fuzzy clustering method for image segmentation based on variable-length intelligent optimization algorithm. In: Cai, Z., Hu, C., Kang, Z., Liu, Y. (eds.) ISICA 2010. LNCS, vol. 6382, pp. 329–337. Springer, Heidelberg (2010)
11. Haralick, R.M.: Textural features for image classification. IEEE Transactions on Systems, Man, and Cybernetics SMC-3(6), 610–621 (1973)
12. Haralick, R.M.: Statistical and structural approaches to texture. Proceedings of the IEEE 67(5), 786–804 (1979)
13. Hassanien, A.E., Abraham, A., Peters, J.F., Schaefer, G., Henry, C.: Rough sets and near sets in medical imaging: A review. IEEE Transactions on Information Technology in Biomedicine 13(6), 955–968 (2009)
14. Henry, C.: Near set Evaluation And Recognition (NEAR) system. In: Pal, S.K., Peters, J.F. (eds.) Rough Fuzzy Analysis Foundations and Applications, pp. 7–1 – 7–22. CRC Press, Taylor & Francis Group (2010), http://wren.ee.umanitoba.ca
15. Henry, C.J.: Near Sets: Theory and Applications. Ph.D. thesis, University of Manitoba, CAN (2010), https://mspace.lib.umanitoba.ca/handle/1993/4267
16. Henry, C.J.: Perceptual indiscernibility, rough sets, descriptively near sets, and image analysis. In: Peters, J.F., Skowron, A. (eds.) Transactions on Rough Sets XV. LNCS, vol. 7255, pp. 41–121. Springer, Heidelberg (2012)
17. Henry, C.J.: Metric free nearness measure using description-based neighbourhoods. Mathematics in Computer Science 7(1), 51–69 (2013)
18. Henry, C.J., Peters, J.F., Hettiarchachichi, R., Ramanna, S.: Content-based image retrieval using a metric free nearness measure. In: Proceedings of the 15th IASTED International Conference on Signal and Image Processing, pp. 374–381 (2013)
19. Henry, C.J., Ramanna, S.: Maximal clique enumeration in finding near neighbourhoods. In: Peters, J.F., Skowron, A., Ramanna, S., Suraj, Z., Wang, X. (eds.) Transactions on Rough Sets XVI. LNCS, vol. 7736, pp. 103–124. Springer, Heidelberg (2013)
20. Henry, C.J., Ramanna, S.: Signature-based perceptual nearness. Application of near sets to image retrieval. Mathematics in Computer Science 7(1), 71–85 (2013)
21. Henry, C.J., Ramanna, S., Levy, D.: Quantifying nearness in visual spaces. Cybernetics and Systems 44(1), 38–56 (2013)
22. Henry, C.J., Smith, G.: Proximity system. Tech. rep., Computational Intelligence Laboratory, University of Manitoba (2012), uM CI Laboratory Technical Report No. TR-2012-021

23. Hergenhahn, B.R.: An Introduction to the History of Psychology. Wadsworth Publishing, Belmont (2009)
24. Maji, P., Pal, S.K.: Maximum class separability for rough-fuzzy C-means based brain MR image segmentation. In: Peters, J.F., Skowron, A., Rybiński, H. (eds.) Transactions on Rough Sets IX. LNCS, vol. 5390, pp. 114–134. Springer, Heidelberg (2008)
25. Mallat, S.: A Wavelet Tour of Signal Processing. Academic Press, California (1999)
26. Mallat, S., Zhong, S.: Characterization of signals from multiscale edges. IEEE Transactions on Pattern Analysis and Machine Intelligence 14(7), 710–732 (1992)
27. Małyszko, D., Stepaniuk, J.: Standard and fuzzy rough entropy clustering algorithms in image segmentation. In: Chan, C.-C., Grzymala-Busse, J.W., Ziarko, W.P. (eds.) RSCTC 2008. LNCS (LNAI), vol. 5306, pp. 409–418. Springer, Heidelberg (2008)
28. Małyszko, D., Stepaniuk, J.: Fuzzified probabilistic rough measures in image segmentation. In: Kim, T.-h., Pal, S.K., Grosky, W.I., Pissinou, N., Shih, T.K., Ślęzak, D. (eds.) SIP/MulGraB 2010. CCIS, vol. 123, pp. 78–86. Springer, Heidelberg (2010)
29. Małyszko, D., Stepaniuk, J.: Probabilistic rough entropy measures in image segmentation. In: Szczuka, M., Kryszkiewicz, M., Ramanna, S., Jensen, R., Hu, Q. (eds.) RSCTC 2010. LNCS, vol. 6086, pp. 40–49. Springer, Heidelberg (2010)
30. Małyszko, D., Stepaniuk, J.: Rough fuzzy measures in image segmentation and analysis. In: Pal, S.K., Peters, J.F. (eds.) Rough Fuzzy Analysis Foundations and Applications, pp. 11-1–11-25. CRC Press, Taylor & Francis Group (2010) ISBN 13: 9781439803295
31. Małyszko, D., Stepaniuk, J.: Rough entropy hierarchical agglomerative clustering in image segmentation. In: Peters, J.F., Skowron, A., Chan, C.-C., Grzymala-Busse, J.W., Ziarko, W.P. (eds.) Transactions on Rough Sets XIII. LNCS, vol. 6499, pp. 89–103. Springer, Heidelberg (2011)
32. Mrózek, A., Mrózek, L.: Rough sets in image analysis. Foundations of Computing and Decision Sciences F18(3-4), 268–273 (1993)
33. Mushrif, M., Ray, A.K.: Color image segmentation: Rough-set theoretic approach. Pattern Recognition Letters 29(4), 483–493 (2008)
34. Naimpally, S.A.: Near and far. A centennial tribute to Frigyes Riesz. Siberian Electronic Mathematical Reports 6, A.1–A.10 (2009)
35. Naimpally, S.A.: Proximity Approach to Problems in Topology and Analysis. Oldenburg Verlag, München (2009) ISBN 978-3-486-58917-7
36. Naimpally, S.A., Peters, J.F.: Topology with Applications.Topological Spaces via Near and Far. World Scientific, Singapore (2013)
37. Naimpally, S.A., Warrack, B.D.: Proximity spaces. Cambridge Tract in Mathematics, vol. 59. Cambridge University Press, Cambridge (1970)
38. Orłowska, E.: Semantics of vague concepts. Applications of rough sets. Tech. Rep. 469, Institute for Computer Science, Polish Academy of Sciences (1982)
39. Orłowska, E.: Semantics of vague concepts. In: Dorn, G., Weingartner, P. (eds.) Foundations of Logic and Linguistics. Problems and Solutions, pp. 465–482. Plenum Pres, London/NY (1985)
40. Pal, S.K., Mitra, P.: Multispectral image segmentation using rough set initialized em algorithm. IEEE Transactions on Geoscience and Remote Sensing 11, 2495–2501 (2002)
41. Pal, S.K., Peters, J.F.: Rough Fuzzy Image Analysis: Foundations and Methodologies. CRC Press, Boca Raton (2010)

42. Pal, S.K., Shankar, B.U., Mitra, P.: Granular computing, rough entropy and object extraction. Pattern Recognition Letters 26(16), 401–416 (2005)
43. Pavel, M.: Fundamentals of Pattern Recognition. Marcel Dekker, Inc., NY (1993)
44. Pawlak, Z.: Classification of objects by means of attributes. Tech. Rep. PAS 429, Institute for Computer Science, Polish Academy of Sciences (1981)
45. Pawlak, Z.: Rough sets. International Journal of Computer and Information Sciences 11, 341–356 (1982)
46. Pawlak, Z., Peters, J.F.: Jak blisko (how near). Systemy Wspomagania Decyzji I, 57–109 (2002)
47. Pawlak, Z., Skowron, A.: Rudiments of rough sets. Information Sciences 177, 3–27 (2007)
48. Peters, J.F.: Classification of objects by means of features. In: Proceedings of the IEEE Symposium Series on Foundations of Computational Intelligence (IEEE SCCI 2007), pp. 1–8 (2007)
49. Peters, J.F.: Near sets. General theory about nearness of objects. Applied Mathematical Sciences 1(53), 2609–2629 (2007)
50. Peters, J.F.: Near sets. Special theory about nearness of objects. Fundamenta Informaticae 75(1-4), 407–433 (2007)
51. Peters, J.F.: Classification of perceptual objects by means of features. International Journal of Information Technology & Intelligent Computing 3(2), 1–35 (2008)
52. Peters, J.F.: Tolerance near sets and image correspondence. International Journal of Bio-Inspired Computation 1(4), 239–245 (2009)
53. Peters, J.F.: Corrigenda and addenda: Tolerance near sets and image correspondence. International Journal of Bio-Inspired Computation 2(5), 310–318 (2010)
54. Peters, J.F.: How near are Zdzisław Pawlak's? In: Skowron, A., Suraj, Z. (eds.) Rough Sets and Intelligent Systems - Professor Zdzisław Pawlak in Memoriam, pp. 545–568. Springer, Berlin (2013)
55. Peters, J.F.: Local near sets. pattern discovery in proximity spaces. Mathematics in Computer Science 7(1), 87–106 (2013)
56. Peters, J.F.: Near sets: An introduction. Mathematics in Computer Science 7(1), 3–9 (2013)
57. Peters, J.F., Borkowski, M.: K-means indiscernibility relation over pixels. In: Tsumoto, S., Słowiński, R., Komorowski, J., Grzymała-Busse, J.W. (eds.) RSCTC 2004. LNCS (LNAI), vol. 3066, pp. 580–585. Springer, Heidelberg (2004)
58. Peters, J.F., Naimpally, S.A.: Applications of near sets. Notices of the American Mathematical Society 59(4), 536–542 (2012)
59. Peters, J.F., Ramanna, S.: Affinities between perceptual granules: Foundations and perspectives. In: Bargiela, A., Pedrycz, W. (eds.) Human-Centric Information Processing Through Granular Modelling, pp. 49–66. Springer, Berlin (2009)
60. Peters, J.F., Wasilewski, P.: Foundations of near sets. Info. Sci. 179(18), 3091–3109 (2009)
61. Peters, J.F., Wasilewski, P.: Tolerance spaces: Origins, theoretical aspects and applications. Information Sciences 195, 211–225 (2012)
62. Poincaré, H.: Science and Hypothesis. The Mead Project, Brock University (1905), L. G. Ward's translation
63. Polkowski, L.: Rough Sets. Mathematical Foundations. Springer, Heidelberg (2002)
64. Ramanna, S., Peters, J.F.: Nearness in associated rough sets: Case study in image analysis. In: Peters, G., Lingras, P., Slezak, D., Yao, Y. (eds.) Selected Methods and Applications of Rough Sets in Management and Engineering, pp. 62–73. Springer, Berlin (2011)

65. Sen, D., Pal, S.K.: Generalized rough sets, entropy, and image ambiguity measures. IEEE Transactions on Systems, Man, and Cybernetics - Part B 39(1), 117–128 (2009)
66. Smeulders, A.W.M., Worring, M., Santini, S., Gupta, A., Jain, R.: Content-based image retrieval at the end of the early years. IEEE Transactions on Pattern Analysis and Machine Intelligence 22(12), 1349–1380 (2000)
67. Sossinsky, A.B.: Tolerance space theory and some applications. Acta Applicandae Mathematicae: An International Survey Journal on Applying Mathematics and Mathematical Applications 5(2), 137–167 (1986)
68. Treisman, A.M., Gelade, G.: A feature integration theory of attention. Cognitive Psychology 12(1), 97–136 (1980)
69. Wolfe, J.M.: Guided search 2.0 a revised model of visual search. Psychonomic Bulletin & Review 1(2), 202–238 (1994)
70. Yu, Y., Mann, G.K.I., Gosine, R.G.: A goal-directed visual perception system using object-based top-down attention. IEEE Transactions on Autonomous Mental Development 4(1), 87–103 (2012)
71. Zeeman, E.C.: The topology of the brain and the visual perception. In: Fort, K.M. (ed.) Topoloy of 3-manifolds and selected topices, pp. 240–256. Prentice Hall, New Jersey (1965)

Rough Sets and Matroids

Victor W. Marek[1] and Andrzej Skowron[2]

[1] Department of Computer Science,
University of Kentucky, Lexington, KY 40506
marek@cs.uky.edu
[2] Institute of Mathematics, The University of Warsaw
Banacha 2, 02-097 Warsaw, Poland
skowron@mimuw.edu.pl

Abstract. We prove the recent result of Liu and Zhu [1] and discuss some consequences of that and related facts for the development of rough set theory.

Keywords: rough set, matroid.

1   Introduction

The goal of this note is to provide a proof of the recent statement by Liu and Zhu [1] and look at some properties of rough sets related to Liu and Zhu realization [2] that rough sets [3–7] relate to one of classical structures of combinatorics and computer science, namely matroid. The importance of that result is that it allows to tie various rough set methods to *greedy algorithms* that succeed when underlying combinatorial structure is defined by matroid [8, 9]. This allows for developments of algorithms for finding properties of maximal and minimal sets in various classes of sets (see also Propositions 2 and 3 below.)

2   Preliminaries

Below we introduce basic notions used in this paper. Generally, we assume that the reader is familiar with the notion of rough sets of [3–5].

2.1   Rough Sets

Any pair $(U, \sim)$, where U i a finite set and $\sim$ is an equivalence relation in U is called an *approximation space*. We denote by $[x]$ the set $\{y \in U : x \sim y\}$. We call sets of the form $[x]$, *monads* (or *elementary granules* [5]). Monads of an equivalence relation $\sim$ form a partition of the set U. Given a set $X \subseteq U$, the sets $\underline{X}$ and $\overline{X}$ are defined as $\bigcup\{[x] : [x] \subseteq X\}$, and $\bigcup\{[x] : [x] \cap X \neq \emptyset\}$, respectively. The sets $\underline{X}$ and $\overline{X}$ are called the *lower approximation* of X (relative to $\sim$) and the *upper approximation* of X (relative to $\sim$), respectively. The set $BN(X) = \overline{X} \setminus \underline{X}$ is called the boundary region of X (relative to $\sim$). If $BN(X) = \emptyset$ then X is *crisp* (relative to $\sim$), otherwise X is *rough* (relative to $\sim$). Pawlak, in [3], established the basic properties of these operations. We assume that the reader is familiar with these properties.

J.F. Peters and A. Skowron (Eds.): Transactions on Rough Sets XVII, LNCS 8375, pp. 74–81, 2014.

2.2   Matroids

Matroids are one of basic structures studied by combinatorists [9, 8]. Matroids occur in many areas of Mathematics and Computer Science as a common generalization of concepts such a collection of independent sets in a linear space and of the cycle-free sets in a graph. Matroids are closely related to issues in combinatorial optimization because of relationship between greedy algorithms and matroids.

Formally, a matroid over a set U is a nonempty family $\mathcal{M}$ of subsets of U satisfying the following conditions:

1. $\emptyset \in \mathcal{M}$.
2. Whenever $X \in \mathcal{M}$, and $Y \subseteq X$, then $Y \in \mathcal{M}$.
3. (Steinitz Exchange Principle) Whenever $X, Y \in \mathcal{M}$ and $|X| < |Y|$ then for some $y \in Y \setminus X$, $X \cup \{y\} \in \mathcal{M}$.

By a parameterized matroid over an index set I we mean a family of matroids $\langle \mathcal{M}_i \rangle_{i \in I}$. In our case the set I will be the powerset of U, $\mathcal{P}(U)$.

3   Matroids Generated by Approximation Spaces

In this section we give a proof of the result of Liu and Zhu [1] on the parametric matroid associated with an approximation space defined by an equivalence relation $\sim$ over a finite set U.

Definition 1. *Given an approximation space $(U, \sim)$ we define for a set $Y \subseteq U$ the family of sets $\mathcal{M}_Y$ as*

$$\{A \subseteq U : \underline{A} \subseteq Y\}. \tag{1}$$

Then, we prove

Theorem 1 (Liu and Zhu). *Let $(U, \sim)$ be an approximation space. Then for every subset $Y \subseteq U$, $\mathcal{M}_Y$ is a matroid.*

Proof: The family $\mathcal{M}_Y$ is closed under subsets because whenever $B \subseteq A$ and $A \in \mathcal{M}_Y$, then, by the definition, $\underline{A} \subseteq Y$. Since $B \subseteq A$, we have $\underline{B} \subseteq \underline{A}$, thus $\underline{B} \subseteq Y$, i.e. $B \in \mathcal{M}_Y$. Therefore the first two conditions on matroid hold for $\mathcal{M}_Y$.

We will now show the exchange property for $\mathcal{M}_Y$. To this end, let A, B be two sets, $A, B \in \mathcal{M}_Y$, $|A| < |B|$. We need to find $x \in B \setminus A$ so that $A \cup \{x\} \in \mathcal{M}_Y$.

Our argument consists of two cases.

Case 1. Some $x \in B \setminus A$ has the property that $[x] = \{x\}$. That is, for $y \neq x$, $y \not\sim x$. We claim that for that x, $A \cup \{x\} \in \mathcal{M}_Y$.

Since $[x] = \{x\}$ and $x \notin A$, we have

$$\underline{A \cup \{x\}} = \underline{A} \cup \{x\}$$

Now, $\underline{A} \subseteq Y$ (because $A \in \mathcal{M}_Y$), and also $x \in Y$ because $\underline{B} \subseteq Y$ and $\{x\} = [x] \subseteq \underline{B} \subseteq Y$. Thus $\underline{A} \cup \{x\} \subseteq Y$, and so $A \cup \{x\} \in \mathcal{M}_Y$.

Case 2. No $x \in B \setminus A$ has the property that $[x] = \{x\}$. We will now assume that for no $x \in B \setminus A$, $A \cup \{x\} \in \mathcal{M}_Y$ and show that this leads to the contradiction.

Let us look at an arbitrary $x \in B \setminus A$. Under our assumption ($A \cup \{x\} \notin \mathcal{M}_Y$), it must be the case that $\underline{A \cup \{x\}}$ is strictly bigger than $\underline{A}$ (because if $\underline{A \cup \{x\}} = \underline{A}$ then as $\underline{A} \subseteq Y$, then, since $\underline{A} \subseteq Y$, $\underline{A \cup \{x\}} \subseteq Y$, so $A \cup \{x\} \in \mathcal{M}_Y$, a contradiction.)

But what is $\underline{A \cup \{x\}}$?

There are two possibilities:

1. $\underline{A \cup \{x\}} = \underline{A}$, or
2. $\underline{A \cup \{x\}} = \underline{A} \cup [x]$.

Since the first possibility has already been eliminated, it must be the case that $\underline{A \cup \{x\}} = \underline{A} \cup [x]$. But this means that for all y such that $y \neq x$, $y \sim x$, the element y must belong to A.

Moreover, since x was an arbitrary element of $B \setminus A$, it must be the case that whenever $x \in B \setminus A$, $y \neq x$, $y \sim x$ then $y \in A$.

Next, we ask if it is possible that for some $x, y \in B \setminus A$, $x \neq y$, $x \sim y$. We claim that this is impossible. Indeed let us assume that for some $x, y \in B \setminus A$, $x \neq y$, $x \sim y$, then $[x] = [y]$ and $[x] \setminus \{x\} \subseteq A$ and $[y] \setminus \{y\} \subseteq A$. Then $y \in [x] \setminus \{x\}$, i.e. $\{y\} \subseteq [x] \setminus \{x\}$. Therefore

$$[y] = ([y] \setminus \{y\}) \cup \{y\} \subseteq ([y] \setminus \{y\}) \cup ([x] \setminus \{x\}) \subseteq A$$

contradicting the fact that $y \notin A$.

Now, for every $x \in B \setminus A$ let us select an element y_x so that:

1. $y_x \sim x$
2. $y_x \in A$.

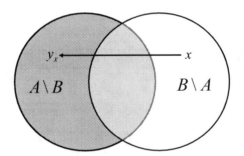

Fig. 1. Mapping $B \setminus A$ into $A \setminus B$

Figure 1 illustrates the fact that $B \setminus A$ can be injected into $A \setminus B$.

We observe that there is such mapping $x \mapsto y_x$ because we are in Case 2. Also, the mapping $x \mapsto y_x$ is an injection, i.e., $x_1 \neq x_2$ implies $y_{x_1} \neq y_{x_2}$. But, of course y_x belongs to $A \setminus B$ (because $y_x \sim x$ and $y_x \notin B$). Therefore we now have an injection of $B \setminus A$ into $A \setminus B$. But then, $|B \setminus A| \leq |A \setminus B|$. This contradicts the fact that $|A| < |B|$ and completes the proof.   □

The matroid $\mathcal{M}_Y$ is called the *matroid defined by Y and the approximation space* $(U, \sim)$.

4   Properties of Parameterized Matroids of Approximation Spaces, and Their Characterization

For a given approximation space $(U, \sim)$ we consider a *parameterized matroid* associated with the approximation space $(U, \sim)$ assuming that $\mathcal{M}^\sim = \{\mathcal{M}_Y : Y \in \mathcal{P}(U)\}$, where $\mathcal{M}_Y$ is a matroid defined by Y and the approximation space $(U, \sim)$, and $\mathcal{P}(U)$ is the powerset of U. Instead $\mathcal{M}^\sim$ we also write $\mathcal{M}$, for short.

We now show the following fact.

Proposition 1. *Let $\mathcal{M}$ be a parameterized matroid associated with the approximation space $(U, \sim)$. Then*

$$\underline{Z} = \underline{X} \text{ iff } \mathcal{M}_Z = \mathcal{M}_X \text{ for any } X, Z \in \mathcal{P}(U). \tag{2}$$

Proof: First assume that $\underline{Z} = \underline{X}$. Then if $A \in \mathcal{M}_Z$, then $\underline{A} \subseteq Z$ and thus $\underline{A} \subseteq \underline{Z}$. Therefore $\underline{A} \subseteq \underline{X}$, thus $\underline{A} \subseteq X$, so $A \in \mathcal{M}_X$. Therefore $\mathcal{M}_Z \subseteq \mathcal{M}_X$. But if $A \in \mathcal{M}_X$, then $\underline{A} \subseteq X$. Then $\underline{A} \subseteq \underline{X} = \underline{Z} \subseteq Z$. This completes implication $\Rightarrow$.

Conversely, let us assume that $\mathcal{M}_Z = \mathcal{M}_X$. We want to show that $\underline{Z} = \underline{X}$. If $\underline{Z} \neq \underline{X}$ then there is Y such that $Y = \underline{Y}$, $Y \neq \emptyset$ and $Y \subseteq \underline{Z}$, $Y \cap X = \emptyset$, or $Y \subseteq \underline{X}$, $Y \cap Z = \emptyset$.

We consider the first case, the other is similar. For that set Y, $Y \in \mathcal{M}_Z$ since $Y \subseteq \underline{Z} \subseteq Z$ and $Y = \underline{Y}$. But $Y \cap X = \emptyset$ so $\underline{Y} \cap X = \emptyset$, a contradiction.   □

Definition 2. *Let $(U, \sim)$ be an approximation space. Let $X \subseteq U$ be a crisp set, i.e., $X = \underline{X}$. We define $\mathcal{D}_X$ as the collection of all monads M such that $M \cap X = \emptyset$.*

We observe that the elements of $\mathcal{D}_X$ are pairwise disjoint and nonempty. Moreover, the union of all monades from $\mathcal{D}_X$ is equal to the lower approximation of $U \setminus X$, i.e., $\underline{U \setminus X} = \bigcup \mathcal{D}_X$. Let us assume that $X \neq U$. Then the family $\mathcal{D}_X$ possesses *selectors*, i.e., sets S such that $S \subseteq \bigcup \mathcal{D}_X$ and for all $D \in \mathcal{D}_X$, $|S \cap D| = 1$. We can now present the description of bases of matroids in $\mathcal{M}$.

Proposition 2 (Truszczynski). *Let $(U, \sim)$ be an approximation space, and let $\mathcal{M}$ be its parameterized matroid. Then for every set $X \neq U$ such that $X = \underline{X}$, the bases for $\mathcal{M}_X$ are precisely the sets of the form $U \setminus S$, where S is a selector for $\mathcal{D}_X$.*

Proof: A base B of $\mathcal{M}_X$ is an inclusion-maximal set in $\mathcal{M}_X$. This means that for any $x \notin B$, $B \cup \{x\}$ does not belong to $\mathcal{M}_X$, that is $\underline{B \cup \{x\}}$ is strictly larger than $\underline{B}$. But $\underline{B} = \underline{X}$. Thus $B \cup \{x\}$ contains at least one more monad M. This means that all the remaining elements of the monad M are already in B. But as this monad M was arbitrary among those not included in $\underline{X}$, we have that B is of the form $U \setminus S$ where S is a selector for the family $\{M \in U/\sim : M \cap \underline{X} = \emptyset\}$. The converse implication is obvious.   □

One consequence of Proposition 2 is that one can use a greedy algorithm to compute a maximal weight set roughly equivalent to a given set X.

To make this claim precise, let us say that subsets X and Y of U are *roughly equivalent* if and only if $\underline{X} = \underline{Y}$ and $\overline{X} = \overline{Y}$ [4]. The following property characterizes roughly equivalent sets $X, Y \subseteq U$: $[x] \subseteq X$ if and only if $[x] \subseteq Y$, for all $x \in U$.

A *weight function* on the set U is any function $wt : U \to \mathbf{R}^+$, where $\mathbf{R}^+$ is the set of all positive reals. The weight of a set $Z \subseteq U$ is equal to $\sum_{z \in Z} wt(z)$.

Our task now is, given $X \subseteq U$ to find a roughly equivalent to X set Y of maximum weight. Each basis of $\mathcal{M}_X$ is roughly equivalent to X and by Proposition 2 all we need to do is to find a selector for $\mathcal{D}_X$ of minimal weight. But such selector can be found by choosing in each element $[z]$ of $\mathcal{D}_X$ a single element of least possible weight (we observe that such element does not need to be unique).

Another class of sets associated with rough sets is that of *representative sets*[1].

A set $X \subseteq U$ is representative if $\overline{X} = U$, that is for every $x \in U$, there is $y \in X$ so that $x \sim y$. The parameterized matroid $\mathcal{M}^\sim$ associated with the approximation space $(U, \sim)$ determines a class of special representative sets. Specifically, let $X \subseteq U$ be a crisp set in an approximation space $(U, \sim)$, i.e., $X = \underline{X}$. Then we characterize the minimal representative sets including X that belong to the matroid $\mathcal{M}_X$ as follows.

Proposition 3. *Let $X \subset U$ be a crisp set in an approximation space $(U, \sim)$, i.e., $X = \underline{X}$. Then the minimal representative sets including X belonging to the matroid $\mathcal{M}_X$ are precisely the sets of the form $X \cup S$ where S is a selector for $\mathcal{D}_X$.*

Given Proposition 3, a greedy algorithm can be used to find the minimal representative set of minimal weight.

We now list a number of properties of the parameterized matroid $\mathcal{M}^\sim$.

Proposition 4. *1. For every $X \in \mathcal{P}(U)$, $X \in \mathcal{M}_X$.*
2. For every $X \in \mathcal{P}(U)$, $\mathcal{M}_X = \mathcal{M}_{\underline{X}}$.
3. For all $X, Y \in \mathcal{P}(U)$, $X \subseteq Y$ implies $\mathcal{M}_X \subseteq \mathcal{M}_Y$.
4. For every $X \in \mathcal{P}(U)$, $\underline{X}$ is the $\subseteq$-least set in $\mathcal{M}_X \setminus \bigcup \{\mathcal{M}_Y : \mathcal{M}_Y \subset \mathcal{M}_X\}$.
5. The family $\{\underline{X} : X \in \mathcal{P}(U)\}$ forms a Boolean algebra.
6. For all $X, Y \in \mathcal{P}(U)$ if $Y \in \mathcal{M}_X$ than $Y \setminus \underline{X} = \emptyset$.
7. For all $X, Y, Z \in \mathcal{P}(U)$ if $Y = \underline{X} \cup Z$ and $\overline{Z} = \emptyset$ then $Y \in \mathcal{M}_X$.
8. If $X \subseteq Z \subseteq Y$ and $\mathcal{M}_X = \mathcal{M}_Y$ then $\mathcal{M}_X = \mathcal{M}_Z$.

Points (1)–(8) are almost obvious, except possibly (4).

But the same points provide a key to the answer to the following question:

Given a parameterized matroid

$$\mathcal{N} = \{\mathcal{N}_X : X \in \mathcal{P}(X)\},$$

---

[1] Note that in [4, 5] such sets are called externally or totally undefinable relative to a given approximation space. Such sets were also used by Pawlak in investigating the notion of rough truth [10].

when there exists an approximation space $(U, \approx)$ so that

$$\mathcal{N} = \mathcal{M}^{\approx}.$$

Specifically, we will formulate seven abstract conditions, corresponding to points (1)–(7) above and show that under these conditions, indeed the parameterized matroid is determined by an approximation space $(U, \approx)$ that is determined by a parameterized matroid $\mathcal{N}$.

So let $\mathcal{N} = \langle \mathcal{N}_X \rangle_{X \in \mathcal{P}(U)}$ be a parameterized matroid. We formulate *conditions* (A)-(F) that $\mathcal{N}$ needs to satisfy.

(A) For all $X \in \mathcal{P}(U)$, $X \in \mathcal{N}_X$.

(B) For all $X, Y \in \mathcal{P}(U)$, $X \subseteq Y$ implies $\mathcal{N}_X \subseteq \mathcal{N}_Y$.

(C) For all $X \in \mathcal{P}(U)$, the family

$$\mathcal{N}_X \setminus \bigcup \{ \mathcal{N}_Y : \mathcal{N}_Y \subset \mathcal{N}_X \},$$

possesses a $\subseteq$-least element, further referred as $[X]$.

(D) The family $\{[X] : X \in \mathcal{P}(U)\}$ forms a Boolean Algebra, further referred as $\mathcal{B}_{\mathcal{N}}$, or simply $\mathcal{B}$.

(E) For all $X \in \mathcal{P}(U)$, $\mathcal{N}_X = \mathcal{N}_{[X]}$.

(F) For all $X, Y \in \mathcal{P}(U)$, if $Y \in \mathcal{N}_X$ then $[Y \setminus [X]] = \emptyset$.

(G) For all $X, Y, Z \in \mathcal{P}(U)$, if $Y = [X] \cup Z$ and $[Z] = \emptyset$ than $Y \in \mathcal{N}_X$.

Once $\mathcal{N}$ is a parameterized matroid satisfying conditions (A)-(F), we define a relation $\approx$ in U by setting:

$x \approx y$ if and only if there is an atom A of $\mathcal{B}$ such that $x \in A$ and $y \in A$.

It is easy to see that (under conditions (A)-(G), in particular condition (D), we have

Proposition 5. $x \approx y$ *if and only if*

for every $X \subseteq U, x \in [X]$ if and only if $y \in [X]$.

One can also observe the following fact:

Proposition 6. *Let $\mathcal{N} = \langle \mathcal{N}_X \rangle_{X \in \mathcal{P}(U)}$ be a parameterized matroid satisfying conditions (A)-(G). Then for any $Y \subseteq U$ we have $[Y] = \underline{Y}$, where $\underline{Y}$ is the lower approximation of Y in the approximation space $(U, \approx)$ and $[Y] \in \mathcal{B}$.*

Proof: Let us assume $x \in [Y]$. Then from Proposition 5 we have $y \in Y$ for $y \approx x$. Since $[Y] \subseteq Y$, we obtain $[x]_{\approx} \subseteq Y$, i.e., $x \in \underline{Y}$.

Now let us assume $x \in \underline{Y}$, i.e., $[x]_{\approx} \subseteq Y$. Suppose that $x \notin [Y]$. Then by Proposition 5 we have $[x]_{\approx} \subseteq U \setminus [Y]$. Hence, $[x]_{\approx} \subseteq Y \setminus [Y]$, a contradiction with (F) (where we take $X = Y$). $\square$

We now show the main result of this section.

Proposition 7. *Let $\mathcal{N} = \langle \mathcal{N}_X \rangle_{X \in \mathcal{P}(U)}$ be a parameterized matroid. Then $\mathcal{N}$ is a parameterized matroid defined by some approximation space, i.e., $\mathcal{N} = \mathcal{M}^{\sim}$ for some approximation space $(U, \sim)$ if and only if $\mathcal{N}$ satisfies conditions (A)-(G) above.*

Proof: By Proposition 4, if $\mathcal{N}$ is a parameterized matroid for an approximation space, then $\mathcal{N}$ satisfies conditions (A)–(G).

Conversely, if $\mathcal{N}$ satisfies conditions (A)-(G), then we show that $\mathcal{M}^{\approx} = \mathcal{N}$. That is we show that for every $X \subseteq U$, $\mathcal{N}_X = \mathcal{M}^{\approx}_X$.

First, assume $Y \in \mathcal{M}^{\approx}_X$. The set $Y \setminus [Y]$ is sparse w.r.t. $\approx$, i.e., $\underline{Y \setminus [Y]} = \emptyset$, where the lower approximation is relative to the approximation space $\overline{(U, \approx)}$. By proposition 6 we obtain $[y \setminus [Y]] = \emptyset$. By condition (G), $[X] \cup (Y \setminus [Y]) \in \mathcal{N}_X$. But $\underline{Y} \subseteq X$. Hence, by Proposition 6 $[Y] \subseteq X$. Therefore $[Y] \subseteq [X]$. But then, $Y = [Y] \cup (Y \setminus [Y]) \subseteq [X] \cup (Y \setminus [Y]) \in \mathcal{N}_X$, as desired.

Conversely, let $Y \in \mathcal{N}_X$. By Proposition 6 we need only to show that $[Y] \subseteq X$. But $Y \in \mathcal{N}_X$ means (see (F)) that $[Y] \subset [X]$ or $[Y] = [X]$. In either case, as $[X] \subseteq X$, we have $[Y] \subseteq X$, that is $\underline{Y} \subseteq X$, by Proposition 6. Hence $Y \in \mathcal{M}^{\approx}_X$. This completes the argument.   □

5   Conclusions

In the paper we have presented some relationships between rough set theory and matroid theory. We plan to explore possibilities of application of heuristics based on combinatorial optimization developed in matroid theory to algorithmic problems in rough set theory.

Acknowlegements. The authors acknowledge valuable discussions with Professor Miroslaw Truszczynski. Andrzej Skowron was supported by the Polish National Science Centre grant DEC-2012/05/B/ST6/03215 as well as by the Polish National Centre for Research and Development (NCBiR) under the grant SYNAT No. SP/I/1/77065/10 in frame of the strategic scientific research and experimental development program: "Interdisciplinary System for Interactive Scientific and Scientific-Technical Information" and the grant No. O ROB/0010/03 /001 in frame of the Defence and Security Programmes and Projects: "Modern engineering tools for decision support for commanders of the State Fire Service of Poland during Fire & Rescue operations in buildings".

References

1. Liu, Y., Zhu, W.: Parametric matroid of rough set. arXiv: 1209.4975, 1–15 (2012)
2. Tanga, J., Shea, K., Minb, F., Zhu, W.: A matroidal approach to rough set theory. Theoretical Computer Science 471, 1–11 (2013)
3. Pawlak, Z.: Rough sets. International Journal of Computer and Information Sciences 11, 341–356 (1982)
4. Pawlak, Z.: Rough Sets: Theoretical Aspects of Reasoning about Data. System Theory, Knowledge Engineering and Problem Solving, vol. 9. Kluwer Academic Publishers, Dordrecht (1991)
5. Pawlak, Z., Skowron, A.: Rudiments of rough sets. Information Sciences 177, 3–27 (2007)
6. Pawlak, Z., Skowron, A.: Rough sets: Some extensions. Information Sciences 177, 28–40 (2007)

7. Pawlak, Z., Skowron, A.: Rough sets and Boolean reasoning. Information Sciences 177, 41–73 (2007)
8. Edwards, J.: Matroids and the greedy algorithm. Mathematical Programming 1, 127–136 (1971)
9. Oxley, J.: Matroid Theory. Oxford University Press, Oxford (2006)
10. Pawlak, Z.: Rough logic. Bulletin of the Polish Academy of Sciences, Technical Sciences 35, 253–258 (1987)

An Efficient Approach for Fuzzy Decision Reduct Computation

P.S.V.S. Sai Prasad and C. Raghavendra Rao

School of Computer and Information Sciences, University of Hyderabad
Hyderabad, Andhra Pradesh, India
{saics,crrcs}@uohyd.ernet.in

Abstract. Fuzzy rough sets is an extension of classical rough sets for feature selection in hybrid decision systems. However, reduct computation using the fuzzy rough set model is computationally expensive. A modified quick reduct algorithm (MQRA) was proposed in literature for computing fuzzy decision reduct using Radzikowska-Kerry fuzzy rough set model. In this paper, we develop a simplified computational model for discovering positive region in Radzikowska-Kerry's fuzzy rough set model. Theory is developed for validation of omission of absolute positive region objects without affecting the subsequent inferences. The developed theory is incorporated in MQRA resulting in algorithm Improved MQRA (IMQRA). The computations involved in IMQRA are modeled as vector operations for obtaining further optimizations at implementation level. The effectiveness of algorithm(s) is empirically demonstrated by comparative analysis with several existing reduct approaches for hybrid decision systems using fuzzy rough sets.

Keywords: Fuzzy rough sets, Hybrid decision systems, Reduct, Quick Reduct, Fuzzy decision reduct.

1 Introduction

Rough sets [38], developed by Prof. Z. Pawlak [39], has emerged as an important soft computing paradigm being applied for several data mining and machine learning applications [20,40]. Feature selection using reduct based on rough set principles is extensively employed in several application domains [5,20,35]. Rough sets provide a non invasive data mining approach for knowledge discovery in databases (KDD) [11,50]. The process of knowledge discovery in a given decision system primarily consists of reduct computation as the preprocessing step for dimensionality reduction. But classical rough sets are applicable to decision (or information) systems with qualitative attributes.

Hybrid decision systems contain a mixture of qualitative and quantitative attributes and occur frequently in real world decision systems. The classical definitions of rough sets are based on an indiscernibility relation, which is an equivalence relation. Hence under indiscernibility relation using a quantitative attribute, two objects will be unrelated even though they have near values on

J.F. Peters and A. Skowron (Eds.): Transactions on Rough Sets XVII, LNCS 8375, pp. 82–108, 2014.
© Springer-Verlag Berlin Heidelberg 2014

a real-values scale. A reduct computed thus would contain primary key like attributes and leads to classifiers with less generalization capacity.

Hence classical rough sets cannot be applied directly to hybrid decision systems for reduct computation. Traditionally indiscernibility relation using quantitative attribute was defined after discretization. The process of discretization converts a quantitative attribute into a qualitative attribute. A discretization algorithm places cuts in the domain of quantitative attribute and divides the continuous domain into non overlapping intervals (bins). Two objects having values in the same interval are assigned the same symbolic label. Optimal way of finding cuts was proved to be NP-Hard [31]. Hence heuristic based sub optimal discretization algorithms were used [57,58] and a widely used approach which is based on rough set principles and boolean reasoning was given in [33].

Discretization approach provided a computationally effective way for dealing with hybrid decision systems using classical rough sets. But information loss is inevitable in any discretization approach. Slowinski et al., in [54], reported that the primary issue in data preprocessing for rough set analysis is with the uncertainty arising out of discretization of quantitative attributes. Based on repeated experiments using several discretization approaches, it was observed that no single discretization approach is appropriate for all decision systems [4]. In [18] it is empirically established that decision systems having attributes conforming to statistically highly skewed distributions or having high peaks are resulting in higher classification errors irrespective of discretization method used. It is also observed in [46] that discretization proves to be ineffective and leads to more information loss and results in significant classification error in inconsistent decision systems (having overlapping feature space of objects belonging to different decision classes). Approaches like soft cuts [32] and fuzzy discretization [46] were introduced to represent the ambiguity present in the neighborhood region of crisp cuts. These approaches were proved to be effective in reducing the classification error for inconsistent decision systems [35,37].

The above mentioned improvements for discretization were intended at building efficient classifiers and not oriented at developing reduct computation approaches for hybrid decision systems. Two fundamental issues of rough sets, as stated in [67] are, representations of indiscernibility/discernibility and attribute reduction based on indisernibility/discernibility. Hence attribute reduction approaches for hybrid decision systems were formulated by defining new representations of indiscernibility relation.

Alternative approaches for reduct computation in hybrid decision systems (without relying on discretization) were evolved by generalizing the classical rough set theory. Generalization was primarily meant to relax the equivalence properties of indisernibility relation and using similarity relation which are more general binary relations [14,65]. These approaches lead to significant advancements in rough set theory both from theoretical and application view points. Similarity relations [25,34,56] allow a limited degree of variability in the quantitative attribute values and the degree is determined by a threshold value. The similarity class of an object is defined as the collections of related objects under similarity relation [55]. The rough approximations are defined based on similarity classes of objects. In particular, tolerance rough sets [41,28,51,53] are based on reflexive and symmetric tolerance relations and reduct of hybrid decision system under tolerance rough sets is named as tolerance reduct. Determining appropriate threshold value is essential in tolerance rough sets for obtaining effective feature selection and classifier performance. Some of the approaches for threshold determination are using genetic algorithm [24] and EM algorithm [12].

Under tolerance relation, an object is related to all objects at varying distance as long as the threshold requirement is satisfied. A more semantically apt representation was obtained by bringing in fuzzy context. The fuzzy similarity relation [13] defines the degree of relatedness between any two objects based on the quantitative attribute values. Fuzzy similarity relation results in representation of approximate equality or graded indisernibility [7] of objects based on quantitative attribute values. Among the generalizations of rough sets for reduct computation [9,10,15,27,30,42,64] fuzzy rough sets [9,10] evolved to be an important generalization of rough sets. Fuzzy rough sets are extensively used for reduct computation in hybrid decision systems.

Fuzzy rough sets and rough fuzzy sets were introduced by Dubois and Prade [9,10] which have become standard approaches. The approximation of a crisp set in a fuzzy approximation space is called as a fuzzy rough set, and the approximation of a fuzzy set in a crisp approximation space is called as rough fuzzy set [21]. The modifications of this standard approach has been addressed in [17,45,61]. Radzikowska and Kerre [45] has given a generalized approach for Dubois-Prade fuzzy rough set model in the form of fuzzy rough set, defined by an implication and a triangular norm (t-norm) . Hu et al. [17] developed a new fuzzy rough set model using fuzzy generalized definitions of subset and intersection of sets in classical notions of lower and upper approximations. Wang et al. [61] developed a fuzzy rough set model based on thresholds on similarity degree for inclusion into lower and upper approximations. Qian et al. [62] has given a comparative study of these fuzzy rough set models. Qian et al. [62] established that Hu's,Wang's models are derivable from Dubois-Prade's model whereas Dubois-Prade's model is derivable from Radzikowska-Kerry's model. Thus Radzikowska-Kerry's model is a generalized model in the class of fuzzy rough set models.

Research has been advanced in fuzzy rough set models and also in reduct computation (feature selection) using these models in hybrid decision systems. Fuzzy rough feature selection (FRFS) was introduced by Jensen et al. [19] based on Dubois-Prade's fuzzy rough set model. In FRFS approach every quantitative attribute is represented by a set of linguistic variables (fuzzy sets) resulting in fuzzy partition of the quantitative attribute [54]. The reduction algorithm was of positive region based heuristic approach similar to quick reduct algorithm [5] called as fuzzy rough quick reduct algorithm. The computational efficiency of fuzzy rough quick reduct algorithm is improved using the concept of compact computational domain by Bhatt et al. [3]. Later Jensen et al. [22] proposed reduct algorithms like B-FRFS, L-FRFS, FDM which are computationally more efficient than FRFS approach. Hu et al. [16] provided a reduct algorithm based on fuzzy conditional entropy. Hu et al. [17] developed FAR-VPFRS algorithm (Forward attribute reduction based on variable precision fuzzy rough model) using Hu's Fuzzy Rough Set model [17] which is a positive region heuristic based algorithm. Qian et al. [44] has given an improved feature selection algorithm based on forward approximation (FA-FSCE). FA-FSCE algorithm incorporates positive region removal (positive approximation) in an iteration of reduct finding process using Hu's fuzzy rough set model and fuzzy conditional entropy. Cornelis et al., [7] introduced the concept of fuzzy decision reduct and provided a modified quick reduct algorithm (MQRA) for finding fuzzy decision super reduct based on Radzikowska-Kerry's fuzzy rough set model. MQRA is an extended version of classical quick reduct algorithm [5] for hybrid decision systems and is a positive region heuristic based algorithm.

Finer and apt knowledge representation in hybrid decision system is achieved monotonically from classical rough sets to tolerance rough sets to fuzzy rough sets. The effective enhancement of representation incurs more computational and space complexities. Reduct computation using fuzzy rough sets incurs higher time and space complexities compared to classical rough sets. Slezak et al. in [52] established that fuzzy similarity relation can be approximated by an averaging of several random discretizations of quantitative attribute. Hence a complex problem of reduct computation using fuzzy rough sets, especially for large decision systems, can be approximated well by reduct computations on several randomly discretized decision systems which can be performed on distributed systems in parallel. In [49] authors have provided an alternative approach which evolves a reduct computed on discretized decision system as a seed for MQRA algorithm for finding fuzzy decision super reduct, making computation of fuzzy decision super reduct viable for large hybrid decision systems. The present study aims at effective fuzzy decision reduct computation by reducing practical space and time complexities of MQRA algorithm. It is expected that the proposed method provides solution with less time and space complexities than MQRA for moderate hybrid decision systems whereas a hybrid of this with seed based approach [49] as a preprocessor will further improve the performance even for large hybrid decision systems.

In classical rough set reduct finding approaches based on sequential forward selection using heuristics such as gamma measure, information entropy measure [2,26,43,47,48,68] an optimization principle known as positive region removal or positive approximation was found to be very effective in reduction of space and time complexities. Using this principle, obtained positive region objects in an iteration are omitted for further iterations without affecting the subsequent inferences. The approach is evident in classical rough sets as positive region equivalence classes are disjoint from non positive region equivalence classes in the partition induced by indiscernbility relation. But in fuzzy rough set approach computing positive region membership of an object requires computations with respect to every object of the universe. This necessitates theoretical validation of removal of objects which belong absolutely in positive region in fuzzy rough set model based reduct computation approaches. There exists generalized rough set models like variable precision rough set model [69] in which monotonic criteria of positive region membership of an object is not satisfied [66]. Hence it is required to establish that in the generalized model of fuzzy rough sets monotonic criteria of positive region membership holds which is the precondition for positive region removal aspect. The positive region removal (positive approximation) was used in [44] for Hu's fuzzy rough set model. Present study establishes theoretically positive region removal aspect in Radzikowska-Kerry's fuzzy rough set model.

This paper addresses mainly three aspects. The first is developing a simplified computational model for Radzikowska-Kerry's fuzzy rough set model and addressing the positive region removal issues. Second is to incorporate the simplified computational model and positive region removal in MQRA resulting in Improved MQRA (IMQRA). The third part involves modeling the computations in IMQRA with vector based operations so that further computational efficiency is obtained. The Matlab implemented versions of MQRA, IMQRA are referred as MQRA_MW, IMQRA_MW respectively. The computational efficiency of MQRA_MW, IMQRA_MW is established empirically by performing comparative analysis with existing FSCE and FA-FSCE algorithms [44] and the algorithms FRFS, B-FRFS, L-FRFS, FDM [22].

The paper is organized as follows. Section 2 gives the overview of fuzzy decision reduct for hybrid decision system and gives MQRA algorithm. Section 3 provides the theoretical basis for the improvements to MQRA algorithm and gives IMQRA algorithm. Section 3 also discusses the implementation aspects for benefitting the Matlab environment and analyzes the complexity of the algorithm. Section 4 demonstrates the proposed concepts using a simple example. Experimental Results are provided in section 5. Analysis of results is dealt in section 6.

2   Overview of Fuzzy Decision Reduct

Let $HDT = \left(U, C^h = C^c \cup C^n, \{d\}, \{V_{a_c}, f_{a_c}\}_{a_c \in C^c \cup \{d\}}, \{V_{a_n}, f_{a_n}\}_{a_n \in C^n} \right)$ be a hybrid decision system. Here U is the set of objects, C^c is set of qualitative (categorical) attributes and C^n is set of quantitative (numerical) attributes and C^h is the set of hybrid conditional attributes comprising of C^c and C^n. For every qualitative attribute $a_c \in C^c \cup \{d\}$, V_{a_c} denotes set of finite domain of values of attribute a_c and $f_{a_c} : U \to V_{a_c}$ is a mapping of assigning a symbol in V_{a_c} to every object in U. For every quantitative attribute $a_n \in C^n$, V_{a_n} has a range, a finite interval on real line with normalized range between 0 and 1 and $f_{a_n} : U \to V_{a_n}$ is a mapping assigning a value in V_{a_n} to every object in U. Notation $a_n(x)$ in place of $f_{a_n}(x)$ for quantitative attributes, $a_c(x)$ in place of $f_{a_c}(x)$ for qualitative attributes being used in the rest of the paper for simplicity. This paper deals with decision systems and hence decision attribute d is taken as qualitative attribute. There are approaches for computation of reduct for hybrid regression systems where in d is a quantitative attribute [7] and will be explored in our future work.

This section provides associated definitions, terminology, concepts for fuzzy rough sets and fuzzy decision reduct, for completeness, based on [6,7,45]. In classical rough sets, the approximation space is defined as (U, R), where U is the set of objects and R is the indiscernbility relation defined on U satisfying the properties of an equivalence relation. A concept $A \subseteq U$ is approximated with the knowledge of R as lower and upper approximations. An object $y \in U$ belongs to lower approximation of A ($R \downarrow A$) iff the equivalence class of y with respect to R is contained in A and belongs to upper approximation of A ($R \uparrow A$) iff the equivalence class of y with respect to R has nontrivial intersection with A. Hence,

$$y \in R \downarrow A \iff [y]_R \subseteq A. \tag{1}$$

$$y \in R \uparrow A \iff [y]_R \cap A \neq \phi. \tag{2}$$

The equivalent forms of Eq. (1), Eq. (2) are [6] given in Eq. (3), Eq. (4).

$$y \in R \downarrow A \iff (\forall x \in U)\,((x, y) \in R \Rightarrow x \in A). \tag{3}$$

$$y \in R \uparrow A \iff (\exists x \in U)\,((x, y) \in R \wedge x \in A). \tag{4}$$

2.1   Fuzzy Rough Set Theory

Similarity relation has been introduced [9,10,15,27,30,64] for addressing the issues associated with discretization instead of indiscernbility relation through fuzzification. The theory developed in this direction is known as fuzzy rough set theory. In fuzzy rough set theory the similarity relation is a fuzzy tolerance relation or a fuzzy equivalence relation. A fuzzy relation R on $U \times U$ is said to be a fuzzy tolerance relation if,

$$R(x, x) = 1 \ \forall x \in U. \tag{5}$$

$$R(x, y) = R(y, x) \ \forall x, y \in U. \tag{6}$$

R becomes a fuzzy equivalence relation with an additional requirement of Γ-transitivity using a given t-norm Γ. The Γ-transitivity property is

$$\Gamma(R(x, y), R(y, z)) \leq R(x, z) \ \forall x, y, z \in U. \tag{7}$$

In modeling fuzzy rough set based system for HDT, the fuzzy tolerance or equivalence relation is defined as R_{a_n} ($\forall a_n \in C^n$). Hence $R_{a_n}(x, y)$ for objects x, y denote the degree of similarity between x and y using the attribute values of a_n. If a_c is a qualitative attribute then the resulting similarity relation is taken as the classical indiscernbility relation and hence is defined as,

$$R_{a_c}(x, y) = \begin{cases} 1 & if \quad a_c(x) = a_c(y), \\ 0 & if \quad a_c(x) \neq a_c(y). \end{cases} \tag{8}$$

The similarity relation is extended for a set of attributes using a specified t-norm Γ, i.e. given $B \subseteq C^h \cup \{d\}$, fuzzy tolerance or equivalence relation is extended as,

$$R_B(x,y) = \Gamma \left(\underbrace{R_a(x,y)}_{a \in B} \right) \ \forall x, y \in U. \tag{9}$$

In classical rough sets, concept to be approximated is a subset of U corresponding to an equivalence class associated with a decision value. In fuzzy rough set theory the concept to be approximated is generalized into a fuzzy set A. Each object x has a degree of membership into concept A with value $A(x)$. In HDT, $R_{d,y}$ of each object y defines a fuzzy concept where,

$$R_{d,y}(x) = R_d(x,y) \ \forall x \in U. \tag{10}$$

Definition of $R_{d,y}(R_d\text{-foreset})$ is general and not restricted only to decision variable. It is to be noted that as d is a qualitative attribute, $R_{d,y}$ represents the characteristic function of set of y's decision class objects.

Radzikowska-Kerry's Fuzzy Rough Set Model. In a fuzzy rough set model, the concepts of lower and upper approximation of a fuzzy set A defined on U are defined using a fuzzy Γ-equivalence relation (or fuzzy tolerance relation) R in U. In Dubois-Prade's fuzzy rough set model [9,10], given a fuzzy Γ-equivalence relation (or a fuzzy tolerance relation) in U, the lower and upper approximations of a fuzzy concept A are defined $\forall y \in U$ as,

$$\mu_{\underline{R}(A)}(y) = \inf_{x \in U} \max\left(1 - R(x,y), A(x)\right). \tag{11}$$

$$\mu_{\overline{R}(A)}(y) = \sup_{x \in U} \min\left(R(x,y), A(x)\right). \tag{12}$$

Eq. (11) and Eq. (12) are fuzzy generalized versions of lower and upper approximations in classical rough set model given in Eq. (3) and Eq. (4) using Kleene-Dienes Implication [1] and minimum t-norm [63]. Radzikowska-Kerry's model [45] generalizes Dubois-Prade's model by a pair of implication $\Im$ and t-norm Γ as,

$$R \downarrow A(y) = \inf_{x \in U} \Im\left(R(x,y), A(x)\right). \tag{13}$$

$$R \uparrow A(y) = \sup_{x \in U} \Gamma\left(R(x,y), A(x)\right). \tag{14}$$

The above definitions treat lower and upper approximations as fuzzy sets and for each object its degree of membership into lower and upper approximation is defined. Similarly positive region is defined. Given $B \subseteq C^h$ the fuzzy B-positive region of an object $y \in U$ is defined as,

$$POS_B (y) = \left(\underset{x \in U}{\cup} R_B \downarrow R_{d,x} \right) (y). \tag{15}$$

Hence the positive region of object y is the maximum membership degree of lower approximation of R_B into R_d-foreset of each object in U. As the decision attribute is qualitative, using proposition-1 given in [7], the fuzzy B-positive region computation requires computing lower approximation of $R_{d,y}$ only. Hence Eq. (15) simplifies to,

$$POS_B (y) = R_B \downarrow R_{d,y} (y). \tag{16}$$

Fuzzy Decision Reduct. Reduct for HDT using the definitions of lower approximation and positive region given in section 2.1.1 is called as fuzzy decision reduct. Fuzzy decision reduct [7] is defined as given below.

Definition 1. *Let M be monotonic function from $\wp (U) \rightarrow [0,1]$ ($\wp (U)$ denotes power set of U) such that $M (C^h) = 1$. A set of attributes B subset of C^h is said to be a fuzzy M-decision super reduct to degree α if $M (B) \geq \alpha$. It is called a fuzzy M-decision reduct to degree α if $\forall B^* \subset B$, $M (B^*) < \alpha$.*

The measure M which satisfies the required properties and in extension to the classical rough set based gamma measure is,

$$\gamma_B = \frac{|POS_B|}{|POS_{C^h}|}. \tag{17}$$

where $|POS_A| = \sum_{x \in U} POS_A (x)$ for $A \subseteq C^h$.

2.2 Modified Quick Reduct Algorithm

Quick reduct algorithm (QRA) for classical rough sets, given by Chouchoulas et al. [5], is a sequential forward selection (SFS) based algorithm using positive region heuristic for arriving at a super reduct for a given decision system. In QRA the recommended reduct B is initialized to empty set. In each iteration an attribute is included into B which gives maximum increase in gamma measure. The algorithm terminates when $\gamma_B (\{d\}) = \gamma_{C^h} (\{d\})$. Modified quick reduct algorithm (MQRA) [7] is formulated for finding fuzzy M-decision super reduct. For completeness, the algorithm is reproduced from [7] in Algorithm 1 using the notations of this paper.

The time complexity of computing $M (B)$ in HDT is $O \left(|U|^2 \right)$ based on Eq. (17). Hence the time complexity of MQRA is $O \left(|C^h|^2 |U|^2 \right)$.

Quick Reduct algorithm has been modified for giving better computational performance in IQuickReduct algorithm (IQRA) [47]. Inspired to improve MQRA in that direction the following section provides schematic development of an algorithm Improved Modified Quick Reduct (IMQRA).

Algorithm 1. Modified Quick Reduct

Input :

$HDT = \left(U, C^h = C^c \cup C^n, \{d\}, \{V_{a_c}, f_{a_c}\}_{a_c \in C^c \cup \{d\}}, \{V_{a_n}, f_{a_n}\}_{a_n \in C^n} \right),$

Measure M,

Degree α

Output: Fuzzy M−decision super reduct B

$B = \phi$

repeat

    $T = B$

    $best = -1$

    foreach $a \in C^h - B$ **do**

        if $M\left(B \cup \{a\}\right) \geq best$ **then**

            $T = B \cup \{a\}$

            $best = M\left(B \cup \{a\}\right)$

        end

    end

    $B = T$

until $M\left(B\right) \geq \alpha$

Return B.

---

3   Improvements to Modified Quick Reduct Algorithm

IQRA described in [47] works similar to QRA but differs in situations where there is no increase in gamma measure with any of the available attributes. This ambiguous situation is handled by using other heuristics to determine the attribute to be included into recommended reduct B. The heuristic followed in IQRA is based on variable precision rough set (VPRS) [69] calculations. If a positive gamma gain is found the attribute leading to the maximum gain is included into B and objects which are in positive region are removed which decreases the computational time for future iterations.

Occurrence of ambiguous situations in MQRA is a rarity due to $M(B)$ being a continuous value which increases till the required $M\left(C^h\right)$ is obtained. Hence improvements to MQRA are aimed at obtaining positive region removal aspect in MQRA. Validation of positive region removal aspect is essential in fuzzy rough set models for proving the monotonic criteria of positive region membership and also that omission of objects does not affect the subsequent inferences. The later is essential as computation of positive region membership of an object involves fuzzy similarity values of the object with all members of U.

In Section 3.1 necessary theory is developed for simplified computation model for lower approximation and positive region in Radzikowska-Kerry's fuzzy rough set model for decision systems and validation of positive region removal in MQRA algorithm using Radzikowska-Kerry's fuzzy rough set model. Section 3.2 contains the proposed algorithm IMQRA and Section 3.3 describes the modeling of IMQRA computations using vector based operations suitable for environments such as Matlab [29]. Section 3.4 contains the time and space complexity analysis of IMQRA.

3.1   Theoritical Foundations

The Radzikowska-Kerry's fuzzy rough set model described in Section 2 is characterized by a fuzzy tolerance or an Γ-equivalence relation R, an implication $\Im$, a t-norm Γ and the definition of M. The discussion in our approach assumes that R is a fuzzy tolerance relation, M is the gamma measure defined in Eq. (17).

For any object $y \in U$ let $U_1(y)$ denote the set of objects belonging to the decision class of y and let $U_2(y)$ denote the remaining objects belonging to other decision classes. The natural negation generated by an implication $\Im$ according to [1] is as given below.

Definition 2. *Let $\Im$ be any fuzzy implication. The natural negation of $\Im$, denoted by $N_\Im$, is given by $N_\Im(x) = \Im(x,0)\ \forall x \in [0,1]$.*

Given an implication $\Im$, Lemma 1 derives the simplified formula for positive region POS_B.

Lemma 1. *For any $y \in U$ and $B \subseteq C^h$, using fuzzy tolerance relation R_B and implication $\Im$, if $U_2(y) \neq \phi$ then*

$$POS_B(y) = \min_{x \in U_2(y)} N_\Im(R_B(x,y))$$

else $POS_B(y) = 1$ where $N_\Im$ is the natural negation generated by $\Im$.

Proof. Using Eq. (16),

$$POS_B(y) = R_B \downarrow R_{d,y}(y)$$
$$= \inf_{x \in U} \Im(R_B(x,y), R_d(x,y))$$
$$= \min_{x \in U} \Im(R_B(x,y), R_d(x,y)) \text{(since U being finite, inf is equal to}$$

min)

$$= \min\left(\min_{x \in U_1(y)} \Im(R_B(x,y), R_d(x,y)),\ \min_{x \in U_2(y)} \Im(R_B(x,y), R_d(x,y))\right)$$
$$= \min\left(\min_{x \in U_1(y)} \Im(R_B(x,y), 1),\ \min_{x \in U_2(y)} \Im(R_B(x,y), 0)\right) \text{(since d being quali-}$$

tative using Eq. (8))

$$= \min\left(\min_{x \in U_1(y)} 1,\ \min_{x \in U_2(y)} N_\Im(R_B(x,y))\right) \text{(Using neutrality of truth property}$$

[59] $\forall x \in [0,1]\ \Im(x,1) = 1$ and Definition 2)

$$= \min\left(1,\ \min_{x \in U_2(y)} N_\Im(R_B(x,y))\right).$$

Hence if $U_2(y) \neq \phi$

then

$$POS_B(y) = \min_{x \in U_2(y)} N_\Im(R_B(x,y))$$

else

$$POS_B(y) = 1. \qquad \blacksquare$$

From Lemma 1 it can be observed that only $N_\Im$ instead of $\Im$ is needed in computing lower approximation and positive region while using Radzikowska-Kerry's

$(\Im, \Gamma)$-fuzzy rough set model for decision systems. As [1,59] $N_\Im$ is a fuzzy negation, an useful conclusion is that for decision systems Radzikowska-Kerry's fuzzy rough set model is completely defined by a pair of fuzzy negation N and t-norm Γ and can be specified as (N, Γ)-fuzzy rough set model. This observation facilitates experimenter in varying only the fuzzy negation for different experiments of reduct computation using Radzikowska-Kerry's model as experimenting with different implications generating the same natural negation has no impact in the outcome.

Confining to similarity between objects defined over quantitative attributes, in practice, strong (and hence continuous) involutive [59] fuzzy negations are preferred and used. Considering standard negation $N(x) = 1 - x$ which is often used, the simplified equation for positive region based on Lemma 1 is given in Eq. (18).

$$POS_B (y) = R_B \downarrow R_{d,y} (y) = \begin{cases} \min\limits_{x \in U_2(y)} (1 - R_B (x, y)) & if \ U_2 (y) \neq \phi \\ 1 & Otherwise \end{cases} \quad (18)$$

It is observed that Eq. (18) coincides with lower approximation obtained in Dubois-Prade's model [8] for approximating a crisp set. In [62], while comparing fuzzy rough set models, it is established that Radzikowska-Kerry's model is equal to Dubois-Prades model by using a pair of Kleene-Dienes's implication $\Im_{KD}$ and minimum t-norm Γ_M. In view of the above discussion, for decision systems using (N, Γ)-fuzzy rough set model it is further stated that, Radzikowska-Kerry's model becomes equal to Dubois-Prade's model by using a pair of standard negation and minimum t-norm Γ_M.

In rest of the paper, Eq. (18) is used for computation of $POS_B (y)$. To derive a formula similar to positive region removal monotonic criteria for $POS_B (y)$ is required. Lemma 2 proves the monotonic property of $POS_B (y)$.

Lemma 2. *If $B_1 \subseteq B_2 \subseteq C^h$, then for any $y \in U$, $POS_{B_1} (y) \leq POS_{B_2} (y)$.*

Proof. If $U_2 (y) = \phi$ then trivially $POS_{B_1} (y) = POS_{B_2} (y) = 1$ using Lemma 1. Because of usage of t-norm in construction of R_B in Eq. 9 ,
$R_{B_2} (x, y) \leq R_{B_1} (x, y)$ for all $x \in U$
$\Rightarrow 1 - R_{B_1} (x, y) \leq 1 - R_{B_2} (x.y) \quad \forall x \in U$
$\Rightarrow 1 - R_{B_1} (x, y) \leq 1 - R_{B_2} (x.y) \quad \forall x \in U_2 (y)$
$\Rightarrow \min\limits_{y \in U_2(x)} (1 - R_{B_1} (x, y)) \leq \min\limits_{y \in U_2(x)} (1 - R_{B_2} (x, y))$
$\Rightarrow POS_{B_1} (y) \leq POS_{B_2} (y)$ using Eq. 18. ∎

If the degree of dissimilarity of y with all objects in $U_2 (y)$ is '1' then $POS_B (y)$. In which case object y belongs absolutely into positive region. Let $ABSOLUTE\_POS_B$ denote collection of all such absolute members of positive region, i.e.,

$$ABSOLUTE\_POS_B = \{y \in U / POS_B(y)=1\} \quad (19)$$

Here the possibility of removal of $ABSOLUTE\_POS_B$ objects without affecting calculations of positive region for remaining objects is investigated. Let $HDT' = \left(U', C^h = C^n \cup C^c, \{d\}, \{V_a, f_a\}_{a \in C^h \cup \{d\}} \right)$ denote the decision system after removal of $ABSOLUTE\_POS_B$ objects in HDT, i.e., $U' = U - ABSOLUTE\_POS_B$. Lemma 3 follows from the definition of HDT'.

Lemma 3. *For an object $y \in U'$,*
$U_1'(y) = U_1(y) - ABSOLUTE\_POS_B$
$U_2'(y) = U_1(y) - ABSOLUTE\_POS_B$.

Lemma 4 establishes an useful property of objects of HDT'.

Lemma 4. *For an object $y \in U'$, $U_2'(y) \neq \phi$.*

Proof. Assume in contrary, $U_2'(y) = \phi$.
$U_2'(y) = \phi \Rightarrow U_2(y) \subseteq ABSOLUTE\_POS_B$(Using Lemma 3)
$\Longrightarrow R_B(x,y) = 0 \quad \forall x \in U_2(y)$
$\Longrightarrow POS_B(y) = 1$
$\Longrightarrow y \in ABSOLUTE\_POS_B$
which is a contradiction as $y \in U' \Longleftrightarrow y \in U - ABSOLUTE\_POS_B$. ∎

Let $B^* \supset B$. Let POS_{B^*}, POS'_{B^*} denote positive region with respect to HDT, HDT' respectively.

Theorem 1. $POS_{B^*}(y) = 1 \quad \forall y \in ABSOLUTE\_POS_B$ and
$POS'_{B^*}(y) = POS_{B^*}(y) \quad \forall y \in U - ABSOLUTE\_POS_B$.

Proof. Let $y \in ABSOLUTE\_POS_B$. Using Lemma 2, since $B \subset B^*$ we have,
$POS_{B^*}(y) \geq POS_B(y) = 1$. Hence, $POS_{B^*}(y) = 1$.
Let $y \in U - ABSOLUTE\_POS_B$, (i.e., $y \in U'$)
Using Lemma 4, $U_2'(y) \neq \phi \Rightarrow U_2(y) \neq \phi$.
Using Eq. (18),
$$POS_{B^*}(y) = \min_{x \in U_2(y)} (1 - R_{B^*}(x,y))$$
$$= \min \left(\min_{x \in U_2'(y)} (1 - R_{B^*}(x,y)), \min_{x \in U_2(y) \cap ABSOLUTE\_POS_B} (1 - R_{B^*}(x,y)) \right)$$
$$= \min \left(\min_{x \in U_2'(y)} (1 - R_{B^*}(x,y)), \min_{x \in U_2(y) \cap ABSOLUTE\_POS_B} (1) \right)$$
$$\left(\begin{array}{c} \because x \in U_2(y) \cap ABSOLUTE\_POS_B \\ \Rightarrow y \in U_2(x) \wedge x \in ABSOLUTE\_POS_B \\ \Rightarrow R_B(x,y) = 0 \Rightarrow R_{B^*}(x,y) = 0 \end{array} \right)$$
Hence, $POS_{B^*}(y) = \min_{x \in U_2'(y)} (1 - R_{B^*}(x,y)) = POS'_{B^*}(y)$. ∎

Theorem 1 establishes the redundancy of $ABSOLUTE\_POS_B$ objects in the computation of POS_{B^*} for $U - ABSOLUTE\_POS_B$ objects and also that for any $B^* \supset B$ it is the case that $ABSOLUTE\_POS_B \subseteq ABSOLUTE\_POS_{B^*}$. The presence or removal of objects belonging to $ABSOLUTE\_POS$ does not alter subsequent ambiguity resolution process. Hence in MQRA, objects which belong to absolute positive region in an iteration continues to remain in absolute positive region and does not affect the positive region computations for remaining objects in the further iterations.

Assume that absolute positive region objects are removed after each iteration of MQRA. Let $\{B_i\}$ be the sequence of recommended reduct sets in forward selection based incremental reduct algorithm, in particular MQRA, such that $B_{i+1} - B_i$ is a singleton set representing the attribute selected in $i + 1^{th}$ iteration where $B_0 = \phi$. Similarly let $\{HDT_i\}$ denote the sequences of hybrid decision systems, and let $\{U_i\}$ denote the set of objects of HDT_i used in i^{th} iteration of MQRA satisfying, $HDT_1 = HDT$, $U_1 = U$ and for $i > 1$, $U_i = U - ABSOLUTE\_POS_{B_{i-1}}$. Theorem 1 validates the computations involved in each iteration of MQRA working only on non absolute positive region objects U_i. Hence the following corollary,

Corollary 1. *The results of MQRA are unaffected by incorporating absolute positive region removal such that the computation of positive region of objects are performed in i^{th} iteration using Theorem 1 such that $B^* = B_i$, $B = B_{i-1}$ with $B_0 = \phi$ and $HDT' = HDT_i$, $HDT = HDT_{i-1}$.*

The removal of absolute positive region objects in each iteration reduces the computational complexity and space complexity of MQRA. The equation for gamma measure becomes,

$$\gamma_{B^*} = \frac{|ABSOLUTE\_POS_B| + |POS'_{B^*}|}{|POS_{C^h}|} \tag{20}$$

In Eq. (20) taking the individual terms will give rise to Eq. (21) useful for implementation perspective. Denoting $\frac{|ABSOLUTE\_POS_B|}{|POS_{C^h}|}$ as the gamma component arising out of absolute positive region objects as $\gamma_{ABSOLUTE}$, Eq. (20) becomes,

$$\gamma_{B^*} = \gamma_{ABSOLUTE} + \gamma'_{B^*} \tag{21}$$

Section 3.2 gives the algorithm Improved Modified Quick Reduct (IMQRA) based on the results obtained in this section.

3.2   Improved Modified Quick Reduct Algorithm (IMQRA)

IMQRA algorithm (given in Algorithm 2) is MQRA algorithm with incorporation of absolute positive region removal at each iteration of MQRA. Eq. (21) is used for updating of gamma measure.

3.3   Modeling of IMQRA Computations Using Vector Operations

The implementation of IMQRA is done in Matlab [29] environment. Using Eq. (16) for computation of positive region implies that the computations need to be

---

Algorithm 2. Improved Modified Quick Reduct

Input :

$HDT = \left(U, C^h = C^c \cup C^n, \{d\}, \{V_{a_c}, f_{a_c}\}_{a_c \in C^c \cup \{d\}}, \{V_{a_n}, f_{a_n}\}_{a_n \in C^n} \right),$

R: Fuzzy similarity relation

N: Fuzzy Negation

Γ: t-norm

α: Degree

Output: B: Fuzzy M- Super reduct with degree α

Compute $R_{a,x}$ for all $a \in C^h \cup \{d\}$ and $x \in U$.

Compute $POS_{C^h}, |POS_{C^h}|$.

$B = \phi$

$\gamma_B = 0$

$ABSOLUTE\_POS_B = \phi,$

$\gamma_{ABSOLUTE} = 0$

$\gamma'_B = 0$

while $\gamma_B < \alpha$ **do**

$\quad$ $T = B,$

$\quad$ $best = -1$

$\quad$ **foreach** $a \in C^h - B$ **do**

$\quad\quad$ **if** $\gamma'_{B \cup \{a\}} > best$ **then**

$\quad\quad\quad$ $T = B \cup \{a\}$

$\quad\quad\quad$ $best = \gamma'_{B \cup \{a\}}$

$\quad\quad$ **end**

$\quad$ **end**

$\quad$ $B = T$

$\quad$ $\gamma_B = \gamma_{ABSOLUTE} + \gamma'_B$

$\quad$ Update $ABSOLUTE\_POS_B, \gamma_{ABSOLUTE}.$

$\quad$ Construct HDT' by removal of $ABSOLUTE\_POS_B$ objects

end

Return B.

---

done with each object in isolation. In this section theory is developed so that the required computations for positive region are modeled as matrix and sub matrix operations to obtain computational gains using Matlab like environments.

The implementation level aspects of IMQRA has been demonstrated by using following fuzzy similarity relation R_{a_n} for any quantitative attribute $a_n \in C^n$ given in Eq. (22) as defined in [7] (examples of more fuzzy similarity relations can be found in [24]) and its equivalent form is provided in Eq. (23). Generalized Lukasiewicz t-norm [63] given in Eq. (24) is used for applying Eq. (9).

$$R_{a_n}(x,y) = \max\left(\min\left(\frac{a_n(x) - a_n(y) + \sigma_{a_n}}{\sigma_{a_n}}, \frac{a_n(y) - a_n(x) + \sigma_{a_n}}{\sigma_{a_n}}\right), 0\right) \forall x, y \in U \tag{22}$$

where σ_{a_n} is the standard deviation of $a_n (.)$ of quantitative attribute a_n.

$$R_{a_n} (x, y) = \begin{cases} 1 & if \ a_n (x) = a_n (y), \\ 1 - \frac{|a_n(x) - a_n(y)|}{\sigma_a} & if \ |a_n (x) - a_n (y)| < \sigma_{a_n}, \\ 0 & if \ |a_n (x) - a_n (y)| \geq \sigma_{a_n}. \end{cases} \qquad (23)$$

$$\Gamma (x_1, x_2, \cdots, x_n) = \max \left(0, \left(\sum_{i=1}^{n} x_i \right) - n + 1 \right) \ x_i \in [0, 1]. \qquad (24)$$

As specified earlier, standard fuzzy negation is used for positive region computation as per the result of Lemma 1. The fuzzy similarity matrix for R_a for $a \in C^h \cup \{d\}$ is represented as a matrix of dimensions $|U| \times |U|$ representing $R_a (x, y) \ \forall x, y \in U$. It is to be noted that row of R_a corresponding to an object $x \in U$ represents the fuzzy set $R_{a,x} (R_{a,x} - foreset)$. Computation of R_a either by using Eq. (22) for quantitative attributes or by using Eq. (8) for qualitative attributes and computation of R_B using Eq. (24) with collection of $R_a \ (a \in B)$ matrices are implemented as matrix operations.

The task of evaluating $POS_B (x) \ \forall x \in U$ is based on another matrix named as Implicator value matrix $\Im_B$. $\Im_B$ contains for any pair of objects $x, y \in U$ the value of $\Im (R_B (x, y), R_d (x, y))$ using Eq. (18). Let $(D_1, \cdots, D_{|V_d|})$ represent the decision equivalence classes of d using indiscernbility relation IND. For any $i \in 1 \cdots |V_d|$, and for any $x, y \in D_i$ we have $U_1 (x) = U_1 (y) = D_i$ and $U_2 (x) = U_2 (y) = U - D_i$. Hence the nature of computations involved for implicator values using Eq. (18) is same for all objects belonging to a decision class. Hence computing $\Im_B$ can be done in stages with each decision class objects D_i for i from 1 to $|V_d|$ and by performing simultaneous operations for objects of D_i at once using sub matrix operations. For each D_i, a portion of matrix $\Im_B$ is constructed using two operations given below.

$$\Im_B (D_i, D_i) = 1 \qquad (25)$$

$$\Im_B (D_i, U - D_i) = 1 - R_B (D_i, U - D_i) \qquad (26)$$

After constructing $\Im_B$ in $|V_d|$ stazes using two operations in Eq. (25), Eq. (26) , the required $POS_B (y)$ for all $y \in U$ is constructed by applying minimum operation on each column of $\Im_B$ resulting in a row vector of $POS_B (y)$ for all $y \in U$. In any iteration of IMQRA, matrix R_B contains the extended similarity matrix of selected B attributes. The memory occupied by individual similarity matrices of $a \in B$ are recovered as and when attribute a is included into B. The removal of $ABSOLUTE\_POS_B$ objects from decision system resulting in reduced decision system is achived by removing the rows and columns corresponding to new $ABSOLUTE\_POS_B$ objects obtained in an iteration from current similarity matrices R_B and R_a for all $a \in C^h - B$.

Hence the process of computing similarity matrices, positive region are modeled as either matrix or sub matrix based vector operations. Algorithms MQRA,

IMQRA implemented in Matlab environment using these vector operations are named as MQRA_MW, IMQRA_MW respectively.

3.4   Analysis of IMQRA Algorithm

The space complexity of IMQRA algorithm is $O\left(\left|C^h\right||U|^2\right)$ upto first iteration of IMQRA used for storing fuzzy similarity matrices R_a for $a \in C^h \cup \{d\}$, R_B, and the implicator value matrix $\Im_B$. In each of the remaining iterations of IMQRA, space complexity reduces as the memory occupied by similarity matrix of attribute selected into recommended reduct B is released. The space complexity further reduces in each iteration as the objects belonging to $ABSOLUTE\_POS_B$ are removed resulting in smaller size similarity matrices. Hence using the notations developed for Corollary 1 the space complexity of ith iteration of IMQRA is $O\left(\left(\left|C^h\right| - i + 1\right)|U_i|^2\right)$.

The time complexity for computing $POS_{\{a\}}$ is $O\left(|U|^2\right)$ in the first iteration for each attribute $a \in C^h$ and hence time complexity of first iteration of IMQRA becomes $O\left(\left|C^h\right||U|^2\right)$. Owing to removal of absolute positive region objects, the time complexity of ith iteration of IMQRA is $O\left(\left(\left|C^h\right| - i + 1\right)|U_i|^2\right)$ using the notations developed for Corollary 1. Hence the theoritical time complexity of IMQRA is $O\left(\left|C^h\right|^2|U|^2\right)$ but in practice it will be much smaller.

4   Illustration

This section contains an illustration of the concepts proposed using a simple decision system consisting of quantitative conditional attributes. Consider the decision system given in Table 1.

Table 1. Example Decision System

Object	A	B	C	D	E	d
x1	3.59	3.52	2.86	0.76	1.30	1
x2	1.97	8.31	7.57	0.54	5.69	2
x3	2.51	5.85	7.54	5.31	4.69	2
x4	6.16	5.50	3.80	7.79	0.12	1
x5	4.73	9.17	5.68	9.34	3.37	1

The standard deviation values (σ) of conditional attributes are needed for fuzzy tolerance relation given in Eq. (22). The σ values for A, B, C, D, E attributes are 1.6981, 2.2750, 2.1411, 4.0084, 2.3134 respectively. The similarity matrix for attribute A is given below summarizing the fuzzy tolerance relation values using A attribute.

$$R_{\{A\}} = \begin{bmatrix} 1 & 0.099 & 0.417 & 0 & 0.2757 \\ 0.099 & 1 & 0.682 & 0 & 0 \\ 0.417 & 0.682 & 1 & 0 & 0 \\ 0 & 0 & 0 & 1 & 0.1579 \\ 0.2757 & 0 & 0 & 0.1579 & 1 \end{bmatrix}$$

4.1   Positive Region Computation Using Conventional Approaches

In this section computation of $POS_{\{A\}}$ is demonstrated using the theory given in section 2. Considering y as $x1$, required fuzzy sets $R_{\{A\},x1}$, $R_{d,x1}$ for applying Eq.(16) to compute $POS_{\{A\}}(x1)$ are depicted in Table 2.

Table 2. Fuzzy sets $R_{\{A\},x1}$, $R_{d,x1}$

	x1	x2	x3	x4	x5
$R_{\{A\},x1}$	1	0.099	0.417	0	0.2757
$R_{d,x1}$	1	0	0	1	1

Using Kleene-Dienes's implication [1], the computation of $POS_{\{A\}}(x1)$ is given below.

$$POS_{\{A\}}(x1) = R_{\{A\}} \downarrow R_{d,x1}(x1)$$
$$= \inf_{x \in U} \Im \left(R_{\{A\}}(x, x1), R_{d,x1}(x) \right)$$
$$= \inf_{x \in U} \max \left(1 - R_{\{A\}}(x, x1), R_{d,x1}(x) \right)$$
$$= \min \left(1, 1 - R_{\{A\}}(x2, x1), 1 - R_{\{A\}}(x3, x1), 1, 1 \right)$$
$$= \min \left(0.901, 0.583, 1, 1 \right)$$
$$= 0.583.$$

Similarly positive region membership values for other objects are computed as,

$$POS_{\{A\}}(x2) = 0.901, \ POS_{\{A\}}(x3) = 0.583,$$
$$POS_{\{A\}}(x4) = 1.0, \ POS_{\{A\}}(x5) = 1.0.$$

4.2   Positive Region Computation Using Vector Operations

The computation of $POS_{\{A\}}$ using vector operation modeling developed in Section 3 is demonstrated in this section. The natural negation generated by Kleene-Diene's implication is standard negation. Hence based on the theory developed in Section 3, computing $POS_{\{A\}}$ using Eq. (25), Eq. (26) will result in the same results obtained in Section 4.1. The equivalence classes of decision attribute d are $D_1 = \{x1, x4, x5\}$ and $D_2 = \{x3, x4\}$. Implication values matrix $\Im_{\{A\}}$ is initially taken as a zero matrix. Steps involved in applying Eq. (25), Eq. (26) for D_1, D_2 objects as submatrix operations are given in Table 3.

Table 3. Implication Value Matrix $\Im_{\{A\}}$ Computation

Step	State of $\Im_{\{A\}}$
$\Im_{\{A\}}(D_1, D_1) = 1$ and $\Im_{\{A\}}(U - D_1, D_1) = 1 - R_{\{A\}}(U - D_1, D_1)$	$\begin{bmatrix} 1 & 0 & 0 & 1 & 1 \\ 0.901 & 0 & 0 & 1 & 1 \\ 0.583 & 0 & 0 & 1 & 1 \\ 1 & 0 & 0 & 1 & 1 \\ 1 & 0 & 0 & 1 & 1 \end{bmatrix}$
$\Im_{\{A\}}(D_2, D_2) = 1$ and $\Im_{\{A\}}(U - D_2, D_2) = 1 - R_{\{A\}}(U - D_2, D_2)$	$\begin{bmatrix} 1 & 0.901 & 0.583 & 1 & 1 \\ 0.901 & 1 & 1 & 1 & 1 \\ 0.583 & 1 & 1 & 1 & 1 \\ 1 & 1 & 1 & 1 & 1 \\ 1 & 1 & 1 & 1 & 1 \end{bmatrix}$

Column wise minimum values in resulting $\Im_{\{A\}}$ are the required $POS_{\{A\}}$ membership values. Hence $POS_{\{A\}}$ obtained using vector operation approach is,
$POS_{\{A\}}(x1) = 0.583, POS_{\{A\}}(x2) = 0.901,$
$POS_{\{A\}}(x3) = 0.583, POS_{\{A\}}(x4) = 1.0, POS_{\{A\}}(x5) = 1.0.$
It can be seen that the result is same as obtained in Section 4.1.

4.3   Illustration of IMQRA_MW Algorithm

The input parameters for IMQRA_MW algorithm are specified in Section 3.3. The first step in IMQRA_MW algorithm is to compute similarity matrices for all conditional attributes. The similarity matrix for C^h is computed using attribute similarity matrices by applying Eq. (9) using generalized Lukasiewicz t-norm. The resulting R_{C^h} is given below.

$$R_{C^h} = \begin{bmatrix} 1 & 0 & 0 & 0 & 0 \\ 0 & 1 & 0 & 0 & 0 \\ 0 & 0 & 1 & 0 & 0 \\ 0 & 0 & 0 & 1 & 0 \\ 0 & 0 & 0 & 0 & 1 \end{bmatrix}$$

Computing POS_{C^h} as demonstrated in Section 4.2 results in,
$POS_{C^h}(x1) = 1, POS_{C^h}(x2) = 1, POS_{C^h}(x3) = 1,$
$POS_{C^h}(x4) = 1, POS_{C^h}(x5) = 1.$ Hence, $|POS_{C^h}| = 5.$

In the first iteration of IMQRA_MW the gamma measure for each conditional attribute is evaluated. As an example consider the evaluation of $\gamma_{\{A\}}$. Using $POS_{\{A\}}$ evaluated in Section 4.2, $\gamma'_{\{A\}} = \dfrac{|POS_{\{A\}}|}{|POS_{C^h}|} = \dfrac{4.067}{5} = 0.8134.$ Initially, $\gamma_{ABSOLUTE} = 0$ and hence it follows that $\gamma_{\{A\}} = 0.8134$ using Eq. (21). Similar computation result in gamma measure for other attributes as $\gamma_{\{B\}} = 0.4127, \gamma_{\{C\}} = 0.924, \gamma_{\{D\}} = 0.8687, \gamma_{\{E\}} = 0.8282$. Hence attribute

C is added into recommended reduct B. Since $POS_C(x1) = 1, POS_C(x2) = 0.8827, POS_C(x3) = 0.8687, POS_C(x4) = 1, POS_C(x5) = 0.8687$, objects $\{x1, x4\}$ are added to $ABSOLUTE\_POS_B$ and $\gamma_{ABSOLUTE} = 0.4$.

The memory allotted for similarity matrix of C attribute is released and rows and columns corresponding to $ABSOLUTE\_POS_B$ objects are removed from similarity matrix of recommended reduct B and similarity matrices of the available attributes $\{A, B, D, E\}$. In the next iteration all computations are performed on the reduced hybrid decision system and attribute A is added into recommended reduct B as it gives the maximum gamma gain. The positive region memberships in the reduced HDT i.e., HDT' by $\{C, A\}$ is $POS_{\{C,A\}}(x2) = 1, POS_{\{C,A\}}(x3) = 1, POS_{\{C,A\}}(x5) = 1$. Hence, $\gamma'_{\{C,A\}} = \frac{3}{5} = 0.6$ and using Eq. (21), $\gamma_{\{C,A\}} = 1$. As the terminating criteria of IMQRA_MW is satisfied algorithm returns $\{C, A\}$ as the fuzzy decision super reduct for HDT.

5   Experiments and Results

The experiments are conducted in Intel Core 2 Duo CPU @2 GHz with 2 GB RAM under Fedora 10 (Linux) operating system in Matlab environment [29]. The hybrid decision systems used for experimental analysis are described in Table 4. All the data sets are from UCI Machine Learning repository [60] excepting web dataset (from [23]) and Olitos (from [36]).

Table 4. Details of Hybrid Datasets

Dataset	Objects	Conditional Features			Decision Classes
		Total	Quantitative	Qualitative	
Image Segmentation	2310	19	16	3	7
Sonar, mines vs. rocks	208	60	60	0	2
Wisconsin diagnostic breast cancer (WDBC)	569	30	30	0	2
Cleveland	297	13	5	8	5
Glass	214	9	9	0	6
Ionosphere	230^a 351^b	34	33	1	2
Web	149	2556	2556	0	5
Wine	178	13	13	0	3
Heart	270	13	5	8	2
Olitos	120	25	25	0	4

^{a} Size of the dataset reported in [22]
^{b} Size of the dataset used in the present
study from online repository [60]

It is to be particularly noted that the same reducts (for example with spam-base dataset) and reduct lengths are obtained with the results for MQRA reported in [7]. The difference in results for MQRA_MW, IMQRA_MW with

MQRA in [7] is in terms of computational time. But in [7] results there is no reporting of computational time used in obtaining reduct. Hence comparative analysis of proposed algorithms with the results given in [7] for MQRA algorithm could not be performed.

5.1 Experiments with IMQRA_MW Algorithm

The FA-FSCE algorithm given in [44] uses absolute positive region object removal (forward acceleration) in Hu's fuzzy rough set model [17] and FA-FSCE algorithm without positive region removal is given as FSCE. Hence FSCE is comparable to MQRA_MW algorithm and FA-FSCE is comparable to IMQRA_MW algorithm. A comparative analysis is done for IMQRA_MW and MQRA_MW algorithms with the results reported for FSCE algorithm and FA-FSCE algorithm in [44]. Table 5 contains the results (reduct size, computational time in seconds) for FSCE and FA-FSCE algorithms reported in [44] along with the results obtained for MQRA_MW and IMQRA_MW algorithms. Table 6 gives the percentage of computational gain obtained by MQRA_MW and IMQRA_MW algorithms.

Table 5. Comparison Results of IMQRA_MW and MQRA_MW with FA-FSCE and FSCE algorithms

Dataset	FSCE		FA-FSCE		MQRA_MW		IMQRA_MW	
	Reduct size	Time	Reduct size	Time	Reduct size	Time	Reduct size	Time
Image Segmentation	17	1258.0468	17	900.5781	16	41.75	16	17.4
Sonar, Mines vs Rocks	41	300.5625	41	50	5	0.57	5	0.69
WDBC	27	228.9218	27	171.875	6	1.7257	6	1.3469

Table 6. Computational gain percentages of MQRA_MW and IMQRA_MW algorithms over FSCE and FA-FSCE algorithms

Dataset	MQRA_MW over FSCE	MQRA_MW over FA-FSCE	IMQRA_MW over FSCE	IMQRA_MW over FA-FSCE	IMQRA_MW over MQRA_MW
Image Segmentation	96.68	95.36	98.62	98.07	58.32
Sonar, Mines vs Rocks	99.81	98.86	99.77	98.62	-21.05
WDBC	99.25	99.00	· 99.41	99.22	21.95

In [22] Jensen et al. have described new approaches for fuzzy rough feature selection in improvement to the earlier Fuzzy-Rough Feature selection (FRFS) algorithm [19]. The algorithms described in [22] are fuzzy boundary-region-based FS (B-FRFS), fuzzy lower-approximation-based FS (L-FRFS), and fuzzy discernibility-matrix based algorithm (FDM). All these algorithms are forward selection based quick reduct like algorithms. Hence a comparative analysis is done with MQRA_MW, IMQRA_MW algorithms with these algorithms. The results reported in [22] along with the results obtained for MQRA and IMQRA algorithms are given in Tables 7 and 8. Table 7 gives the results of reduct lengths whereas Table 8 contains the computational time in seconds.

Table 7. Reduct Size Comparison with FRFS and associated algorithms

Dataset	FRFS	B-FRFS	L-FRFS	FDM	MQRA_MW and IMQRA_MW
Cleveland	11	9	9	9	8
Glass	9	9	10	9	9
Heart	11	8	8	8	7
Ionosphere	11	9	9	8	7
Olitos	10	6	6	6	5
Web	24	20	21	18	20
Wine	10	6	6	6	5

Table 8. Computational Time (in seconds) Comparison with FRFS and associated algorithms

Dataset	FRFS	B-FRFS	L-FRFS	FDM	MQRA_MW	IMQRA_MW
Cleveland	24.11	8.78	3.32	10.68	0.37	0.5
Glass	1.61	3.3	1.53	4.88	0.14	0.28
Heart	11.84	3.61	2.17	8.77	0.24	0.31
Ionosphere	61.8	8.53	3.77	17.54	1.08	0.84
Olitos	11.2	1.29	0.72	4.07	0.17	0.25
Web	5642.65	949.69	541.65	1782.69	89.62	62.17
Wine	1.42	1.69	0.97	4.6	0.11	0.23

6   Analysis of Results

The computational gain percentages obtained in IMQRA_MW over FA-FSCE, FSCE and MQRA_MW algorithms is given in Fig. 1 and over FRFS and associated algorithms is given in Fig. 2. The size of data sets used in each iteration of IMQRA_MW for the data sets used for FA-FSCE, FSCE comparative analysis is given in Fig. 3 and the same with respect to FRFS associated algorithms is given in Fig. 4. From Fig. 1 it is observed that IMQRA_MW has achieved over 98% computational gain over FSCE, FA-FSCE algorithms. Similarly from Fig. 2 it is observed that IMQRA_MW has achieved over 60-99% computational gains over FRFS, B-FRFS, L-FRFS, FDM algorithms. The highly significant

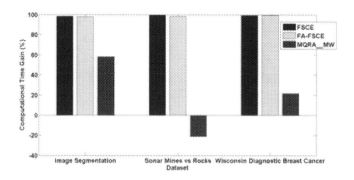

Fig. 1. Computational Gain Percentage of IMQRA_MW over FSCE, FA-FSCE, MQRA_MW algorithms

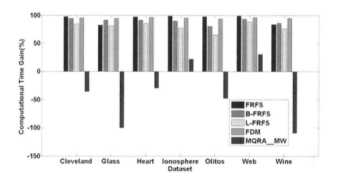

Fig. 2. Computational Gain Percentage of IMQRA_MW over FSCE, FA-FSCE, MQRA_MW algorithms

computational gains illustrate the importance of vector based implementation and absolute positive region removal incorporated in IMQRA_MW.

It is observed that the vector implementation alone has turned out to be significantly effective in the computational gain even in MQRA_MW. In comparison of computational gains between MQRA_MW, IMQRA_MW mixed results can be noticed. The gain due to removal of absolute positive region is sensitive to the nature of the data sets, i.e., the quantum of absolute positive region in the initial iterations as can be seen from Fig. 3 and 4. Identifying the absolute positive region objects and filtering is an overhead in IMQRA_MW and lead to little bit of inferior performance of IMQRA_MW for smaller data sets. But for large decision systems such as Image Segmentation and Web data sets the overhead is significantly compensated by gains obtained in absolute positive region removal and in these two data sets IMQRA_MW has achieved significant gains over MQRA_MW.

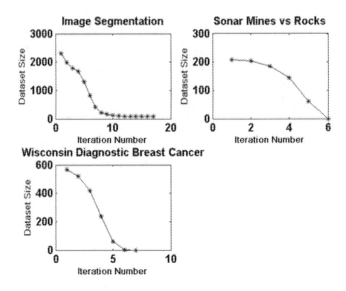

Fig. 3. Reduction in Data set Size in IMQRA_MW iterations in FA-FSCE comparative Experiment

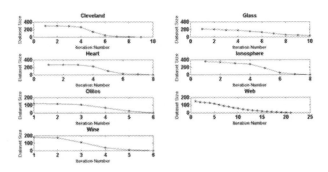

Fig. 4. Reduction in Data set Size in IMQRA_MW iterations in FRFS comparative Experiment

7   Conclusion

The study aims at efficient implementation of fuzzy decision reduct for hybrid decision systems using fuzzy rough set theory. Simplified computational model for positive region computation in Radzikowska's fuzzy rough set model is developed. It is identified that Radzikowska-Kerry's fuzzy rough set model for decision systems can be specified using a pair of fuzzy negation and t-norm instead of an implication and t-norm. Theoretical validation for redundancy of presence of absolute positive region objects for computation of positive region for remaining objects is provided. These improvements are incorporated into MQRA as algorithm Improved MQRA (IMQRA). The computations in MQRA, IMQRA are modeled as vector based operations. It is observed that MQRA_MW and IMQRA_MW have achieved significant computational gains over several existing approaches for reduct computation using fuzzy rough set theory.

Present study is limited to decision systems where decision attribute is qualitative. It is proposed to extend the study to the systems where decision attribute is quantitative. It is also proposed to extend to build ensemble systems by considering few fuzzy similarity relations instead of confining to one.

Acknowledgments. Authors would like to express their gratitude to the reviewers for their constructive comments for quality enhancement of the paper.

References

1. Baczynski, M., Jayaram, B.: S- and R- implications: A state-of-the-art survey. Fuzzy Sets and Systems 159(14), 1836–1859 (2008)
2. Bazan, J.G., Nguyen, H.S., Nguyen, S.H., Synak, P., Wroblewski, J.: Rough Set Algorithms in Classification Problem. In: Polkowski, L., Tsumoto, S., Lin, T.Y. (eds.) Rough Set Methods and Applications. STUDFUZZ, vol. 56, pp. 49–88. Physica-Verlab GmbH, Heidelberg (2000)
3. Bhatt, R.B., Gopal, M.: On the compact computational domain of fuzzy-rough sets. Pattern Recognition Letters 26, 1632–1640 (2005)
4. Blajdo, P., Grzymala-Busse, J.W., Hippe, Z.S., Knap, M., Mroczek, T., Piatek, L.: A Comparison of Six Approaches to Discretization—A Rough Set Perspective. In: Wang, G., Li, T., Grzymala-Busse, J.W., Miao, D., Skowron, A., Yao, Y. (eds.) RSKT 2008. LNCS (LNAI), vol. 5009, pp. 31–38. Springer, Heidelberg (2008)
5. Chouchoulas, A., Shen, Q.: Rough Set aided Keyword Reduction for Text categorization. Applied Artificial Intelligence 15, 843–873 (2001)
6. Cornelis, C., Cock, M.D., Radzikowska, A.M.: Fuzzy Rough Sets: from theory into practice. In: Pedrycz, W., Skowron, A., Kreinovich, V. (eds.) Handbook of Granular Computing, pp. 533–552. John Wiley and Sons (2008)
7. Cornelis, C., Jensen, R., Hurtado, G., Slezak, D.: Attribute selection with fuzzy decision reducts. Information Sciences 180(2), 209–224 (2010)
8. Dubois, D., Prade, H.: Similarity versus Preference in Fuzzy Set-Based Logics. In: Incomplete Information: Rough Set Analysis. STUDFUZZ, vol. 13, pp. 441–461. Physica-Verlag, HD (1998)
9. Dubois, D., Prade, H.: Rough fuzzy sets and Fuzzy Rough Sets. Int. J. General Systems 17(2-3), 191–209 (1990)
10. Dubois, D., Prade, H.: Putting fuzzy sets and rough sets together. In: Slowiniski, R. (ed.) Intelligent Decision Support, pp. 203–232. Kluwer Academic, Dordrecht (1992)
11. Duntsch, I., Gediga, G., Nguyen, H.S.: Rough set data analysis in the KDD process. In: Proceedings of IPMU, Madrid, Spain, pp. 220–226 (2000)
12. Fazayeli, F., Wang, L., Mandziuk, J.: Feature Selection Based on the Rough Set Theory and Expectation-Maximization Clustering Algorithm. In: Chan, C.-C., Grzymala-Busse, J.W., Ziarko, W.P. (eds.) RSCTC 2008. LNCS (LNAI), vol. 5306, pp. 272–282. Springer, Heidelberg (2008)

13. Greco, S., Matarazzo, B., Słowiński, R.: Fuzzy Similarity Relation as a Basis for Rough Approximations. In: Polkowski, L., Skowron, A. (eds.) RSCTC 1998. LNCS (LNAI), vol. 1424, pp. 283–289. Springer, Heidelberg (1998)
14. Henry, C.J., Ramanna, S.: Parallel computation in finding near neighbourhoods. In: Yao, J., Ramanna, S., Wang, G., Suraj, Z. (eds.) RSKT 2011. LNCS, vol. 6954, pp. 523–532. Springer, Heidelberg (2011)
15. Hu, Q.H., Xie, Z.X., Yu, D.R.: Fuzzy probabilistic approximation spaces and their information measures. IEEE Transactions on Fuzzy systems 14, 191–201 (2006)
16. Hu, Q.H., Yu, D.R., Xie, Z.X.: Information preserving hybrid data reduction based on fuzzy rough techniques. Pattern Recognition Letters 27(5), 414–423 (2006)
17. Hu, Q.H., Yu, D.R., Xie, Z.X.: Hybrid attribute reduction based on a novel fuzzy-rough model and information granulation. Pattern Recognition 40, 3509–3521 (2007)
18. Ismail, M.K., Ciesielski, V.: An Empirical Investigation of the Impact of Discretization on Common Data Distributions. In: Proc. of HIS-2003 on Design and Application of Hybrid Intelligent Systems, pp. 692–701. IOS Press (2003)
19. Jensen, R., Shen, Q.: Fuzzy Rough attribute reduction with application to web categorization. Fuzzy Sets and Systems 141(3), 469–485 (2004)
20. Jensen, R., Shen, Q.: Rough Sets, their Extensions and Applications. International Journal of Automation and Computing 4(3), 217–228 (2007)
21. Jensen, R., Shen, Q.: Computational Intelligence and Feature Selection: Rough and Fuzzy Approaches. IEEE (2008)
22. Jensen, R., Shen, Q.: New approaches to fuzzy-rough feature selection. IEEE Transactions on Ruzzy Systems 17(4), 824–838 (2009)
23. Jensen's repository of datasets, http://users.aber.ac.uk/rkj/datasets/index.php
24. Kretowski, M., Stepaniuk, J.: Selection of objects and attributes, a tolerance rough set approach. In: Proceedings of the Poster Session of Ninth International Symposium on Methodologies for Intelligent Systems, Zakopane Poland, pp. 169–180 (1996)
25. Lin, T.: Neighborhood systems and approximation in database and knowledge base sys-tems. In: Proceedings of the 4th International Symposium on Methodologies for Intelligent Systems (1989)
26. Liu, Y., Xiong, R., Chu, J.: Quick Attribute Reduction Algorithm with Hash. Chinese Journal of Computers 32(8), 1493–1499 (2009)
27. Liu, W.-N., Yao, J., Yao, Y.: Rough approximations under level fuzzy sets. In: Tsumoto, S., Słowiński, R., Komorowski, J., Grzymała-Busse, J.W. (eds.) RSCTC 2004. LNCS (LNAI), vol. 3066, pp. 78–83. Springer, Heidelberg (2004)
28. Marcus, S.: Tolerance Rough Sets, Cech topologies, learning processes. Bull. Polish Academy of Sciences, Technical Sciences 42(3), 471–487 (1994)
29. Matlab, http://www.mathworks.com
30. Nanda, S., Majumdar, S.: Fuzzy Rough Sets. Fuzzy Sets and Systems 45, 157–160 (1992)
31. Nguyen, H.S.: Discretization Problem for Rough Sets Methods. In: Polkowski, L., Skowron, A. (eds.) RSCTC 1998. LNCS (LNAI), vol. 1424, pp. 545–552. Springer, Heidelberg (1998)
32. Nguyen, H.S.: On Exploring Soft Discretization of Continuous Attributes. In: Rough Neural Computing: Techniques for Computing with Words, Cognitive Technologies, pp. 333–350. Springer (2003)

33. Nguyen, H.S., Skowron, A.: Quantization of Real Value Attributes, Rough Set and Boolean Reasoning Approach. In: Proceedings of the 2nd Annual Joint Conference on Information Sciences, pp. 34–37 (1995)
34. Nieminen, J.: Rough tolerance equality. Fundamenta Informaticae 11(3), 289–296 (1988)
35. Ningler, M., Stockmanns, G., Schneider, G., Dressler, O., Kochs, E.F.: Rough Set-Based Classification of EEG-Signals to Detect Intraoperative Awareness: Comparison of Fuzzy and Crisp Discretization of Real Value Attributes. In: Tsumoto, S., Słowiński, R., Komorowski, J., Grzymała-Busse, J.W. (eds.) RSCTC 2004. LNCS (LNAI), vol. 3066, pp. 825–834. Springer, Heidelberg (2004)
36. Olitos Dataset website at
    http://michem.disat.unimib.it/chm/download/datasets.htm#olit
37. Paul, S., Maji, P.: Fuzzy Discretization for Rough Set Based Gene Selection Algorithm. In: Proceedings of EAIT, pp. 317–320. IEEE (2011)
38. Pawlak, Z.: Rough Sets. International Journal of Computer and Information Science 11, 341–356 (1982)
39. Pawlak, Z.: A treatise on rough sets. In: Peters, J.F., Skowron, A. (eds.) Transactions on Rough Sets IV. LNCS, vol. 3700, pp. 1–17. Springer, Heidelberg (2005)
40. Pawlak, Z., Grzymala-Busse, J., Slowinski, R., Ziarko, W.: Rough Sets. Communications of ACM 38(11), 89–95 (1995)
41. Peters, J.F., Wasilewski, P.: Tolerance spaces: Origins, theoretical aspects and applications. Information Sciences 195, 211–225 (2012)
42. Peters, J.F., Ramanna, S.: Feature Selection: Near Set Approach. In: Raś, Z.W., Tsumoto, S., Zighed, D.A. (eds.) MCD 2007. LNCS (LNAI), vol. 4944, pp. 57–71. Springer, Heidelberg (2008)
43. Qian, Y., Liang, J., Pedrycz, W., Dang, C.: Positive approximation: An accelerator for attribute reduction in rough set theory. Artificial Intelligence 174(9-10), 597–618 (2010)
44. Qian, Y., Li, C., Liang, J.: An Efficient Fuzzy-Rough Attribute Reduction Approach. In: Yao, J., Ramanna, S., Wang, G., Suraj, Z. (eds.) RSKT 2011. LNCS, vol. 6954, pp. 63–70. Springer, Heidelberg (2011)
45. Radzikowka, A.M., Kerre, E.E.: A comparative study of Fuzzy Rough Sets. Fuzzy Sets and Systems 126, 137–155 (2002)
46. Roy, A., Pal, S.K.: Fuzzy discretization of feature space for a rough set classifier. Pattern Recognition Letters 24(6), 895–902 (2003)
47. Sai Prasad, P.S.V.S., Rao, C.R.: IQuickReduct: An Improvement to Quick Reduct Algorithm. In: Sakai, H., Chakraborty, M.K., Hassanien, A.E., Ślęzak, D., Zhu, W. (eds.) RSFDGrC 2009. LNCS, vol. 5908, pp. 152–159. Springer, Heidelberg (2009)
48. Sai Prasad, P.S.V.S., Raghavendra Rao, C.: Extensions to iQuickReduct. In: Sombattheera, C., Agarwal, A., Udgata, S.K., Lavangnananda, K. (eds.) MIWAI 2011. LNCS, vol. 7080, pp. 351–362. Springer, Heidelberg (2011)
49. Sai Prasad, P.S.V.S., Rao, C.R.: Seed based fuzzy decision reduct for hybrid decision systems. In: Proceedings of FUZZ-IEEE, pp. 1–8. IEEE (2013), doi:10.1109/FUZZ-IEEE.2013.6622535
50. Skowron, A.: Rough Sets in KDD. In: Proceedings of the 16th World Computer Congress, Beijing, China, pp. 1–14 (2000)
51. Skowron, A., Stepaniuk, J.: Tolerance approximation spaces. Fundamenta Informaticae 27(2-3), 245–253 (1996)

52. Ślęzak, D., Betliński, P.: A Role of (Not) Crisp Discernibility in Rough Set Approach to Numeric Feature Selection. In: Hassanien, A.E., Salem, A.-B.M., Ramadan, R., Kim, T.-h. (eds.) AMLTA 2012. CCIS, vol. 322, pp. 13–23. Springer, Heidelberg (2012)
53. Ślęzak, D., Wasilewski, P.: Granular Sets – Foundations and Case Study of Tolerance Spaces. In: An, A., Stefanowski, J., Ramanna, S., Butz, C.J., Pedrycz, W., Wang, G. (eds.) RSFDGrC 2007. LNCS (LNAI), vol. 4482, pp. 435–442. Springer, Heidelberg (2007)
54. Slowinski, R., Stefanowski, J.: Handling various types of Uncertainty in the Rough Set Approach. In: Proceedings of RSKD, pp. 366–376. Springer, Heidelberg (1993)
55. Slowinski, R., Vanderpooten, D.: Similarity relation as a basis for rough approximations. Advances in Machine Intelligence & Soft Computing, Dept. of Electrical Engineering, Duke University, Durham, North Carolina, USA, 17–33 (1997)
56. Slowinski, R., Vanderpooten, D.: A Generalized Definition of Rough Approximations Based on Similarity. IEEE Transactions on Knowledge and Data Engineering 12(2), 331–336 (2000)
57. Su, C.-T., Hsu, J.-H.: An Extended Chi2 Algorithm for Discretization of Real Value Attributes. IEEE Trans. on Knowledge and Data Engineering 17(3), 437–441 (2005)
58. Tian, D., Zeng, X., Keane, J.: Core-generating approximate minimum entropy discretization for rough set feature selection in pattern classification. International Journal of Approximate Reasoning 52, 863–880 (2011)
59. Tick, J., Fodor, J.: Fuzzy implications and inference processes. In: Proceedings of International Conference on Computational Cybernetics, pp. 105–109. IEEE (2005)
60. UCI Machine Learning Repository,
    http://archive.ics.uci.edu/ml/datasets.html
61. Wang, X.Z., Tsang, E.C.C., Zhao, S.Y., Chen, D.G., Yeung, D.S.: Learning fuzzy rules from fuzzy samples based on rough set technique. Information Science 177, 4493–4514 (2007)
62. Wei, W., Liang, J., Qian, Y.: A comparative study of rough sets for hybrid data. Information Sciences 190, 1–16 (2012)
63. Wikipedia on t-norm, http://en.wikipedia.org/wiki/T-norm
64. Yao, Y.Y.: Combination of rough and fuzzy sets based on α-level sets. In: Lim, T.Y., Cercone, N. (eds.) Rough Sets and Data Mining: Analysis for Imprecise Data, pp. 301–321. Kluwer Academic Publishers, Boston (1997)
65. Yao, Y.Y.: Relational interpretation of neighborhood operators and rough set approximation operators. Information Sciences 111(1-4), 239–259 (1998)
66. Zhao, Y., Luo, F., Wong, S.K.M., Yao, Y.: A general definition of an attribute reduct. In: Yao, J., Lingras, P., Wu, W.-Z., Szczuka, M.S., Cercone, N.J., Ślęzak, D. (eds.) RSKT 2007. LNCS (LNAI), vol. 4481, pp. 101–108. Springer, Heidelberg (2007)
67. Zhao, Y., Yao, Y., Luo, F.: Data analysis based on discernibility and indiscernibility. Information Sciences 177(22), 4959–4976 (2007)
68. Zhong, N., Dong, J.: Using Rough Sets with Heuristics for Feature Selection. Journal of Intelligent Information Systems 16, 199–214 (2001)
69. Ziarko, W.: Variable precision rough set model. Journal of Computer and System Sciences 46, 39–59 (1993)

Rough Sets in Economy and Finance

Mariusz Podsiadło[1] and Henryk Rybiński[2]

[1] Misys plc
[2] Institute of Computer Science,
Warsaw University of Technology
Nowowiejska 15/19, 00-665 Warsaw, Poland
mariusz.podsiadlo@misys.com,
hrb@ii.pw.edu.pl

Abstract. The Rough Set Theory makes it possible to represent and infer knowledge from incomplete or noisy data, and has attracted much focus of the research community and applications have been found in a wide range of disciplines where knowledge discovery and data mining are indispensable. This paper provides a detailed review of the currently available literature covering applications of rough sets in the economy and finance. The classical rough set model and its important extensions applied to the economic and financial problems in crucial areas of risk management (business failure, credit scoring), financial market prediction, valuation and portfolio management are described, showing that the rough set theory is an interesting and increasingly popular method employed alongside traditional statistical methods, neural networks and genetic algorithms to support resolution of the most difficult problems in economy and finance.

Keywords: soft computing, rough sets, artificial intelligence, risk management, stock market prediction, credit scoring.

1 Introduction

Financial ecosystems are characterized by large amounts of noisy and incomplete information and inherent uncertainty of any forward looking predictions. On the other side, any wrong decision can have potentially catastrophic consequences for the economic well-being of individuals, institutions, nations or even the whole world, as recently shown by the demise of Lehmann Brothers and the Eurozone crisis.

It is therefore not surprising that academia, financial industry and its regulatory bodies have ever since been looking for methods able to analyze historical and real time data and infer reliable observations, applicable to a wide variety of problems, ranging from macroeconomic crisis prevention to stock market movement prediction and optimal composition of investment portfolios.

Statistics and probability theory traditionally form the basis for formal methods used in economy and finance for data analysis and prediction. The applications range from a self-contained technical analysis of historical stock data

J.F. Peters and A. Skowron (Eds.): Transactions on Rough Sets XVII, LNCS 8375, pp. 109–173, 2014.
© Springer-Verlag Berlin Heidelberg 2014

series[1], which tries to discover buy/sell signals by looking for repeatable trend patterns of price, trading volume and related statistical measures, to sophisticated pricing models, which use complex stochastic processes to model the future uncertainty of price movements. The strong mathematical foundation allows us to deliver predictive models being sophisticated enough to capture many aspects of the modeled reality with statistically measurable precision of the results.

However, the generated knowledge has to be considered within the limits of the underlying model of reality and associated assumptions, be it the type of statistical distribution, a pricing model or stochastic process. The multivariate models, typical for the financial environment driven by multiple causal factors, grow complex together with the number of independent variables, and their non-linear correlations, which makes the analysis difficult to follow by anyone but field experts. The unknown causal relationships, potentially present in the analyzed data set, are replaced by, usually simplifying, assumptions of the applied model, making any conclusions vulnerable to the reality check.

One prominent example was the *Value at Risk* (VaR) framework, popularized in 90's by the J.P. Morgans RiskMetrics service and accepted as the standard financial risk management methodology to measure the impending risk of portfolio loss. The framework is based on statistical principles so its predictive quality depends from its calibration (i.e. statistical distribution, simulation horizon, simulation method, etc.). This model risk was evident during the crisis started in 2007, where many financial institutions, having an operational VaR framework in place, failed or were at a brink of bankruptcy. It seemed that the models did not consider the major causal effects, which resurfaced and have been shaping the market since 2007[2].

Consequently, despite the dominance of statistical methods, attempts to employ data discovery and soft computing models in order to unearth and account for complex data characteristics and relationships have a long track record in economy and finance. Multiple soft computing methodologies, like neural networks and genetic algorithms, have gained the required sophistication and maturity and been successfully applied to various problems ranging from credit scoring and banking crisis prediction to portfolio management, prediction of market movements and financial derivatives pricing[3]. A comprehensive discussion of the most popular soft computing models and methods used in finance can be found in Bahrammirzaee [6], Kovalerchuk and Vitayev [76], Dymova [42], Chen at al. [19], and Kumar and Ravi [103].

---

[1] Technical Analysis can be applied to any financial data time series, not only stock movements. For a comprehensive treatment of this topic see Murphy [92].

[2] The overreliance on the normal distribution assumption ignoring extreme events ('fat tails') was seen as one reason for the failure (see [130]).

[3] Derivative is a financial instrument, which value is dependent on pricing of one or more underlying financial instruments, including other derivatives. The dependency can be complex, non-linear and have no closed-form solution.

The *Rough Set* theory, describing a way to represent and infer knowledge from incomplete or noisy data, is also increasingly used in the financial domain due to the following advantages (Dimitras *et al.* [39]; Greco *et al.* [49]):

- The classical rough set model considers only the original data as its universe of discourse and does not need any additional information or calibration, unlike probability in statistics or grade of membership in the fuzzy set theory (Dubois and Prade [41]).
- The rough sets model is able to consider both quantitative and qualitative attributes.
- It is possible to generate decision rules using natural language (if then rules).
- The model is able to remove redundant information from the input data set, generating consistent and general decision rules, which are free from noise.
- The generated decision rules are supported by factual information, as every rule is based on a subset of the underlying data set.

This paper contains an up-to-date representative review of rough set applications in the economy and finance domain.

Google Scholar returns currently more than 70000 references to the phrase *"rough set"*, with over 15000 ($>$ 20%) references quoted in the last 4 years. The number of publications focused on the application of the rough set theory in the financial domain is also increasing very quickly, with a great interest in hybrid models: Google Scholar returned 12000 hits, when looking for the phrase *"rough set" finance*, including more than 4500 references ($>$ 37%) from the last 4 years. Earlier attempts to describe the research activity in this area can be found in Mrózek and Skarbek [91] as well as Tay and Shen [132]. A brief discussion of the rough set usage in the economic context, together with an introduction to the *fuzzy variable precision rough set* model is also provided in Zhang *et al.* [155].

This paper is organized as follows: Section 2 provides a brief overview of the classical rough set model and its most relevant extensions, with references to the literature covering the respective topics in more detail. Section 3 describes the various application areas of rough sets in economics and finance based on an extensive discussion of the currently available empirical studies using the classical rough set theory, its multiple extensions as well as their combinations with other methods (hybridization). The concluding remarks, describing the current state and promising areas for further research, are given in Section 4.

2   Rough Sets Overview

The concept of rough sets was introduced by Pawlak [97,98] as an extension of the set theory, which allows to define an approximate classification of the given set of data objects (data universe) in presence of data vagueness.

The basic assumption is that every object present in the considered data universe can be described using only the available associated information. For example, information associated with a portfolio of financial assets forms a data universe, where each individual transaction represents an object described by properties of the used financial instrument and associated characteristics, transaction size, value, trade date, etc.

Depending on the granularity of the information, some objects in the universe cannot be differentiated from each other and appear to be the same. Using the above example and ignoring the trade date, transaction size and value causes the transactions using the same financial instrument to be indiscernible. Such groups of indiscernible objects form elementary sets (called *equivalence classes*) within the considered universe, also referred as *crisp* or precise sets. This indiscernibility relation forms the mathematical basis of rough sets.

Any union of elementary sets is also a crisp set. Any other set is a rough (imprecise) approximation of a class, i.e. it contains objects which cannot be surely classified as belonging to the set or not. Such a boundary region cannot be precisely classified using the available information.

Crisp sets do not have any boundary region. Therefore, the rough set theory describes vague concepts using two crisp concepts:

- the *lower approximation*, which consists of all objects belonging to the given class (elementary set) with certainty, and
- the *upper approximation*, which contains objects possibly belonging to the considered class.

Following Pawlak and Skowron [100], one can observe the following about the rough set approach:

- availability of formal methods for hidden data pattern discovery,
- ability to propose optimal set of data and descriptive attributes (data reduction),
- quantifiable assessment of data/attribute significance,
- ability to derive decision rules from the observed data set,
- easy-to-understand formulation of rules,
- straightforward interpretation of obtained results,
- suitability of many of its algorithms for parallel processing.

Basic concepts of the rough set theory and their extensions are described in the following sections. For a more detailed discussion of rough sets refer to Komorowski *et al.* [74], Pawlak and Skowron [99], and Shen and Jensen [115].

2.1 Information and Decision Systems

The data universe is represented as a data table constructed from the individual objects (rows) and their attributes (columns).

The data table is called an information system (Pawlak [96]), defined as the tuple $I = (\mathbb{U}, A)$, where $\mathbb{U}$ is a non-empty finite set of objects (the data universe) and A is a non-empty finite set of attributes such that $a : \mathbb{U} \to V_a$ for every $a \in A$. The set V_a contains all values the attribute a can take.

If the data classification is known, then the information system turns to be a decision system defined as the tuple $S = (\mathbb{U}, C \cup D)$, where $C \cup D = A$, $C, D \subseteq A$

and $C \cap D = \emptyset$. In this context C is a set of conditional attributes (features) of all objects belonging to $\mathbb{U}$, and D is a set of decision attributes, describing the elementary set (decision class), the given object belongs to[4]. The decision attribute may take several values though binary outcomes are rather frequent.

2.2 Indiscernibility Relation

The data table representing an information system may contain redundant information due to the presence of multiple indiscernible objects. Rough Set Theory identifies redundant objects using the notion of an equivalence (indiscernibility) relation $IND(B)$:

$$IND(B) = \{(x, y) \in \mathbb{U}^2, B \subseteq A : \forall a \in B a(x) = a(y)\} \tag{1}$$

If $(x, y) \in IND(B)$ then objects x and y belong to the same partition of $\mathbb{U}$ ($\mathbb{U}/IND(B)$ or $\mathbb{U}/B$), i.e. x and y are indiscernible from each other using attributes from B and belong to the same B-equivalence class or B-elementary set, denoted as $[x]_B$, $x \in \mathbb{U}$.

2.3 Set Approximation and Boundaries

The basic concept of rough sets is that of a set approximation.

Let $I = (\mathbb{U}, A)$ be an information system. Considering $X \subseteq \mathbb{U}$ and $B \subseteq A$, one can approximate set X using only attributes contained in B by defining the following operations on X:

- $\underline{B}(X) = \{x \in \mathbb{U} : [x]_B \subseteq X\}$, called B-*lower approximation* of X, which is a crisp set of all objects in $\mathbb{U}$, which can be *surely* classified as belonging to X using the attributes set B.
- $\overline{B}(X) = \{x \in \mathbb{U} : [x]_B \cap X \neq \emptyset\}$, called B-*upper approximation* of X, which is a crisp set of all objects in $\mathbb{U}$, which can be *possibly* classified as belonging to X using the attributes set B.

Based on the above operations, the following sets can be defined:

- $BND(X) = \overline{B}(X) - \underline{B}(X)$, called a B-*boundary region* of X, which is a set of all objects in $\mathbb{U}$, which cannot be surely classified as belonging or not belonging to X using the attributes set B.
- $NEG(X) = \mathbb{U} - \overline{B}(X)$, called B-*negative region* is a set of all objects of $\mathbb{U}$, which with certainty cannot be classified as belonging to X using the attributes set B.

---

[4] From this perspective rough sets can be seen as a classification framework using the supervised learning approach.

A set with $BND(X) \neq \emptyset$ is called a rough set with its approximation accuracy measured by the following coefficient:

$$\alpha_B(X) = \frac{|\underline{B}(X)|}{|\overline{B}(X)|} \tag{2}$$

Other approximation measures are also available (Greco *et al.* [52], Ziarko [159]).

2.4   Attribute Dependency in Decision Systems

Given a decision system $S = (\mathbb{U}, C \cup D)$, it is important to induce dependencies between conditional and decision attributes. The dependency level is described by the coefficient $k (0 \leq k \leq 1)$, defined as follows:

$$k = \gamma(C, D) = \frac{|POS_C(D)|}{|\mathbb{U}|}, \tag{3}$$

where $POS_C(D)$ is called a positive region and defined as:

$$POS_C(D) = \bigcup_{(X \in \mathbb{U}/D)} \underline{C}(X) \tag{4}$$

If $k = 1$ then D depends totally on C, otherwise D depends only partially on C, with $k = 0$ denoting a lack of dependency of D from C.

2.5   Reducts and Discernibility Matrix

The attribute dependency measure k, described in the previous section, provides a way to identify the set of conditional attributes R, which are relevant for the induction of the given set of decision attributes D, i.e. $\gamma(R, D) > 0$. Consequently, the decision system can be optimized to carry only these conditional attributes from set C, which are necessary to identify decision attributes without any loss of the dependency information. The remaining attributes can be removed as their absence will not weaken the classification dependencies.

A set of conditional attributes R, meeting the condition:

$$\gamma(R, D) = \gamma(C, D), |R| < |C|; \gamma(R - \{a\}, D) \neq \gamma(R, D), \forall a \in R \tag{5}$$

is called a *reduct* in the context of D.

There can be many reducts available in the decision system for the given decision set D. The set of attributes present in all such reducts is called a *core* in the context of set D. Being an intersection of all reducts, the core attribute set is present in all reducts of the decision system $S = (\mathbb{U}, C \cup D)$. Thus, removing any of the core attributes introduces more classification uncertainty into the analyzed dataset. However, it is possible to have a core being an empty set.

To find all reducts, the classical rough set theory proposes a methodology based on the discernibility relation and Boolean reasoning applied to the decision system $S = (\mathbb{U}, C \cup D)$. A symmetric $|\mathbb{U}| \times |\mathbb{U}|$ discernibility matrix is created, which contains entries c_{ij}, defined as follows:

$$c_{ij} = \{a \in C : a(x_i) \neq a(x_j), i, j = 1, \ldots, |\mathbb{U}| \tag{6}$$

Therefore, each entry c_{ij}, contains attributes with different values for objects x_i and x_j. Based on the discernibility matrix, a boolean discernibility function of m boolean variables $a_1^*, \ldots, a_m^*$ (corresponding to the m attributes stored in the given entry c_{ij} of the discernibility matrix) is defined:

$$f_S(a_1^*, \ldots, a_m^*) = \wedge\{\vee c_{ij}^* : 1 \leq j \leq i \leq |\mathbb{U}|, c_{ij} \neq \emptyset\}, \tag{7}$$

where $c_{ij}^* = \{a^* | a \in c_{ij}\}$

Having the set of all prime implicants of the discernibility function f_S allows finding all reducts of S (Komorowski et al. [74], Kryszkiewicz [79], Pawlak and Skowron [100]).

The problem of finding a minimal reduct (i.e. having the smallest number of attributes) is NP-hard (Skowron and Rauszer [121]). However, there is a fair amount of research devoted to the design of heuristic algorithms delivering multiple good quality reducts in acceptable time for classical rough sets (Kumar [80], Wróblewski [138], Jensen and Shen [64]) and their extensions (Yao and Zhao [145], Cornelis et al. [28]).

2.6 Extensions of Rough Set Theory

The rough set approach was criticized for its requirement to discretize data (Koczkodaj et al. [72] but see also Grzymala-Busse and Ziarko [57]), strict definition of lower and upper approximations (no fuzziness) and sole reliance on the available data to induce knowledge about the real world.

Consequently, the rough set theory has been studied intensively since their introduction and numerous extensions have been proposed, which alleviate many of the above-mentioned limitations and allow the use of rough sets beyond the pure supervised learning-based data classification domain (Pawlak and Skowron [99], Shen and Jensen [115]).

One of the first extensions was the *Variable Precision Rough Set* model (VPRS) proposed by Ziarko (Ziarko [159]), which allows relaxation of the upper and lower approximations by introducing classification thresholds. The main motivation for this extension was to provide the notion of classification uncertainty, which allows to define a degree of inclusion between two non-empty subsets X, Y of $\mathbb{U}$. This is an important extension, as it allows the rough set model to provide information similar to the confidence level known from statistical models widely used in finance.

In a similar fashion *Tolerance Rough Sets* replace the equivalence relation used by the classical rough set theory by another eligible tolerance (Skowron and Stepaniuk [122,123]) or similarity relation (Yao and Wong [141]). The approach allows one to use any suitable similarity measure in the similarity relation definition for each attribute, and relax transitive, reflexive and symmetric properties of the classical rough set equivalence classes (Kawasaki *et al.* [69]).

Another interesting extension of the classical rough set theory introduced the notion of a preferential attribute order, pairwise comparison table (Greco *et al.* [48]) and resulting *Dominance-based Rough Sets* (Greco *et al.* [49,50,52]), which defines a dominance (outranking) relation to be used in place of the equivalence relation. This allows considering the relative importance of ordered domain values when analyzing choice and ranking problems using the rough set approach.

The rough set theory is also seen as a model complimentary to that of fuzzy sets, proposed by Zadeh [153] as the way to express inexact concepts using the notion of a membership function. Due to its ability to deal with vague information the rough set theory was compared to that of fuzzy sets and hybridization of both approaches, known as *Fuzzy Rough Sets*, is the focus of research looking at fuzzy and rough set theories being complementary, rather than competitive, ways to tackle vague information (Dubois and Prade [41], Radzikowska and Kerre [102], Cornelis *et al.* [27]).

Other hybridizations include the usage of rough sets with neural networks and genetic algorithms, being the most popular soft computing methods used in finance. A more detailed description of individual extensions is given in section 3, where their usage in the context of economy and finance is discussed.

3   Application of Rough Sets in Economy and Finance

The application of rough sets in economy and finance has its roots in classification problems using classical rough set theory for the purpose of identifying failed businesses or predicting a threat of bankruptcy or merger/acquisition. A related field of research investigates the use of rough sets in selection of strong indicators for financial crises and resulting better insight in mechanisms of their appearance.

The second important application domain is that of predictive models for stock market movements and associated trading strategies. A related active research area is the use of rough sets in the active portfolio management.

All the above-mentioned areas of research observe a growing sophistication of approaches ranging from straight applications of the classical rough set theory to hybrid combinations of rough sets with a multitude of other knowledge discovery models. This section provides a detailed discussion of these approaches based on references to the existing empirical research and applications of rough sets in the economy and finance domains.

3.1   Risk Management

The financial risk management, understood as the way to discover and predict the possibility of financial losses, business failures and financial crises, has an obvious and prominent importance for economic ecosystems, as mentioned in Section 1.

Among many types of risk, the valuation of a solvency risk or the assessment of the business bankruptcy likelihood is of great interest to the financial sector in particular, and the public in general. The ability to correctly classify a transaction counterparty or security issuer (be it a commercial company or a sovereign) according to its bankruptcy likelihood (credit scoring) allows one to correctly price it into the transaction or reject it altogether if the risk is deemed to be too large. The plausible assessment of financial health is also crucial in the heavily regulated financial services sector, where mere rumors about solvency problems may cause a deep financial stress to the affected financial institution, market panic and liquidity squeeze, resulting in country- or worldwide financial crises. It is therefore not surprising that financial regulatory bodies (e.g. Basel Committee), independent industry watchdogs (e.g. rating agencies), and the financial institutions themselves are constantly looking after improved methods for forward-looking solvency assessment and risk management. Especially credit ratings, being a crisp assessment of creditworthiness[5], are an important factor considered when looking at the financial health of the given party, be it an individual, business, sovereign country or international organization. Public credit ratings provided by specialized internationally recognized rating agencies, the most prominent ones being Moodys, Standard & Poors, and Fitch Ratings, have a large material impact on the market perception of creditworthiness of assessed parties. This influence manifests itself in the direct relationship between the lower rating and increased borrowing cost, the given party has to pay the market, when issuing debt (so-called *credit spread*). The implications of the credit rating reach much deeper, and in case of financial institutions, have a material impact on the way the business is conducted.

However, in the wake of the 2008 mortgage crisis and the following Eurozone sovereign debt crisis, the overreliance on the ratings issued by the private rating agencies has been harshly criticized, as one of the factors contributing to the crisis. The objectivity of the rating agencies has been doubted upon and the fact that the rating methodology was a proprietary knowledge of the respective agencies called for more transparency and limitation of the credit agencies monopoly. Therefore, it is even more important to develop alternative methods for credit assessment as well as propose ways to verify the given credit rating and explain the rationale behind it, without knowing the actual methodology employed by the rating agencies.

Consequently, these problem areas attracted a lot of research attention – an early survey of Dimitras *et al.* [38] showed the usage of methods ranging from the discriminant analysis and logit/probit models to recursive partitioning (tree) algorithms and expert systems. A more recent discussion of empirical research on financial crises and business failures in the banking sector and the artificial intelligence methods proposed for their prediction was given by Demyanyk and Hasa [36].

Not surprisingly, business failure analysis and prediction, looking predominantly after accurate classification methods and the way to replicate the rules

---

[5] See Chen [22] for a description of credit ratings and further references.

used to assign a credit rating by rating agencies, is also actively approached using the rough set theory and its extensions.

An early work of Slowinski and Zapounidis ([124,125]) applied the rough set theory classification approach to the problem of credit scoring for a sample of 39 companies described by a set of 12 conditional attributes, being a mixture of selected financial ratios (quantitative attributes) and descriptive variables (qualitative attributes). Each company in the learning set was assigned a credit score (decision attribute) by a domain expert. The numerical (quantitative) attributes were discretized into qualitative ranges using the experts opinion (i.e. using the best practice with regards to the interpretation of financial ratios). Based on the decision system, the significant set of conditional attributes (reducts) and resulting if-then rules were generated. The authors pointed out the importance of qualitative input variables in the induced decision rules (e.g. management experience), which were not considered by traditional quantitative risk models. The model was applied to a real case data of a Greek bank and verified by a financial expert resulting in a positive opinion as with regards to its usability, even though the size of the data sample was relatively small and no out-of-sample test was conducted.

Slowinski *et al.* [126] continued to research the application of classical rough sets in the business domain by applying it to the problem of predicting a company acquisition, based on a sample of Greek companies. The rough set based prediction delivered a classification accuracy of 100%, 75% and 66.7% for 1, 2 and 3 years before the acquisition, respectively. Objects not matching any of the generated rules have been assigned the 'closest' rules – a distance measure based on the *valued closeness relation* (Mienko *et al.* [90]) has been introduced for this purpose. The other advantages of the rough set approach were also presented, like the ability to deliver the minimal subset of significant attributes (reducts) or the possibility to generate human readable if-then rules. Furthermore, an empirical evidence of advantages of rough sets vs. discriminant analysis, especially their explanatory power, was shown.

The relatively small sample size (60 companies) and somewhat arbitrary selection of conditional attributes (a set of financial ratios) and discretization algorithm provided only a glimpse of a practicable solution but at the same time showed the advantages and potential of the rough set based approach.

Dimitras *et al.* [39] compared the effectiveness of the rough set model with that of discriminant analysis and logit models, applied to the case of Greek companies. In comparison to Slowinski *et al.* [126] and Slowinski and Zapounidis [124] a relatively larger learning and test samples of firms (80 and 36, respectively) across several industrial sectors were created and similarly, an expert judgment used to identify conditional attributes (12 out of 28 financial ratios) and discretize the entry dataset. Similarly to Ruggiero [106], authors reported a better predictability using rules generated with the rough set approach against the results of a discriminant analysis, whereas the effectiveness of rough sets and a logit model were comparable. The effectiveness of the valued closeness relation (VCR) approach to classifying objects not exactly matching the generated rules was also confirmed (VCR helped to correctly classify 60% of previously not classified objects). The availability of the minimal set of most significant attributes (the selected reduct had 5 variables) and resulting compact decision rules were quoted as additional advantages of the rough set approach. The authors admitted however, that the induced rules were mostly relevant to the bank and the expert user, who delivered the data sample and selected decision rules. Using the described approach for another bank was however possible by working with the dataset delivered by the bank (especially, the training sample containing the decision attributes).

Another empirical confirmation of rough set theory suitability for the business failure prediction was delivered by McKee [87]. The author used the rough set methodology to construct a company failure prediction model generated and tested using financial report data for years 1986 – 1988 of a sample of 200 US companies, whereas 100 companies constructed the learning sample and another 100 the testing sample. The information table had 8 conditional variables, being financial ratios selected arbitrarily based on prior research and authors experience. The author used decision class rule strength and valued closeness relation based classification for objects having no exact match to the generated rules. The model accuracy was given as 93% for the learning sample and 88% for the testing sample. The model accuracy was much better than that of the recursive partitioning method developed previously using the ID3 method (McKee [86]), which when applied to the same data sample had an accuracy of 65%.

The, apparently successful, application of the classical rough set model theory to the business failure prediction was verified by McKee in his later empirical research study (McKee [89]), where the data sample from 1990-1997 for 291 companies (146 bankrupt and 145 remaining going concern) was selected. Care was given to the representativeness of the sample and comparativeness of the bankrupt and non-failed companies (matching size and revenues). The conditional variables were again selected based on their theoretical support and appearance in literature. Their discretization used the percentile ranking to create 10 subintervals for each of the selected variables. The input data was used to generate the learning sample of 150 companies and a testing sample of 141 companies. The author used a similar resolution of not exact matches as done in previous experiments (i.e. decision class rule strength and valued closeness relation-based classification).

The model achieved the accuracy on the test sample between 61% and 68%, using the VCR based approach similar to McKee [87] and sets of rules generated by two reducts. In comparison, the auditor signaling rate of 66% was computed on the same data. The author concluded that "the rough set models developed in the current research offer no significant comparative predictive advantage over auditors' current methodologies".

The author subsequently discussed the possible reasons for the achieved low accuracy against previous empirical studies, quoting a more realistic and larger data set, wider industry sector coverage, and calibration of non-deterministic matching strategies. The author concluded that the rough set theory was a worthwhile research candidate when looking for efficient business failure prediction method, especially when looking at data in the boundary region and searching for an optimal set of explanatory variables (reducts).

Bose [15] applied the rough set method to the search for predictive rules about the financial health of internet companies. The used data sample, taken at 30th of June 2001, encompassed 240 internet companies and 24 financial ratios selected based on the relevant literature and their applicability to the internet company specifics, and calculated for the year 2000, the year of the Internet Bubble crisis. The division into healthy and unhealthy companies in the data sample was somewhat arbitrary, i.e. companies with the stock price below 10 cents have been deemed unhealthy. The author did not provide details about the sample selection criteria other than that the selected companies had .com in their name or their business was conducted mostly over the Internet.

The dataset was initially divided into the learning (80%) and test samples (20%). A cross-validation with 10 random sample compositions was used in order to avoid overfitting and assure a general applicability of the generated rules outside of the used data sample.

Thanks to the used software (ROSETTA, Øhrn [94]) the author was able to apply multiple discretization algorithms (Boolean reasoning, Entropy algorithm, equal frequency range, naive algorithm, semi-naive algorithm) and two different ways of computing reducts (a genetic algorithm and Johnsons algorithm) and generating classification rules (standard voting and voting with object tracking). This allowed testing the efficiency of the methods combination on the classification accuracy. The best overall accuracy ratio of 72.1% was achieved using the equal frequency discretization with a genetic algorithm based reduct generation and standard voting classification (i.e. class with the highest certainty factor has been assigned).

The author performed a statistical analysis of the accuracy sensitivity to the number of rules, their length in terms of conditional attributes, as well as to the balance between and size of learning and testing samples. The marginal contribution of individual conditional variables was also verified.

The resulting observation was that the rough set model implemented using the ROSETTA software generated lots of redundant rules, as the reduction of the number of rules by 90% resulted in only 4.3% decrease of the classification accuracy. Out of the over 8000 generated rules, 500 rules were sufficient to generate prediction results having a reasonable accuracy. Neither rules length nor the balance between and size of learning and testing sample did have any statistically significant impact on the model accuracy.

The rough set model efficiency was compared with the one of logistic regression and discriminant analysis. The rough set model performed better than the statistical models, delivering accuracy of 72.1% on the test sample, against 65.2% and 65.8% delivered by the logit model and discriminant analysis, respectively. Another conclusion was that the rough set model was better in identifying the unhealthy companies than the healthy ones, even though their occurrence frequency in the data sample was similar.

A related work and findings were published by Liu and Zhu [84], who applied the rough set theory to the analysis of construction industry in China, using a sample of 16 financial ratios for 296 Chinese construction contracting companies. Similarly to Bose [15], the authors used the ROSETTA software and noted the better performance of the rough set model in classifying negative samples (unhealthy companies) than the positive ones. They have also noted the large number of reducts and redundant rules. The constructed rough set model had 5% to 10% worse classification rate than the other methods tested (decision tree, logistic regression and neural network). However, it had the best prediction accuracy for bad credit and the lowest misclassification error of all tested methods.

Sanchis *et al.* [110] provided a discussion of financial crisis and insurance company insolvency phenomena and proposed explanatory models for both the macro- and microeconomic problem based on the rough set theory. The authors mentioned the dominance relation rough sets, introduced by Greco *et al.* [49], but decided to use the classical rough set theory based on the observation that the considered financial ratios exhibit complex correlations (as observed by McKee and Lensberg [88]), which prevent the applicability of the dominance relation.

The rough set model created to analyze financial crises (model A) employed a mixture of 12 quantitative and qualitative variables representing macroeconomic policies (e.g. central bank independence) and variables (GDP, inflation, etc.) as well as financial ratios data for a set of 79 countries and period of 1981 – 1999. The learning and test samples used data from the period 1981 – 1997 and 1997 – 1999, respectively. Numerical input data was uniformly discretized into quartiles. The generated rough set core matched well the expected set of the most influential economic variables. The model achieved an 80% accuracy ratio on the test sample. The generated classification rules enabled the authors to confirm the importance of the monetary policy in times of crisis, as suggested in literature, but also claim that the policy clarity is more important than its function in terms of the resilience to a financial crisis.

The rough set model created to analyze the insolvency risk of non-life insurance companies (model B) used an input dataset of 17 financial ratios of 72 Spanish insurers (36 solvent, and 36 insolvent) for the period 1983 – 1994. Similarly to model A, the conditional variables were uniformly discretized into quartiles. The model generated 452 reducts, no core and identified 9 redundant conditional attributes.

The generated classification rules exposed the most significant variables, which were in line with literature, with one significant variable being an original observation derived from the model.

The accuracy of the rough set model B was compared with linear and quadratic discriminant functions, showing performance similar to that of linear discriminant analysis and better than that of the quadratic discriminant function.

Another empirical study on rough set application in the financial sector comes from Ruzgar *et al.* [109]. The authors looked at the performance of the Turkish banking sector in years 1995 – 2007 and tested the efficiency of the rough set theory in predicting insolvencies in this sector and time period (i.e. looking for early warning attributes). The time around the year 2000 was a difficult time for the Turkish banking sector (e.g. 6 banks failed alone in 1999). Therefore, the used data reflects well the conditions occurring during a banking crisis in a developing economy.

Altogether, 41 banks and 36 financial ratios (conditional attributes) were included in the sample. The continuous variables were discretized using quartile buckets, similarly to Sanchis *et al.* [110]. Reducts generated using the Johnson's algorithm were considered to be of better quality than the ones generated using the genetic algorithm (others than Bose [15], who used the same ROSETTA software package). A sample of generated rules and their strength was shown but no accuracy statistics or comparison with other predictive methods was provided (although the authors planned to continue their research and publish the comparative analysis).

Based on the results delivered by the rough set model, the low capitalization, asset quality and profitability were pointed out as the strongest discriminant variables in the analyzed data sample. The authors concluded by recommending the rough set method as a promising basis for an early warning system in the banking sector.

Greco *et al.* [53] showed the usability of the rough set model to the generation of decision rules, which describe the rating process, based on the data set containing the rating information. The authors applied the rough set model to the data sample describing the investment risk ranking of 52 countries as complied by the Wall Street Journal using 27 indicator variables. An extension of the rough set model able to cope with incomplete information was also described and used to generate the decision table, as 9.5% of data were missing. Finally, the LERS algorithm (Grzymala-Busse [55]), which was modified to account for missing values, was used to generate decision rules. The average accuracy of the model was relatively low at 61.54% – the small sample size was quoted as the main reason for this. Furthermore, the focus of the study was on the explanatory quality of the decision rules and not their predictive accuracy. Nevertheless, rough sets compared favorably to the multiple criteria decision analysis (MCDA – see Doumpos and Zapounidis [40]), decision tress and neural network based analysis of the same dataset in terms of accuracy but more importantly in terms of the explanatory power of the generated *if-then* rules.

In general, the basic weakness of the so far discussed research seems to be a relative small and biased data sample, relatively arbitrary selection of conditional variables and discretization methods used, as well as overfitting of the learning sample.

It was suggested that a further improvement of the model expressiveness could be achieved by using the *dominance relation* extension of the rough set theory, proposed by Greco *et al.* [49] (although McKee [87] and Sanchis *et al.* [110] decided not to use it, referring to the complexity of economic variables' correlations, for which a dominance relation was not expressive enough). The authors introduced the concept of *Dominance-based Rough Sets* (DRS) and applied it to the same empirical data set used in Slowinski and Zapounidis [124] in order to show the added value of the new approach against the classical rough set method based on the indiscernibility relation.

The use of dominance relation instead of the classical indiscernibility relation, i.e. the ability to define the preferential order of attribute values, resulted in a smaller number of reducts (4 vs. 29) and larger core, what is seen as indicators of a good approximation of the decision system (Pawlak [98]). Furthermore, the system generated a smaller set of compact decision rules with a more predictive power than the classical rough set approach.

Boudreau-Trudel and Zaras [16] applied the dominance rough set model to the problem of sorting investment projects based on a sample of 10 candidate investment projects from the Canadian province of Quebec. The DRS model was compared to the widely used Analytic Hierarchy Process (AHP) method. The initial input was defined using 24 variables describing the companies asking for the credit. The AHP method requires analysis and pairwise comparison of all the attributes to arrive at their weighting and thus resulting score of the analyzed data. In contrary, the DRS approach does not require the analysis but needs a learning sample to generate classification rules. The set of 4 candidate projects was used as the basis for DRS processing and generation of decision rules. The learning sample was created using a pairwise combination of the selected 4 projects, resulting in 6 objects described by the initial 24 conditional variables. The variables were replaced by the outcome an outranking relation applied to the values of the pairwise combined projects. The DRS model generated a reduct consisting only of 2 variables (vs. the full set of 24 initial variables required by AHP) with a 100% quality of approximation. The resulting classification rules allowed ranking all the 10 projects in the order very similar to that of AHP but with much less effort (time spent) thanks to no need for attribute analysis, what was quoted by authors as one of the main advantages of the rough set based approach.

Bioch and Popova [14] introduced a related method of rough set based analysis based on the concept of the monotone discernibility matrix and monotone reduct. The authors showed the applicability of the proposed model to the case of the bankruptcy risk analysis referenced by Greco *et al.* [49] and compared the effectiveness of the proposed approach with this of the dominance based rough sets. Thanks to the limiting assumptions of the monotone reduct the described model was able to generate only 5 decision rules covering the whole input dataset vs. 11 rules generated by the model proposed by Greco *et al.* [49].

Dominance-Based Rough Sets

The proposed extension defines an outranking relation s_a on $\mathbb{U}$ relative to attribute $a, \forall a \in C$, where $s_a(x, y)$ is interpreted as "*x outranks y*", or "*x is at least as good as y*" with respect to the attribute $a \in C$. The set V_a contains all values the attribute a can take. Furthermore, it is assumed that s_a is a strongly complete and transitive binary relation and the universe $\mathbb{U}$ can be partitioned into a classification set $K = \{K_t, t \in T\}, T = \{1, \ldots, n\}$, such that each $x \in \mathbb{U}$ belongs to one and only one $K_t \in K$ (Greco *et al.* [36])[6].

The following ordering relationship can then be observed $\forall x, y \in \mathbb{U}, \forall r, s \in T$ and $B \subseteq C$:

$$[x \in K_r, y \in K_s, r > s] \Rightarrow [s_B(x,y) \text{ and not } s_B(y,x)] \tag{8}$$

That is to say, x dominates y with respect to the attribute set $B \subseteq C$, denoted as $s_B(x,y)$, if $\forall a \in B : s_a(x,y)$.

The relation s_B can then be used to derive dominance sets with respect to $x \in \mathbb{U}$:

$$D_B^+(x) = \{y \in \mathbb{U} : s_B(y,x)\} \tag{9}$$

$$D_B^-(x) = \{y \in \mathbb{U} : s_B(x,y)\} \tag{10}$$

Having the classification set K divided as follows:

$$K_t^+ = \bigcup_{s \geq t} K_s \tag{11}$$

$$K_t^- = \bigcup_{s \leq t} K_s \tag{12}$$

lower and upper rough set approximations can be defined:

$$\underline{B}(K_t^+) = \{x \in \mathbb{U} : D_B^+(x) \subseteq K_t^+\}, \tag{13}$$

$$\overline{B}(K_t^+) = \bigcup_{x \in K_t^+} D_B^+(x) \tag{14}$$

$$\underline{B}(K_t^-) = \{x \in \mathbb{U} : D_B^-(x) \subseteq K_t^-\}, \tag{15}$$

$$\overline{B}(K_t^-) = \bigcup_{x \in K_t^-} D_B^-(x) \tag{16}$$

The model defines the quality of approximation as follows:

$$k = \gamma(B, K) = \frac{|\mathbb{U} - \bigcup_{t \in T} BND_B(K_t^-) \cup \bigcup_{(t \in T)} BND_B(K_t^+)|}{|\mathbb{U}|} \tag{17}$$

Definition of respective boundary regions and approximation accuracy coefficients follows the classical rough set theory.

---

[6] This corresponds to the partition of $\mathbb{U}$ determined by $IND(D)$, i.e. $\mathbb{U}/D$, where D is the decision attributes set D in a decision system $S = (U, C \cup D)$, such that $d : U \to V_d$ for every $d \in D$.

Greco *et al.* [51] introduced a further extension of the Dominance-based Rough Sets (DRS) approach, called *Variable Consistency Dominance-based Rough Sets* (VC-DRS), which reportedly improved the quality of rules induced from the lower approximation set vs. DRS by relaxing the assignment conditions of the lower approximation set via the consistency level parameter allowing a gradual violation of the dominance principle. An alternative extension of the dominance-based rough set, based on maximum likelihood estimation, was proposed by Dembczynski *et al.* [35], and further extended into so called *Stochastic Dominance-based Rough Sets* in Kotlowski *et al.* [75]. Interested readers are referred to these papers for further details.

An alternative interpretation of the dominance relation in the framework of the *fuzzy probabilistic rough set* model based on the coherent partial conditional probability paradigm was proposed by Capotorti and Babanera [17]. The authors applied the model to the credit scoring (probability of default) classification problem, and tested it on a sample of 80 companies from the Italian Umbria region, characterized by a set of 10 conditional variables, denoting their financial health over 2 years, and a binary decision variable D denoting the status of a company (i.e. default $=1$, or not default $= 0$).

As one of the first steps of the proposed method, the entry decision system $S = (\mathbb{U}, C \cup D)$ was analyzed by a domain expert providing an ordered criteria (positive, neutral, negative), which corresponded to value ranges of each conditional variable. The criteria express the expert's view of *a priori* probability of default $P(X)$ of a company in the context of conditional attributes' values. For example, an attribute a values 0-3 were judged to be positive criteria, whereas value 5 was judged to be negative one, with the value of 4 being neutral. Thus, the conditional variable's value set is replaced by the ordered set (positive, neutral, negative).

The same process is repeated for all n conditional variables resulting in a set of n criteria described by the same value set mapped to value sets of the original conditional variables. By introducing a strong assumption of the *conditional exchangeability* of the resulting criteria, i.e. two objects were equivalent if they had a matching count of criteria (positive i, neutral k, negative j), the authors derived a new set of equivalence classes $[x] \equiv F_{ijk}$ with larger granularity than the original ones. However, the new data set was still vulnerable to violations of dominance relations due to the basic definition of the *rough membership function* $\mu_X(x)$ referenced by the authors as:

$$\mu_X(x) = \frac{|X \cap [x]|}{|[x]|} \equiv P(X|[x]) = P(X|F_{ijk}) \tag{18}$$

where $X = \{x \in \mathbb{U} : fx, D = 1\}$

The proposed resolution introduced an alternative membership function based on the notion of a *T-norm* and holding the assumption of conditional exchangeability:

$$\mu(X|F_{ijk}) = \Theta_{pos}^i \Theta_{neg}^j \Theta_0^{(I_{((i+j=0)})} \tag{19}$$

where Θ_{pos}, Θ_{neg} and Θ_0 are *T-conorms* measuring the respective influence of the present positive, negative and neutral criteria on the company default. The authors described a maximum likelihood like procedure used to compute the T-conorms.

Having defined an equivalent *T-norm* and *T-conorms* for the membership function in the set of non-defaulted companies, it was possible to define the positive and negative regions of the resulting rough sets, which respectively classified the included companies as defaulting or healthy based on the available annual balance sheets separately for each year[7]. At that stage the classification accuracy was comparable with that of linear discriminant analysis and logistic regression but worse than that of VC-DRS. Therefore, the two classifications resulting from the annual data available in the 2 years sample were joined to create new positive, negative and boundary regions based on the alignment of equivalence classes and default/not default membership function comparison. The resulting classification accuracy was improved form 64.1% to 77.4%.

Beynon and Peel [10] applied the *Variable Precision Rough Set* model (VPRS), proposed by Ziarko [159], to the problem of business failure prediction based on a sample of 12 financial variables describing the condition of 90 UK companies (45 failed and 45 remaining a going concern) between 1997 and 1998. Similarly to the previous publications, the selection of the conditional variables was based on their occurrence in the expert literature in the context of corporate failures. Instead of an expert based discretization of continuous variables, the FUSINTER discretization method, using the quadratic entropy measure (Zighed *et al.* [160]), was applied. The model was able to correctly classify 91.7% of objects in the learning sample and 70% in the testing sample, even though the selected β-reduct was chosen to illustrate the method rather than be the optimal reduct. The model delivered accuracy comparable with multivariate discriminant analysis, logit and decision tree methods.

Yet another example of the VPRS application in finance was presented in Beynon *et al.* [11]. Here it was used to predict the company profitability based on a textual analysis of chairman statements. The authors analyzed chairman statements of 98 UK non-financial companies, 49 with the highest and the other 49 with the lowest annual profit. The statements were preprocessed to generate 10 conditional variables representing their textual characteristics often used for contextual and readability analysis, e.g. Flesch index[8], average number of words per statement, percentage of good/bad news words and positive/negative key words, etc. The input data was discretized using the minimum entropy method, introduced by Fayyad and Irani [43]. The authors decided to use only two value intervals for all discretized conditional variables, resulting in a binary

---

[7] The boundary region was also defined.

[8] The Flesch index, introduced by Flesch [44], is one of the most popular readability indexes, known as the Flesch Reading Ease test. Based on the average sentence length in words L and average sentence length in syllables S (expressed as syllables per 100 words), the index is defined as follows: $206.835 - (1.015L + 0.846S)$. The lower the index of a passage the more difficult is the passage to understand.

representation of the input information. The pre-processed data and resulting decision rules were tested using the n cross-validation method (80 statements in-sample and 18 out-of-sample) on 1000 generated information systems (decision tables). In 977 cases, a single β-reduct was generated. The predictive accuracy of the VPRS method was high at 90%, with the average number of 3-4 rules and the maximum number of 10 rules being generated (3 and 5 rules cases were the most frequent ones, occurring 702 and 225 times, respectively). Similarly, the most common β-reducts contained 2 to 4 conditional variables.

Based on the analysis of the generated β-reducts the authors concluded that the percentage of bad/good news and percentage of positive key words were the conditional variables having the most impact on the classification quality, with the percentage of bad news being the most selective one.

Even though the described experiment was provided rather for an illustrative purpose (small sample size, simple discretization and no information about the used β apart from the applied selection rule set), the successful application of rough sets to the analysis of textual information shows the importance of seemingly qualitative information for the prediction process and the respective advantages of rough sets due to their ability to process such data alone or in combination with quantitative information. Therefore, the next logical step would be to extend the amount of textual information available to the rough set based classification system – Yu, Wang and Lai [151] described such a combination of text mining and rough sets used to forecast tendency of the oil market movements.

Griffiths and Beynon [54] subsequently proposed a VPRS-based expert system offering a domain expert the ability to fine tune all steps of the rough set based classification process, from selection of β-reducts to partitioning of the data set into training and testing samples and generating resulting decision rules. The system was presented using the case of rules mining based on Moody's ratings (decision variable) and a set of arbitrarily selected 9 financial ratios (conditional variables) for a sample of 435 US, Canada and West European banks. Therefore, the study is a presentation of a VPRS based software using the example from the financial domain rather than an empirical verification of the VPRS ability to predict or classify data from the financial domain.

Variable Precision Rough Sets

The model defines an inclusion degree for two non-empty sets $X, Y \in \mathbb{U}$:

$$c(X, Y) = 1 - \frac{|X \cap Y|}{|X|} \tag{20}$$

Ziarko [136] defines then a desired classification error β, which allows to control the level of uncertainty allowed in the inclusion relation:

$$X \subseteq_\beta Y \; iff \; c(X, Y) \leq \beta, 0 \leq \beta < 0.5 \tag{21}$$

The existence of the upper limit of $\beta < 0.5$ is defined as a majority requirement.

The above-defined notion of inclusion is then used to derive the basic rough set operations:

$$\underline{B}_\beta(X) = \bigcup\{[x]_B \in \mathbb{U}/B : c([x]_B, X) \leq \beta\}, \tag{22}$$

$$\overline{B}_\beta(X) = \bigcup\{[x]_B \in \mathbb{U}/B : c([x]_B, X) < 1 - \beta\}, \tag{23}$$

This allows one to apply the rough set theory to analysis of statistical trends, where data uncertainty (aside of incompleteness) is imminent. Nevertheless, the β parameter has to be specified a priori and has an impact on the effectiveness of the β-reduct generation (Kryszkiewicz [78], Beynon [9]).

Hybridization

Another common way of using rough sets is to combine them with other knowledge discovery or predictive methods in order to benefit from a combined strength of both approaches.

A common construct is the use of the rough set model as the pre-processing stage for other methods, where the input dataset is optimized (reduced) both in terms of the data attributes and data objects by the rough set generating reducts and removing indiscernible object copies.

Hashemi et al. [58] proposed a hybrid rough set and neural network process for prediction of merger and acquisition events in the banking sector. The learning sample of 28 financial ratios from 1992 for more than 200 banks from the US state of Arkansas was used as the input data.

The rough sets process was applied as a preprocessing stage, which delivered an optimized (reduced) dataset for a neural network model. The rough set model reduced the size of the data sample (redundant indiscernible objects) and number of attributes to 18 (2-dimensional reduction of the information system). The hybrid model delivered a classification accuracy of 96% against the standalone neural network's accuracy of 84%. The authors concluded, that the optimization (reduction) of the input data performed using the rough set model significantly improved the predictive power of the combined decision support system.

Ahn et al. [1] also combined the rough set method with neural networks to analyze the predictive power of the combined approach for the case of a business failure. They used a historical dataset of 8 financial ratios for 2400 Korean firms and the period between 1994 and 1997. The selection of ratios and their discretization was based on similar studies and an expert judgment. Similarly to [58] the rough set method was applied to the entry data first, generating decision rules and reduced information system, where redundant attributes and objects have been removed (so-called 2D reduction). However, the authors went further than that and used classification rules generated by the rough set model as the primary classification tool. The reduced information system was then used as the entry dataset for a neural network for samples having no exact classification given by the rough set model (referred as a hybrid model).

¬The hybrid model was tested using a k-fold cross-validation approach with $k = 12$, so that each object in the entry data sample participated in the training and testing sample.

Objects which failed prediction using the rough set approach were fed into the neural network. The resulting prediction accuracy (94%) was better than that of a stand-alone neural network (84%) or discriminant analysis (78.6%), confirming the strength of the combined approach and results of previous comparisons with the discriminant analysis.

Another example of the rough set and neural network combination, utilizing the ability of rough sets to filter out insignificant information was described by Jie *et al.* [65]. The rough set model was applied to a learning sample consisting of 18 financial variables describing 64 companies listed on Shangai and Shenzen Stock Exchange in year 2008. Another sample of 64 company data for 2009 was used as a test sample. The authors selected a reduct consisting of 6 variables to construct the learning data set for the neural network based classifier. The reported out-of-sample accuracy of the model was 97%.

Shuai and Li [116] used a combination of rough sets and worst practice data envelopment analysis (DEA - see Seiford and Zhu [111]) to construct a business failure predicting model for Taiwanese companies. The model used 9 quantitative and 4 qualitative input variables derived from annual reports for years 2003-2004 of 396 firms (352 going concern and 44 failed). The DEA model was used to classify the input sample based on quantitative variables. Rough sets were used to generate classification rules based on qualitative variables only. The authors reported a 100% accuracy of the hybrid system against 82% achieved using only the worst case DEA model.

Zhou and Bai [157] described a hybrid credit scoring assessment system using a combination of rough sets, *support vector machines* (SVM) and genetic algorithms (GA). Similarly to Ahn *et al.* [1], data was first fed into a rough set model, resulting in a horizontal reduction of the data set and generation of classification rules. The output of the rough set processing was then analyzed using discriminant analysis (DA), a BPN neural network with one hidden layer (5-9-1), SVM and a GA-SVM hybrid.

The model was tested using a sample of 330 companies being debtors of a Chinese Construction Bank between 2002 and 2004. The sample was divided into learning and testing sets, containing 80% and 20% of data, respectively. The sample contained healthy, doubtful and failed debtors, with each group constituting 33.3% of the sample size (i.e. 110 companies). The data set was described using a set of 12 variables consisting of financial ratios (e.g. sales income/total assets) and qualitative information (e.g. level of management).

The rough set analysis delivered an optimal reduct consisting of 5 variables, thus resulting in a data reduction vs. the input data set described by 12 variables. The test sample was divided into the subset having the matching classification rule (Group I) and the one without the matching classification rule (Group II). The classification rules generated by the rough set model had the best classification ratio on the holdout sample of Group I (95.51% vs. 91.63% of the second best GA-SVM hybrid model). The proposed GA-SVM hybrid delivered the best classification accuracy of 89.95% on the Group II testing sample.

The overall classification accuracy of the combined rough set and GA-SVM model was compared with the performance of a rough set and neural network and rough set and SVN models. The rough set-GA-SVN model provided the best overall performance of 92.4% vs. 91.6% and 89.3% delivered by rough set-SVN and rough set-BPN models, respectively.

The authors pointed out the effectiveness of the rough set method in its capacity of data reduction and generation of classification rules, even though the used data sample ignored the company's industry and size and the strict application of the rough set model did ignore data not matching any of the generated classification rules. Addressing these issues was left for the future research.

Yao [147] applied a combination of the *neighborhood variable precision rough set* - NVPRS (Hu *et al.* [60]) and support vector machine models to a credit scoring assessment problem, on the example of the Australian and German credit data samples available from the UCI Repository of Machine Learning Databases. The German data set consisted of 1000 entries described by 24 numeric variables whereas the Australian data contained 640 entries described by 6 qualitative and 8 numeric variables.

The input data set was first processed using the neighborhood variable precision rough set, resulting in the horizontal data (feature) reduction. Instead of the standard discretization procedure, the neighborhood rough set model employs the concept of information granules expressed in the form of a δ neighborhood relation, acting as an equivalence relation. The reduced data set was classified using the SVM model and the accuracy of the hybrid model verified using the k-fold cross validation ($k = 10$) method. The hybrid NVPRS-SVM model delivered the classification accuracy of 76.60% for the German data set and 87.52% for the Australian sample, compared to the second best neural network based classifier, with the respective accuracy of 75.20% and 86.83%. Statistical models, being logistic regression and linear discriminant analysis delivered even lower accuracy of 72.40% and 66.00% for German data and 85.70% and 85.20% for Australian data, respectively. The author quoted the ability to work on mixed (quantitative and qualitative) data and lossless feature reduction as important advantages of the used rough set model, having a decisive impact on the performance of the proposed hybrid model.

Yeh *et al.* [148] combined a hybrid rough set-SVM model with the data envelopment analysis (DEA) to create a business failure prediction method. The use of rough sets allowed the authors to consider qualitative information, namely management efficiency, as a model input variable prepared using the DEA method. The data sample of 114 companies (38 failed and 76 going concern) was constructed using information from the Taiwan Stock Exchange and Taiwan Economic Journal for years 2005-2007.

The rough set model was utilized as a pre-processor stage for the SVM model, where generated reducts were used to remove redundant information from the entry data. Consequently, out of the initial 17 quantitative input variables (financial ratios selected by an expert) a reduct consisting of 7 input variables was used to construct the entry data set for SVM analysis. The qualitative management efficiency information, prepared by DEA as the 8th variable was then added to the model, and its impact on the accuracy was tested. The authors also compared the hybrid rough set-SVM (RST-SVM) model with the combination of rough sets and back-propagation neural network (RST-BPN) and claimed better classification results achieved using the rough set-SVM model considering the qualitative management efficiency information prepared using DEA, namely RST-SVM accuracy of 83.33% vs. 86.84% when the DEA variable is used, and RTS-BPN accuracy 78.95% vs. 82.46% with the DEA input variable included. The authors concluded that the inclusion of the management efficiency information, as prepared using the DEA method, improved the classification accuracy of both considered hybrid models, i.e. RST-SVM and RST-BPN, whereas the RST-SVM model delivered the best overall accuracy. Selection of input variables was mentioned as having a significant impact on the performance of the SVM classifier – thus the usefulness of the rough set method, which allows one to identify the input variables influencing the decision variable the most. The authors also confirmed a better efficiency of the above mentioned model against the statistical tools of MDA and logit, although the actual test results were not described.

McKee and Lensberg [88] used previously analyzed data and derived a rough set information model (McKee [89][9]) used as a pre-processor (data reduction) stage for a genetic algorithm based predictor of business failures. The learning and test samples were created in a similar fashion like in McKee [89], with the test sample being slightly larger (144) than previously (141). They reported an increased prediction accuracy of the hybrid approach at 80.3% against the 67% accuracy of the standalone rough set model. The conclusion underscored the improved accuracy of the hybrid model and increased insight and understanding of the correlations between conditional variables thanks to the combined usage of rough sets and genetic algorithms.

Chen [22] proposed a hybrid soft computing model to induct rules, which can be used to derive a credit rating of the given party. The model was based on a combination of feature selection methods, probability based data partitioning, fuzzy sets and rough set theory. Expert knowledge was employed at the entry point to remove outliers, noisy and low quality data. Subsequently, core descriptive attributes were selected by averaging results produced by several popular feature selection algorithms, i.e. Chi-square, ReliefF, Gain Ratio and Info Gain. The resulting core data set was discretized using the linguistic value mapping (fuzzification) based on the cumulative probability distribution approach (CPDA – similarly to Teoh et al. [134], a normal distribution function is used) and triangular fuzzy numbers. The constructed decision table was then used by the rough sets based LEM2 algorithm (Grzymala-Busse [56]) to generate a candidate set

---

[9] The original text referenced the article McKee [89] as *McKee, T.E., 1999b. Rough sets bankruptcy prediction models versus auditor signaling rates (revised manuscript under review in international journal).*

of *if-then* rules describing the credit rating assessment process. The final rule set was produced by discarding rules having a support value below the given threshold.

The model was tested using the ratings dataset of Asian banks derived from the BankScope database maintained by the Bureau Van Dijk company. The entry dataset covered the period between 1993 and 2007 and contained 1327 entries consisting of 38 conditional attributes and one decision attribute, being the credit ratio assigned by the Fitch Ratings agency. The author applied his expert knowledge in order to preprocess the entry data set, by removing attributes deemed to be superfluous or suffering from insufficient data quality. The resulting experimental data set consisted of 18 numeric continuous conditional variables and one decision variable (credit rating). The subsequent feature selection step reduced the number of conditional variables to 16 (a rough set reduct based feature selection did not reduce the number of variables in the experimental data set). The data set was then discretized using the fuzzy sets method and CPDA model based on the normal distribution function. The generated decision table was used by the rough set LEM2 algorithm to produce decision rules, whereas rules with support < 2 were discarded. The procedure was repeated 10 times using the experimental data set randomly split into the learning and testing samples, containing 66.7% and 33.3% of data, respectively. In order to compare the efficiency of the proposed method, the same data samples were used to train reference models, being C4.5, Bayesian Networks, neural networks (MLP), Holte's OneR, logistic regression, and single vector machines with sequential minimal optimization.

The proposed model displayed the best accuracy of 83.84%, utilizing 16 conditional variables. The author mentioned however that the method, as other learning models, is susceptible to the quality of the data sample, oversampling and specifics of the application context (here: credit ratings of Asian banks). Further research was suggested in order to test the approach on data coming from different emerging markets as well as including additional financial information, e.g. auditor opinion.

Yeh *et al.* [149] proposed a combination of the Moody's KMV debt pricing model, random forest and rough set to predict credit rating of analyzed companies based on the current market information in addition to the commonly used company accounting reports, which were necessarily backwards looking. The KMV model was often used in the credit risk area to assess the probability of default of analyzed companies and compute the market value of their debt. The so-called distance-to-default is a measure indicating the health of the analyzed company, i.e. the larger the distance measure the lower the probability of default. The KMV distance-to-default was used together with other financial variables to construct the initial data set. The data set was divided into learning and testing samples using the proportion of 80:20. Subsequently, the Random Forest analysis is used to sort and select the candidate input variables in terms of their significance. The resulting data set forms an information system processed using the rough set model in order to generate classification rules via the LEM2 algorithm.

The empirical test of the proposed hybrid model was performed using a data sample taken from the Taiwan Economic Journal gathered for the period of 2003-2008 for 2470 Taiwanese high-tech companies and consisting of 21 financial variables (debt-equity ratio, return on equity, earnings per share, etc.), as well as the KMV distance-to-default computed by authors. Data was then analyzed using the Random Forest ensemble to sort the candidate conditional variables according to their importance for classification. Following this analysis, the sample data for 18 variables, including the KMV distance-to-default measure (assigned the highest importance by the random forest ensemble) was discretized using an expert opinion and used to construct the decision system for the rough set based predictive classification and decision rule generation. The empirical test reached the best accuracy of 93.4% vs. 90.3% achieved by a rough set only model and 84% of the third best combination of random forest and decision tree (C4.5). Other tested models included pure C4.5, classification and regression tree (CART), support vector machines (SVM) and their combinations with a random forest ensemble (RF), similarly to the proposed combination of RF and rough sets (RS). The performed tests also showed the significance of the KMV distance-to-default measure, which removal caused a drop of accuracy to 73.7% for the proposed RF+RS model. This confirms the advantage of models being able to accept heterogeneous information and inherently select the relevant variables. The authors hinted at this potential by naming the ability to integrate potential non-financial factors as a part of the credit scoring analysis. This may include textual information, like news, auditors or CEOs statements, which analysis using rough set based models was already proposed by some researchers mentioned in this paper.

Xiao *et al.* [139] described an ensemble based bankruptcy risk classification model using the Dempster-Schaefer (D-S) evidence theory as the method used to combine classification results of individual classifiers (the evidence). As advantage of this selection, the ability of the D-S model to assign different significance to the individual classifiers was quoted. However, D-S has problems when fusing conflicting information. One of the methods proposed to minimize the impact of the conflicting information is to assign weights to individual classifiers before applying the D-S based combination (so-called weighted D-S evidence combination). The authors proposed the use of the rough set model to assess the significance (weight) of each individual classifier, quoting the self-sufficiency of the method in terms of the required information, i.e. only the analyzed single prediction output is needed to compute its significance. Consequently, the proposed hybrid model consists of a classifier ensemble, a rough set based result analyzer, assigning significance to the results delivered by individual classifiers, and providing the weighted information to the D-S model for the final synthesis. Normalized outputs of each single classifier (i.e. level of financial distress) applied to the learning sample formed the conditional variables of the input information system for the rough set based weight computation. The actual normalized distress level of each company was used as the input decision attribute. The variables were then discretized to form a decision system for the rough set analysis. The rough set attribute significance was then used to compute the weight of the attribute thus providing the weight of the respective classification method. The weights delivered by the rough set model were then combined with outputs of individual classifiers to compute basic probability assignment matrix within the framework of the D-S theory in order to arrive at the forecasted decision attribute (the level of financial distress).

The model was tested using a sample of 253 Chinese companies listed on Shenzen and Shanghai Stock Exchanges in years 2007-2009. 161 companies were classified as going concern, whereas the remaining 92 companies had a Special Treatment (ST) status assigned by the Chinese supervisor. The authors selected the logistic regression, neural network and support vector machine models as individual classifiers. The input data set consisted of 39 financial variables, including current ratio, earnings per share, etc. The forecasting horizon was set on 2 and 3 years, i.e. financial data describing the company's condition 2 and 3 years ago was used to forecast the current financial distress condition. The initial variable set was further reduced using the t-test of significance and stepwise logistic regression before delivering the data to the individual classifiers. Based on the applied 10-fold cross validation procedure the number of selected conditional variables varied between 2 and 6. The resulting financial distress classification of the 3 individual classifiers (resulting in 3 conditional variables for the rough set analysis) was discretized using the equal distance binning method. The authors decided to use two bins (binary mapping in $\{0,1\}$) for the input values with the threshold set at 0.5 ($a < 0.5 \Rightarrow 0$, otherwise 1). The average out-of-sample prediction accuracy of the proposed hybrid model for the 2 years and 3 years prediction horizon was higher than that of individual classifiers and alternative ensemble systems (majority voting, Bayesian, behavior-knowledge space – BKS) at 87.80% and 69.02%, respectively. Results for the training sample were mixed, with alternative ensemble methods obtaining better accuracy in several individual validation runs[10]. The authors hinted at the coarse mapping between the generated output values and financial distress classification as the possible reason for this inefficiency and pointed it out as the possible future research area.

A summary of the described rough set research focused on the risk management is given in Table 1.

The majority of work was concentrated on predicting the bankruptcy risk using various types of the rough set model itself as well as in combination with other techniques, including neural networks, genetic algorithms, support vector machines and fuzzy sets. Especially, the fuzzy rough set combination seems to provide an inherent way to discretize input continuous variables based on the fuzzy membership function and utilize the rough set's ability to reduce the number of variables and generate decision rules. Another interesting advantage of rough sets was the ability to analyze qualitative information and the potential

---

[10] The BKS based ensemble outperformed the proposed hybrid model also in few validation runs on the testing sample, but its average accuracy was lower.

to perform the analysis of a combined qualitative and quantitative information universe. This is especially important in the financial domain, when many economically relevant risk factors are hidden in the narrative comments provided in financial reports or daily news. The ability to extend the set of analyzed information to include the narratives aside of the standard quantitative set of financial ratios seems to be a very promising way to improve the quality and explanatory power of the forecast. The growing research in this area (Beynon *et al.* [11], Yu, Wang and Lai [151]) confirms its potential.

The presented empirical research used widely different data samples ranging from small focused data sets, where the delivered observations can be perceived as an indication of the described model's basic potential rather than its direct usability, to relatively large samples covering multiple industries, where the practical usefulness of the proposed solution was contested. The common feature was the expert based selection of input conditional variables and geographical locality of data, e.g. Taiwan, Greece, or USA. This allowed authors to use standardized descriptors for individual companies (mostly financial ratios based on accounting statements) or countries coming from one or two well-known sources, but on the other hand did not explore the relationships existing between global players domiciled in different regions – a sample including only peers from a local geography may not reflect the increasingly global market dependencies.

Also, the macroeconomic environment (e.g. state of the economy at the snapshot time or performance of a global benchmark index), which may have an impact on company's results in the particular time period, was not sufficiently considered. In addition, the temporal aspect of data should be explored more thoroughly, e.g. a continuous deterioration of free cash flow over a number of years is a more serious issue than a large loss in one year, which may be due to operational (e.g. a litigation) or macroeconomic issues surrounding the given company.

All authors have used an expert knowledge when selecting the candidate input variables. This necessarily introduced a bias into the resulting analysis, aggravated by the relatively small set of input variables, varying between 2 and 38. While the involvement of an expert seems inevitable due to the almost infinite set of candidate variables, it would be beneficial to work with data sets containing as much information as possible and let the formal methods select the ones, which are important in the given context. However, practical issues, like data availability, quality and comparability make this process difficult. Nevertheless, this seems to be the next necessary step required to deliver predictive systems not only confirming the existing expert knowledge but discovering new useful insights mined from data sets too large and too complex for a single human expert to analyze.

The above-mentioned importance of the selection of proper input variables was acknowledged by many authors and underscored by varying accuracy and sensitivity of results to the data being tested.

Table 1. Summary of Rough Set based methods used in the Risk Management Area

Area	Model	Discretisation Method	Sample Size	Cond. variables	Accuracy	Accuracy vs. others[11,12]	References
Bankruptcy Risk	Rough [13] Sets	Expert	39	12	100%[14]	-	[14], [124], [125]
			116	12	50.0 - 76.3%	>DA >logit	[39]
		Objective	41-291	2 to 36	61.5 - 88.0%	>DA >DT >logit >NN =human	[15], [53], [87], [89], [109], [110]
			296	16	75.0%	<DT <logit <NN	[84]
	Variable Precision Rough Sets	Objective	90-435	10 - 12	70.0 - 90.0%	>DA >DT >logit	[10], [11], [54]
	Dominance Relation Rough Sets	Expert	6 - 39	12 - 24	94.9%	=AHP >RS	[16], [49]
	Hybrid Models	Expert	80 - 2470	8 - 22	77.4 - 97.0%	>DA >DT >logit > NN >SVM >RS	[1], [17], [65], [149], [157]
		Objective	253 - 1327	2 - 24	76.6 - 87.8%	>Bayes >DT >logit >NN >RS >SVM	[22], [88], [139], [147]
		—	396	4	100%[11]	>DEA	[116]
			114	17	86.8%	>RS+SVM >RS+NN >RS+NN+DEA	[148]
Macroeconomic Stability	Rough Sets	Objective	79	12	80.0%	>DA	[110]
Merger and Acquisition	Rough Sets	Objective	60	10	66.7%	>DA	[126]
	Hybrid Models	-	200	28	96.0%	>DA >NN	[58]

---

[11] AHP = Analytic Hierarchy Process; Bayes = naive Bayes classifier; DA = Discriminant Analysis; DEA = Data Envelopment Analysis; DT = Decision Tree; NN = Neural Network; RS = Rough Sets; SVM = Support Vector Machines.

[12] > is better; < is worse; = is comparable.

[13] Bioch and Popova [14] applied the concept of a monotone discernibility matrix.

[14] Learning sample only; Bioch and Popova [14] did not provide this information.

3.2 Financial Time Series Forecasting and Trading Rules Generation

The ability to predict market data movements (especially stock market prices) is a focus of much research for obvious reasons. The financial market participants use predictive tools based either on technical analysis of time series (so called chartists) or rely on a thorough evaluation of the company financial data, its economic environment and leading macroeconomic indicators (fundamental analysis).

The basic assumption is that the historical data contains patterns of behavior, which can be applied to the future price movements, be it a particular repeatable sequence of its historical prices or related market data (the chartist case) or a strong pattern of cause-effect relationships between the company's financial ratios, macroeconomic indicators and the phase of the economic cycle.

The predictive model of market movements is usually coupled with optimized trading rules, which together attempt to deliver optimal trading strategies. This sort of analysis has traditionally employed statistical regression but is also at heart of the knowledge discovery methodologies, and numerous attempts were published, although mostly utilizing neural networks and genetic algorithms (Atsalakis and Valavanis [5]).

The market data time series is represented by numerical variables (price, volume, volatility), which generate huge data volume at the level of single stock transactions (so-called tick data). Yet, data has to be discretized and analyzed over a certain period of time to induce generalized rules incorporating historical patterns. Furthermore, multiple data streams have to be considered in order to derive strong predictive rules. Therefore, the problem of attribute transformation is a relatively important topic in the case of time series – the decision to use moving average aggregation, correlation ratios, etc. has a decisive impact on the quality of the predictive model.

Lin and Tremba [34] examined the theoretical aspects of numerical attribute transformation and applied them to the case of stock market and economic data prediction using the rough set model. They recommended using delay windows, averaging and cumulative aggregation as methods to create input attributes for time series analysis. The linear attribute transformation was demonstrated using a rough set model applied to a sample of daily stock prices in the period 1990-1996, where correlations between the mixture of 10 stock prices and index values, serving as conditional attributes, and the target stock of Applied Materials, serving as the decision variable, were discovered in the form of the generated classification rules.

The daily input data was processed using the column transposition operation to generate 3 intermediary decision tables containing the transformation effect of delay window, averaging and summing, having each 77 attributes (a weekly time window) and 1511, 1487, and 1511 rows, respectively.

Conditional variable data was discretized by rounding the daily percentage change to the nearest integer. Initially, the same discretization algorithm was applied to the decision variable, resulting in 67 decision classes. This approach caused generation of weak rules with low support. The decision variable data was then discretized using an arbitrarily defined value buckets with the cut points being -0.5 and 0.5, dividing the continuous value space into 3 decision classes (falling, no change, rising). The accuracy of the rules generated from the daily data was tested using an out-of-sample data for the period from August 1, 1996 to April 9, 1997. The generated rules achieved an average accuracy of 71%[15] on the test sample.

The authors pointed out the importance and difficulty of deriving the right discretization algorithms, as required by many knowledge discovery tools, including rough sets, when applied to the stock market time series. The trial and error approach has to be used. They recommended an expert user input for the selection of required attribute transformations but also suggested a brute force search otherwise.

In Tremba and Lin [135] the authors applied the previous findings in further practical experiments using the rough set software package DataLogic/R+ (Golan and Edwards [45]) and algorithms performing a linear transformation of the input raw market data already referenced in Lin and Tremba [34]. The authors extended the test data set with monthly data consisting of 3 conditional variables being 3-month moving averages of indexes characterizing the order book condition in the semiconductor industry. The decision attribute was still the (monthly average) price of the Applied Materials stock. The input monthly data set consisted of 72 entries[16].

The input data was processed using the column transposition operation to generate 3 intermediary decision tables containing the transformation effect of delay window, averaging and summing. The transformation of the monthly data resulted in three tables having each 28 attributes (a 7 month time window) and 60 rows.

Conditional variable data was discretized by rounding the daily percentage change to the nearest integer. The decision variable data was discretized using an arbitrarily defined value buckets with the cut points being -0.5 and 0.5, dividing the continuous value space into 3 decision classes (falling, no change, rising). The accuracy of the rules generated from the monthly data was tested using an out-of-sample data for the between August 1, 1996 and December 31, 1996. The generated rules achieved an average accuracy of 88.9%[17] on the test sample.

The authors concluded by confirming the importance of the applied discretization (clustering) method for the effectiveness of the rough set based analysis.

---

[15] Only rules with large support, as defined by the authors, were considered when computing the average accuracy.

[16] The daily data, its processing and results were the same like in Lin and Tremba [34].

[17] Only rules with large support, as defined by the authors, were considered when computing the average accuracy.

They discovered some useful time-dependent predictive rules using this approach but noted the necessity of a trial and error process.

Early works introducing the rough set approach to the stock market prediction included Ziarko *et al.* [158], Golan and Edwards [45] and Golan and Ziarko [46], who attempted to employ the variable precision rough set model to generate stock market movement prediction rules based on the discovery of repeatable dependencies in the past market data. The model was tested using monthly historical market data from the period 1980 – 1990 including 5 major Canadian stocks from the Toronto Stock Exchange, major Canadian and US stock indexes (TSE, DOW, S&P, etc.) and a large set of economic indicators, like GDP, interest rates, unemployment and inflation. Altogether the data set was described by 32 to 40 variables. The raw numeric data has been pre-processed to the delta form (increase/decrease vs. the previous value) and discretized based on intervals defined by a domain expert. Movements of the selected 5 stocks served as decision variables. The authors used the value of $\beta=0.55$. Generalized rules, derived from β-reducts, were verified by domain experts and confirmed as sufficiently descriptive for the characteristics of the selected stocks. The authors pointed out the importance of the input information and rough set parameterization (i.e. roughness and precision *beta*) for the quality of the generated rules. An expanded set of input variables, including dividend information and technical indicators as well as the out-of-sample testing were left for future research.

Bazan *et al.*[8] studied the rough set based discovery of possible predictive correlations between end-of-month values of 15 financial indicators and stock prices at the end of the next month. The study concluded with a relatively weak performance of the method (prediction accuracy of 44%) and highlighted challenges stemming from data selection and gap filling (missing data).

A similar prediction attempt was provided by Baltzersen [7] using the index of the Oslo Stock Exchange as his experimental basis. The work described the methods, which can be used to convert time series into rough set objects, as well as the analysis of computed reducts and rules. Although the reported accuracy was below 50%, the system was able to identify core predictive variables having a strong influence on the movements of the index (interest rate level, gasoline consumption, or currency rates).

Skalko [119] was able to generate trading rules using a variable precision rough set model (implemented by the software package DataLogic/R+ (Golan and Edwards [45]) applied to time series of S&P 100, associated options (put/call) and US Treasury bond yields for the period between October 1987 and December 1994. The rules were applied to the testing data set from the year 1995. Based on the rules, the system generated 9 trades, 7 of them were profitable. Analysis of the rules hinted at the relative importance of interest rate levels and market sentiments for tactical trading.

Herbert and Yao [59] used the classical rough set approach to analyze time series data from the New Zealand Stock Exchange for the period 1991 – 2000. The input data, consisting of open, close, high and low daily prices, was transformed into a series of technical trend indicators, including Moving Average

Convergence/Divergence, Wilder Relative Strength Index, Moving Average over 5- and 12-day periods. The associated closing price of the next day was used to compute the decision attribute for the learning sample, being -1 (relative price drop), 0 (unchanged), 1 (relative price increase). The transformed data set was discretized using the equal frequency algorithm and divided into the learning and validation samples. The authors selected 10 high quality rules out of the 96 rules generated by the rough set model based on 18 reducts obtained. The average prediction accuracy on the holdout sample was 64.7%.

In Yao and Herbert [146] the authors provided further details about the experiments and extended research by testing the so-called neutral zone buffer, i.e. the range of relative daily price changes, which are deemed to be insignificant and thus assign the 0 (unchanged) value to the decision variable. The authors showed the impact of the buffer's size on the distribution and bias of the decision variable, what had a strong impact on the number and quality of generated reducts and rules. The remaining properties of the model were re-confirmed.

Al-Qaheri et al. [2] applied the rough set theory to create a stock prediction model using daily movements of a single banking stock from the Kuwait Stock Exchange for the period 2000-2006 as the analyzed data. The conditional attribute set contained a mixture of price values and trend indicators computed from the raw entry data (moving average, momentum, rate of change, price oscillator, etc.). The attribute selection was based on the domain literature. The boolean reasoning discretization algorithm proposed by Qizhong [101] was used. The data set was divided into the training (75%) and testing sample (25% based on random selection). The authors reported model accuracy of 98% measured on a holdout sample. The rough set model was also compared to the performance of a neural network model applied to the same data sample. The proposed rough set model generated almost 50% less rules than the neural network, with higher accuracy, although details of the neural network model were not provided.

Lee et al. [82,83] applied the rough set theory to the case of trading rules for futures market based on real time data. The authors defined 6 so-called trend groups (e.g. short term raising, long term falling, flat falling, etc.) and assigned 5 best performing technical trend indicators (in terms of their annual average return) within each group as conditional variables for the subsequent rough set processing. In the next step, the input data was discretized, using the equal frequency binning method, and the set of reducts generated. The authors decided to manually select 3 best reducts per trend group consisting of 3 (out of 5) randomly technical indicators and generated resulting decision rules. The trading rules base stored the rules generated using values for the 20 day time window. Each subsequent day provided a feedback to the rule base and the rules were recomputed based on the performance comparison between the rule based on old and new time window. The trend groups were then ranked based on the nearest neighborhood method.

The model was tested using the KOSPI 200 index of the Korean Stock Exchange for the time period between July 1996 and December 2006. The data frequency was empirically selected to be 30 minutes, based on the return rates of the different data samples. The data set was divided into learning and testing samples. Trend periods (long term raising, short term falling, etc.) in the learning and validation samples were identified by futures market experts. The authors reported the maximum average return rate of 12.7% and Sharpe ratio of 1.47 for a trading portfolio based on the set of three trend groups.

Khoza and Marwala [70] applied the classical rough set model to the case of stock market time series prediction. The input data was a mixed set of raw price data (open, high, low, and close) and technical indicators, including rate of change, momentum, disparity and 5-day moving average – altogether 9 variables. The decision variable was a trend indicator based on n consecutive daily close prices.

The model was tested using a 5 year sample of daily movements of the Johannesburg Stock Exchange All Share Index between 1 April, 2006 and 1 April 2011. The data set was divided into the learning and testing sample containing 75% and 25% of data, respectively. The authors applied the time series conversion algorithms described in Baltzersen [7] to preprocess the time series data and used the ROSETTA package to discretize and reduce the input data, and generate forecasting (classification) rules. The authors paid special attention to the discretization method, which is critical for the forecast accuracy of the generated rule set. Consequently, four discretization algorithms, namely equal frequency binning, Boolean reasoning, entropy and naive algorithm, were used and their impact on the forecast quality investigated. The best accuracy was achieved using the equal frequency binning algorithm with 4 bins and this data was used as the basis for further processing. The model generated a reduct, using a genetic algorithm provided ROSETTA, consisting of 6 variables and derived 246 rules based on the reduct. The standard voting rule matching algorithm provided 80.41% forecasting accuracy confirming the applicability of the rough set model to the time series forecasting problem.

Another attempt to use the rough set model to explain stock market movements was presented in Ruizhong [108]. The data sample was constructed from financial data of 50 Chinese companies, which were constituents of the Shanghai Stock Exchange index SSE 50 in 2011 and 2012. Altogether 8 financial variables (P/E ratio, main operating margins, etc.) were selected to form the sample, with current ratio and rate of capital accumulation serving as the decision variable. All 7 conditional variables were discretized using manually selected bins. The decision variable was discretized to 1 if greater than average and 0 otherwise. The learning sample was built using data from 2011, whereas the quarterly data from the first quarter of 2012 was used as the test sample. Subsequently conditional variables with the greatest explanatory power were identified. The most important variable was rate of net return and its correlation with the stock movement trend in the learning and testing sample periods was verified. The author found a statistically significant positive correlation between the two variables, what is confirmed by a common sense. The reported out-of-sample accuracy was 97.8%, which may hint at a sample bias, given the short length and low sample frequency of the observed period.

Kim and Han[71] used the Variable Precision Rough Set model to generate profitable trading rules based on the market timing strategy. The conditional attributes consisted of 9 technical indicators, e.g. relative strength index, momentum, rate of change, Stochastic K%, etc., selected based on a domain expert advice and literature review. Similarly, the required discretization used threshold recommended in the expert literature for the respective technical indicators. The underlying data sample was taken from the daily market data of the Korean Stock Exchange KOSPI 200 index for the period from May 1996 until October 1998. The sample was divided in to a training part (65%) and holdout part (35%) in order to allow a simple rule validation. The used ROSETTA software was configured to generate β-reducts and induced rules for $\beta < 0.5$. The resulting set of 280 reducts had an empty core and authors selected 9 reducts as the basis for the rule induction. The selection of optimal reducts was based on their strength (support) and ability to generate bull market (buy) signals. The authors reported a positive performance of up to the absolute cumulated rate of return of 31.5% and excess rate of return of 48.3% against the buy-and-hold strategy, as measured in the validation period between January 1998 and October 1998.

Recently, AdgaM Group [63,120] applied rough set theory to create the AdgaM Trading Robots based on rough sets. Basically the robots are built up using the following types of classifiers:

- classifiers for the automated adaptive best features selection (out of tens of thousands of features) and
- classifiers for automated synthesis of adaptive Boolean expressions for trading rules generation (especially for Forex market).

For the real money on-line adaptive algorithmic trading on Forex market (EUR/USD) since September 2009 until March 2011 (1,5 year live trading) on OANDA platform (http://www.oanda.com/) gave the results as follows:

- 18% ROI;
- risk volatility 9,5%;
- out of 25 thousands EUR/USD positions 80% were the winning positions.

The back testing was performed on simulated trading using historical data and artificially generated thousands years by Monte Carlo with Forex data. The simulation of emotional Forex dynamics was based on dynamics of live trading data streams.

Hybridization

The stock market and financial time series prediction have been traditionally dominated by neural networks models over the last 20 years (e.g. [143,144,142] and [154]). It is therefore not surprising that much of the initial research activities considering the usage of rough sets in stock market prediction were evolving around the hybrid integration with neural networks.

An early attempt to use a hybrid combination of the rough set model and neural networks as a trading rule generator was reported by Ruggiero [106]. The author had already successfully experimented with the variable precision rough set model when building a trading system applied to the S&P 500 index (Ruggiero [104,105]). The VPRS-based system had a success hit ratio of over 70% when predicting positive index movements within the time window of 5 weeks and was able to predict strong index rallies of more than 2%.

In the hybrid system the rules delivered by the rough set model were used to supervise the learning process of the neural network and correct output errors. The system was able to reduce losses by 25-50% and improved the winner/loser ratio to over 50%.

This work was followed in Ruggiero [107] discussing a hybrid trading system based on the rough set theory, decision trees (C4.5) and neural networks.

Wang and Wang [136] also proposed a hybrid rough set-neural network model for the stock market prediction. Rough sets were used in their classical capacity of extracting decision rules out of the trained neural network model, but did not participated in the core data processing and forecast process, which was built based on a regularized neural network and selection of input variables derived from the underlying time series data, which described the time series trend (length and slope) and the signal to noise ratio. Nevertheless, the ability of rough set to generate human readable classification rules out of the neural network black box model was quoted as the advantage of the proposed hybrid.

There are however many interesting attempts to combine rough sets with other knowledge discovery methods aside of neural network-rough set hybridization – a very promising example are *Fuzzy Rough Sets*. Below we present them briefly.

Fuzzy Rough Sets

Fuzzy Sets were proposed by Zadeh [153] as the way to express inexact concepts by the way of a membership function. Unlike in the case of crisp sets, where an element can belong only to the set or its complement, the fuzzy sets allow an element to belong to more than one set with an associated membership grade. Since its introduction fuzzy sets and fuzzy logic are intensively researched and used (Zimmerman [161]).

The fuzzy rough set model is based on the notion of a fuzzy similarity relation $s(x,y)$[18] and fuzzy equivalence class $F = \mu_{[x]_S}$:

$$\mu_{[x]_S}(y) = \mu s(x,y), B \subseteq A \qquad (24)$$

---

[18] Properties of reflexivity, symmetry, and transitivity must hold.

Consequently, the fuzzy lower and upper approximation sets are defined as follows:

$$\mu_{\underline{S}(X)}(y) = \inf_{\omega \notin X}(1 - \mu_S(y, \omega)) \tag{25}$$

$$\mu_{\overline{S}(X)}(y) = \sup_{\omega \in X}\mu_S(y, \omega) \tag{26}$$

The theoretical aspects of the fuzzy rough sets are described in detail in Yeung et al. [150].

Wang [137] proposed a stock movement predicting system based on the fuzzy rough set model. The ability to cope with the noise inherent in the financial time series was quoted as one of the main advantages of the model – it is able to induce general rules even though the stock price movements in two different periods are of different scale but both should be considered as significant. The author used stock market data, including prices and volume information, from the Taiwanese stock market for the year 2001 (training sample) and the first five months of 2002 with 5 minutes sampling frequency. The input data was pre-processed into hourly intervals and a matching degree measure, defined as a fuzzy equality relation, was used to express the predictive power of the generated rules. The system was calibrated using 180 runs and delivered the prediction accuracy of 93%. The system did not account for correlations between stocks, qualitative information (e.g. political events) and missing data (i.e. no gap filling algorithm has been used) what could further improve the predictive effectiveness. This was left for a future research.

Teoh et al. [134] proposed a stock market predictive model based on a combination of a fuzzy set model, cumulative probability distribution approach (CPDA) and rough set based induction of decision rules. The model was based on the assumption that the input dataset distribution can be described using the cumulative probability distribution function. For the analyzed data sample an assumption of the normal distribution was made, where mean and standard deviation sufficiently describe the distribution. This allowed the model to define the data universe boundaries (for the used test dataset these were the minimum and maximum values of the given dataset plus/minus its standard deviation σ) and discretization intervals (seven intervals have been defined). The model then defined fuzzy sets, based on the computed discretization intervals and a fuzzy membership function, and mapped entry values of a stock index into the fuzzy sets (no expert input is used). The resulting fuzzy logical relations were used to construct a rough set decision table and generate decision rules using the LEM2 algorithm. The model was then able to compute the forecasted numerical outcomes of the prediction based on the assigned fuzzy categories.

The model was tested using data sets from Taiwan Stock Exchange (TAIEX) and New York Stock Exchange (NYSE) for years 1990-1999, containing daily closing prices. The authors verified the efficiency of the hybrid method by comparing it with its individual components (CPDA and rough set models). Following the evaluation method presented in the paper, the hybrid method consistently outperformed both individual models when applied to both, TAIEX and NYSE datasets (as measured by the root mean square error RMSE $= 90.29$ vs. 91.03 and 90.51 for TAIEX, and 45.62 vs. 45.67 and 46.39 for NYSE, respectively), although it did not consider the additional information available in correlated markets (e.g. futures market) and the effect of possible corporate actions (e.g. dividend announcements) and earnings surprises[19]. The proposed model outperformed also two other fuzzy time series models (not using rough sets) proposed by Chen [18] and Yu [152].

Cheng *et al.* [24] proposed a hybrid model combining fuzzy set based discretization methods, rough sets and genetic algorithms for stock market prediction. The input set of conditional variables is constructed of a series of technical indicators (moving average, momentum relative strength index, etc.) derived from the underlying time series data, with the 5 days horizon. Daily price fluctuation (i.e. the difference between todays and previous closing price or todays close and opening price) was selected as the decision variable. The candidate set of conditional variables was pruned using the correlation matrix indicating their significant correlation with the decision variable. The authors subsequently discretized the input data using fuzzy set methods, namely:

- the *Minimize Entropy Principle Approach* (MEPA) in order to partition range values of conditional variables. The partitioned set of values was used to generate fuzzy membership functions and assigned the attribute data to one or more linguistic values (fuzzification). The maximum membership degree was then used to select the associated linguistic value, being the output of the discretization process,
- the CPDA method (as referenced above for Teoh *et al.* [134]) in order to partition the decision variable being the daily stock price fluctuation, following the assumption of its normal distribution. The daily stock movement was discretized into three linguistic values denoting the directional trend of the movement (*up, fair, down*).

The discretized decision table was then processed using the rough set model to generate classification rules. Finally, the generated rule set was optimized for improved predictive accuracy using the genetic algorithm model, which was to introduce resilience against unexpected events not present in the learning sample.

The system provided trading rules and computed the forecast accuracy as well as expected return, which were used to compare the performance of the proposed model against its constituents (rough set and genetic algorithm only) as well as with the rate of return of the buy and hold strategy.

The model was tested using a sample of market data for a stock and index instruments. The input data set was selected to contain 8 technical indicators having a significant correlation with the price movements. The model performance when applied to a single stock forecast was tested using the daily market data of the TSMC stock traded on the Taiwan Stock Exchange between June

---

[19] The surprise effect describes the situation where the earnings unexpectedly missed or sur-passed market expectations (usually quantified by analysts' estimates). What follows is a sudden jump or drop of the stock price.

1999 and May 2000. The data was divided into learning and testing samples, having 10 and 2 months, respectively. The model achieved the accuracy of 55% and financial return of 16 stock units, outperforming the rough set (54.5%, 15 stock units), genetic algorithm (42%, 10 stock units), and buy-and-hold strategy (-15 stock units).

The model performance when applied to a stock index forecast was tested using the daily market data of the Taiwan Stock Exchange index (TAIEX) traded on the Taiwan Stock Exchange between January 2000 and December 2005. The data was divided into 6 annual datasets tested separately. Each annual data sample was divided into learning and testing samples, having 10 and 2 months, respectively. The model achieved the average accuracy of 60.1% and financial return of 950.62 index units, outperforming the rough set (55.3%, 531.93 stock units), genetic algorithm (54.7%, 722.56 stock units), and buy-and-hold strategy (275.04 stock units).

Based on the test results for the index data the authors underscored the positive performance of the model in a trending market regardless of the market trend (bull and bear markets). However, they noted the relative underperformance of the rough set, genetic algorithm and proposed hybrid model in periods of high volatility.

Ang and Quek [3,4] proposed a combined stock prediction and trading system based on the rough set theory and fuzzy neural network hybridization, called it Rough Set-based Pseudo Outer-Product (RSPOP) model. The predictive performance of the hybrid model was compared to several other neural network models (e.g. Kasabov [67]) using an artificially generated time series. The predictive accuracy of the RSPOP model, as measured by the mean square error measure, was found to be better than that of the other models.

Furthermore a trading system was proposed, using a simple rule set based on the moving average trend indicators (moving average convergence/divergence and exponential moving average). The model considers transaction costs when computing the total profit and loss, being used as the actual performance measure for the system.

The system was tested empirically using market data for two stocks from the Singapore Stock Exchange for the period 1980 – 2005, taken from the real economy and banking sector.

The data sample was divided into learning and validation samples. The performance of the proposed model was compared against the strategy using only the moving average indicators and another strategy using data forecasted with the dynamic evolving neural-fuzzy inference system (DENFIS – Kasabov and Song [68]), which had a performance comparable with RSPOP when applied to the artificial data set. The authors concluded by pointing out a lower number and complexity of generated rules, thanks to the application of rough sets, and higher efficiency of prediction of the proposed approach against the DENFIS model but underscored also the heavy dependency of the strategy's success on calibration of the trading module (i.e. moving average parameters).

Tan, Quek and Yow [131] subsequently proposed a trading system based on simple trend indicators (MACD, RSI) and using the rough set based pseudo outer-product model as its input predictor with the goal of extending the prediction horizon to up to five days. Thanks to its ability to reduce redundant information and filter out noise signals, the system was able to deliver better prediction accuracy than the benchmark neural network systems (95.8% vs. 92.87% of ANFIS[20] and 90.53% of RBF[21]) and outperformed the buy-and-hold based trading strategies (the average rate of return 693% vs. 139%) when applied to the data sample of daily prices of 5 stocks, representing different industries, traded on the Stock Exchange of Singapore between January 1991 and December 2004.

Cheng *et al.* [25] described a combination of probabilistic neural networks, variable precision rough sets and decision trees (C4.5) applied to the stock market timing problem. The three classifiers formed an ensemble employing the bagging algorithm to deliver the final classification. Aside of serving as a classifier, rough sets were also used to reduct the set of conditional variables used by the other classifiers. The proposed model was tested on a data sample described by a combination of 13 lagging macroeconomic variables (e.g. gross national product) and monthly market data (e.g. opening and closing price, trade volume) of a weighted Taiwan stock index published by the Taiwan Economic Journal. The sample consisted of 208 monthly data entries for the period between January 1988 and April 2005. The first 144 months was used to create the learning sample with the remaining 64 being the test sample. The decision variable was constructed as a trend indicator with values in $\{-1, 0, +1\}$ denoting the expected loss, zero or gain of the stock index over the Taiwan short term risk free interest rate. The variable was forecasted for 1, 3, 6, and 12 months. The VPRS model generated β-reducts reducing the data sample used by all three classifiers. The hybrid model achieved the classification accuracy between 57.81% for the 1 month forecast horizon and 76.56% for the 12 months forecast horizon.

For performance comparison, standalone rough set and probabilistic neural network models were constructed and tested using the same data sample. The probabilistic neural network and variable precision rough sets models achieved classification accuracy between 46.88% and 60.94% for the 1 month forecast horizon and 71.88% and 75% for the 12 months forecast horizon, respectively. Considering the actual earned income, the hybrid model had the best performance over the 1 month horizon, with the neural network performing the best over longer time horizons, over 3 months. All the tested models were better than the buy & hold strategy. The authors concluded that the combination of rough set based variable selection, decision tree noise reduction and ensemble based generation of trading rules allowed to construct a well performing trading rule generation model.

Kumar *et al.* [81] proposed an interesting combination of the rough set model and wavelet transform analysis. The application of the wavelet transforms allowed

---

[20] Adaptive-Network-Based Fuzzy Inference System.
[21] Radial Basis Function Networks.

generating views of the entry variables time series data at several resolution levels in the time-frequency space. This was to expose deterministic features of time series, possibly hidden at various time-frequency resolution levels and filter out the random noise. The attribute reduction capability of rough sets was their main feature used in the proposed model. The rough set reduct was generated both from the entry data set and all data sets derived using the wavelet transform at different resolution levels, i.e. the entry data set was reduced before the wavelet transform was applied; similarly, the data sets generated via the wavelet transform were first processed using the rough set and the resulting reducts generated the entry data set for a neural network based forecasting engine. The forecasts obtained at different scales were then combined recursively to obtain the final forecast.

The predictive model was tested by generating a 10 step (weeks) forecast of the S&P 500 index closing value. The entry data set consisted of weekly time series of 21 input variables describing the S&P 500 index and related macroeconomic variables, e.g. S&P High/Close, S&P Earnings, NYSE Total Volume, Gold Price, Short/Long-term interest rates, for the period January 1980 to December 1992. The learning sample consisted of 654 data entries, with the remaining 15 entries serving as the testing sample.

The rough set reduct analysis limited the initial conditional variable set to 13 variables. The reduced time series data was subsequently decomposed into 7 detail levels and an approximation (smoothing) level using the Daubechies wavelet (Daubechies [33]), resulting in 8 data sets for each input variable. The decision table for each level was build using each data set and generated a rough set reduct. This rough set processing resulted in further removal of 0 to 2 variables at the given level. The rough set reduct (decision table) of the learning sample was subsequently provided to the neural network forecasting engine, starting its training at the coarsest level, and continuing the training by providing the higher resolution levels until the required trade-off between generalization and overfitting was found based on the associated testing sample (the authors did not describe the used criteria). Based on this recursive cross-validation the 2 lowest resolution levels were dropped.

The resulting forecasting model delivered the best out-of-sample accuracy of 69.23%, as compared to the accuracy of alternative models being a combination of wavelets and neural networks, wavelets and Principal Component Analysis, and neural network only, which delivered the out-of-sample accuracy of 61.54%, 33% and 53.85%, respectively.

Although the authors acknowledged the capability of rough sets to generate forecasting rules, they decided that this ability was still not mature enough (referencing to the 44% accuracy reported by Bazan et al. [8]) and used a neural network based forecasting engine. However, the proposed model is generic enough to replace the used neural network model with a rough set classifier, bearing in mind the research progress and superior forecasting capabilities of various rough set and hybrid models delivered after the work of Bazan et al. [8] and listed in this paper. The advantages of a rough set based forecasting engine, including human readable classification rules, relatively low calibration needs and the ability to consider qualitative information make such an experiment worthwhile.

Huang and Jane [62] proposed a hybrid model based on classical rough sets, moving average autoregressive exogenous analysis (ARX – Bhardwaj and Swanson [13]) and Grey Systems theory (Deng [37]). Several financial ratios describing the financial condition of the analyzed companies are used as entry attributes to the trend prediction task employing a predictive ARX model applied to historical time series of quarterly stock market data extracted from the New Taiwan Economy database for the period of 2003-2007. Entries with missing values were rejected and detected outliers assigned a default value. The forecast is limited to the next quarter or half a year period. Forecasted time series data carrying values for the 10 most significant conditional attributes was subsequently selected for further processing using the GM(1,N) Grey method. The data was grouped into three clusters using the K-means clustering algorithm. The rough set model was then applied to the grouped forecast data and the generated lower approximation set used to generate decision rules. The stocks selected using the derived decision rules were used to form a portfolio, where the weight of individual stocks was computed using a Grey Relation Sequence. The rate of return of the selected portfolio was subsequently verified and, if acceptable, the portfolio was run for the next quarter. Otherwise, the portfolio structure was revisited by applying all the above-mentioned processing steps. The model achieved a cumulative rate of return of 82.45% versus the rate of return of 56.67% achieved by a Grey model GM(1,1) based system over the three years period (from the 2nd quarter 2004 to the end of 2006).

Huang [61] proposed an extension of the hybrid stock market prediction model introduced by Huang and Jane [62] by utilizing the Variable Precision Rough Sets (VPRS), which introduce classification uncertainty into the model, controlled using the precision parameter β. A method for determining a suitable β threshold using the Fuzzy C-Means clustering method (Bezdek [12], Cox [29]) and fuzzy algorithms was proposed.

Performance of both models, i.e. the one using the classical rough set and VPRS, was compared and a better performance of the VPRS based model against the RS based model (the cumulative rate of return over 3 years was 87.98% against 81.80%, respectively) was concluded.

Nair et al. [93] proposed a hybrid decision tree-rough sets model to predict stock market movements within 1 day horizon for the SENSEX index of the Bombay Stock Exchange. The data sample used contained index data (open, high, low, close price and volume) for the period between 3rd of September 2003 and 7th of March 2010. The raw data was used to compute 21 technical indicators describing the trend of volume and price, which served as conditional attributes. The price trend was used as the decision attribute. The attribute set was filtered using the C4.5 decision tree pruning and the resulting data used to generate rough set reducts and decision rules.

The model efficiency was verified using 10-fold cross validation. The classification accuracy of the hybrid system was 90.2% against 88.18% of the standalone rough set model, 77.66% of a simple feed-forward neural network and 73.26% of a naive Bayes classifier.

Shen and Loh [114] proposed an application of rough sets combined with clustering, using self-organizing maps (SOM – Kohonen [73]), to the predictive analysis of financial time series in the context of a market timing strategy. The input historical time series were represented by 7 technical indicators capturing the trend on the underlying stock (e.g. moving average convergence/divergence, stochastic oscillator, price rate of change) and selected based on the expert literature and related research (Shen and Tay [112]). The data was converted into a rough set decision table using the *columnizing* method described by Baltzersen [7]. The technical indicators served as conditional attributes. The decision attribute was defined as a computed variable denoting the predicted direction of the market up to 20 days in the future. The model initially applied the self-organizing map clustering to the input data using the number of target groups equal the number of values of the decision variable. The assigned cluster groups were compared to the decision variable computed for the given object (the decision attribute has been discretized using the equal-frequency-interval method). Based on the comparison a new decision attribute was assigned (mismatch) or the existing attribute was retained (match). The so-called reconstructed decision table was subsequently discretized using the Chi-square method (Tay and Shen [133]) and used to generate reducts and rules based on the classical rough set approach. Self-organizing maps were also used here to modify objects' strength values.

The model performance, in terms of the generated net profit, was compared to the buy-and-hold strategy based on the data sample of historical time series of S&P 500 between 4th of January 1988 and 26th of July 1999. Data from the year 1999 was used as the holdout sample. The system outperformed the buy-and-hold strategy in the learning period but did not bring a better performance in the validation period, as measured by the profit and Sharpe ratios. Depending from the tuning parameters, the proposed system performed worse to comparable with the buy-and-hold strategy, with the buy-and-hold performance benchmark for the holdout period (1999) equal to 163 and the performance of the RoughSOM system varying between -115.40 and 168.70, depending from the selected strength threshold (0.5 to 3.0). The classification accuracy remained stable around 58% regardless of the selected strength threshold and did not have much impact on the profitability of the generated trading signals. The conclusion was that by applying an expert knowledge to the selection of optimal trading rules the performance of the proposed system could be further improved.

The system was also tested using three other future indices, namely MATIF-CAC (French stock index futures), EUREX-BUND (German 10-year government bond) and CBOT-US (United States 30-years government bond) – see Shen [113]. The author reported better performance of the proposed system vs. the buy-and-hold strategy for these indices (643-1021 vs. 539.5, -7.54 to -2.02 vs. -7.47, -6.0 to 3.38 vs. -12.125, respectively). The classification accuracy varied between 51% for the CBOT and 74% for MATIF and remained stable regardless of the selected strength threshold. Similarly to the case of S&P 500, the threshold had a large influence on the profitability of the generated trading signals.

Pai *et al.* [95] applied rough set classification model supported by the *directed acyclic graph support vector machines* (DAGSVM) model to the problem of forecasting foreign exchange rates. The DAGSVM model allows using the SVM method for multiclass classification problems. The input data set consisted of 13 conditional variables being technical indicators (moving average, momentum, stochastic oscillators, etc.) and the decision variable being the rate of change of the daily closing price. The proposed hybrid model discretized entry data into 10 categories using the self-organizing maps (SOM) algorithm. The decision variable was mapped into 4 classes. The discretized training data set was subsequently analyzed using rough sets to generate reducts (8 conditional variables were retained) and resulting classification rules, which were subsequently pruned to improve classification accuracy. The reduced testing data set was also used to calibrate (select parameters) for the DAGSVM model using the *immune algorithm* and *tabu search* (IA/TS) method. The testing sample was first analyzed using a rule set generated by the rough set model. Only in case no rule was matching the entry data set, was it classified using the DAGSVN model. The model was empirically tested using a sample of 600 daily market data for the USD/JPY currency pair gathered in the period 2004-2007. The learning sample used 500 points with the remaining 100 points serving as the testing sample. The classification accuracy using rules pruning threshold of 3 was 68%, which was reported as better than the accuracy of individual classifiers of the proposed hybrid model.

Yu, Wang and Lai [151] proposed an interesting combination of text mining and rough set analysis applied to the problem of the crude oil price movement prediction. The model used a text mining algorithm, extracting factors correlated with the oil price movements from unstructured textual sources and creating a metadata repository. This factual dataset was subsequently analyzed using the rough set model in order to remove noisy/redundant information and generate efficient decision rules. The model effectiveness was compared to that of traditional structured data based prediction methods including multivariate (linear regression. back–propagated neural network) and time series models (random walk, autoregressive integrated moving average). The structured data set was based on the monthly WTI (West Texas Intermediate) crude oil spot price for the period 1970 – October 2004. The authors transformed the raw price quotes into trend indicators to improve performance of the prediction models. The used set of indicators was not described. The learning sample was constructed using the time period 1970 1999. The remaining data was used as the validation sample.

The rough set and text mining (RSTM) model outperformed the other approaches reflecting its access to a wider set of information than the other structured data approaches (86.21% vs. 75.86% delivered by the second best neural network). Furthermore, the model contained a feedback loop, which allowed for continuous self-adjustment and improvement of decision rules.

The majority of referenced work focused on delivering methods to predict trends of financial time series and trading systems employing the forecast to generate profitable trading signals. The proposed models were based on classical and variable precision rough sets as well as their combination with other techniques, including neural networks, fuzzy rough sets, self-organizing maps, decision trees, text mining, and autoregressive time series models. Given the inherent noise and volatility of the raw financial time series, they were usually pre-processed to a more stable (smoothed) representation defined by technical trend indicators (moving average, price rate of change, oscillators, etc.), incremental price changes or wavelet transform scaling. Technical indicators used to transform the raw input time series were selected based on an expert knowledge. The indicators provided inherent statistical properties, which account for the crucial data timeliness. This property was often used by the discussed research applying rough sets directly to the input data in order to generate reducts and forecast rules. Another way to use rough sets was to utilize their strength in generating descriptive rules, based on decision tables created using other techniques, like autoregressive exogenous analysis (ARX) or fuzzy time series models.

A generally interesting extension is the combination of rough sets with textual (unstructured) data mining ([151]), in order to increase the amount of relevant information available to the knowledge discovery system. This intuitively sensible approach deserves further research as knowingly the market is riding on the news.

The presented empirical research used widely different data samples ranging from small sample time series with monthly or quarterly frequency, where the delivered observations can be perceived as an indication of the described model's basic potential rather than its direct usability, to relatively large samples with high frequency data going down to minutes, where the practical usefulness of the proposed solution was contested in the form of a trading system generating intraday trading signals.

The seemingly successful application of rough sets in creation of trading systems based on analysis and forecast of time series with granularity ranging from quarterly to intraday shows the flexibility of the model and its ability to work alongside other soft computing methods in hybrid time series forecasting models towards a near real time trading system. Furthermore, the inherent ability to identify the core set of input variables, allows the user to experiment with a large number of correlated input variables, both quantitative (e.g. technical indicators, wavelets) and qualitative (news).

A summary of the described rough set research focused on the financial time series forecasting and trading systems is given in Table 2. Apart from the forecast accuracy researchers used other performance measures, more suitable for financial success measurement, like Sharpe ratio, rate of return and outright profit/loss, as well as statistical error measures like root mean square error (RMSE). The work referenced in Table 2 is therefore grouped according to the application area, methodology and used performance metrics.

Table 2. Summary of Rough Set based methods used for Financial Time Series Forecasting and Trading Systems

Area	Model	Discretisation Method	Sample Size	Cond. variables	Performance Matrics[22]	Performance vs. Others[23,24]	References
financial time series forecasting	Rough Sets	Expert	15100	7	α = 97.8%	–	[108]
		Objective	1250-2220	3-12	α = 64.7-98.0%	>NN >RS	[2], [34], [59], [70], [135], [146]
			10686	5	RoR=12.7% Sharpe=1.47	–	[82], [83]
		–	–	–	α < 50%	–	[7],[8]
	Variable Precision Rough Sets	Expert	600	32-40	Rules positively verified by an expert	–	[45], [46], [158]
	Hybrid Models	Expert	208-820	3-13	α = 76.6-85.0%	>NN >RS	[25], [136]
		Objective	600-21300	11-13	α = 68 − 93%		[95], [137]
			2428-2824	1	RMSE=45.62-90.29	>CPDA >FS >RS	[134]
		–	669-1625	21	α = 69.2-90.2%	>Bayes >NN >RS	[81], [93]
			406	32	α = 86.2%	>ARIMA >LRM >NN >RW	[151][25]
trading system	Variable Precision Rough Sets	Expert	132000	9	RoR=31.5%	>RW	[71]
		–	–	–	α = 70 − 77%	–	[104], [105], [119]
	Hybrid Models	Objective	5917-6222	10	profit= 385.8 -1659.8%	>MA >NN	[4]
			–	10	RoR= 27.5-28.3%	>Grey >RS	[61], [62]
			1539-2929	7-10	α = 58.54− 95.8%	>GA >NN >RS >RW	[24], [113], [114], [131]
		–	–	–	drawdown reduction= 25%-50% avg winner/loser ratio increase=50%-100% avg trade length reduction=50%-80%	–	[106]

---

[22] α = accuracy; RoR = annual rate of return; RMSE=root mean square error.

[23] ARIMA = auto-regressive integrated moving average; Bayes=naive Bayes classifier; CPDA = cumulative probability distribution approach; FS = fuzzy sets; LRM = linear regression model; MA = simple moving average strategy; NN = neural network; PCA = principal component analysis; RS = rough sets; RW = random walk (buy & hold).

[24] > is better; < is worse; = is comparable.

[25] Rough Sets + Text Mining.

3.3   Active Portfolio Management and Asset Valuation

The active portfolio management attempts to generate returns higher than the reference benchmark by modifying the benchmark portfolio structure (constituents and their weights). For example, an active portfolio manager may overweight some stocks listed in the S&P 500 index, which he believes will perform better than the benchmark. Instrumental for this decision is information about growth and value prospects of candidate assets (rate of return, growth rate, etc.), so undervalued assets can be identified. This is often difficult given either complex behavior of the valued assets or scarcity of available data. Many accepted statistical valuation models for financial assets are based on normalizing assumptions, thus introducing model risk in the valuation. The models have also an inherent difficulty of incorporating qualitative information, which is very often related to the behavioral aspect of the markets (the common assumption of a so-called rational investor). The ability to get a data-inferred valuation or its trend using soft computing techniques is useful in that case, as they are able to work with small samples. The inherent ability of rough sets to analyze qualitative information and identify the relevant information in the input data set allows the analysis to focus only on input variables having a factual impact on the observed outcome.

An early discussion of this topic was provided in Greco et al. [47], where the advantages of the rough set based analysis of portfolio data and related market factors vs. the widely used single- and multifactor statistical risk models were described. The authors compared the data requirements and predictive capabilities of standard statistical models (using the multivariate regression as the example model) with that of a rough set extended with the concepts of Pairwise Comparison Table, being the basis for the Dominance Relation Rough Sets extension. The sample data set consisted of daily prices of an equally weighted portfolio of 22 blue chip Italian companies and 7 market factors (e.g. Italian/US Government bond rate, Italian overnight rate)[26] for the period 1987 – 1992. The time series were converted into their monthly logarithmic delta equivalent as follows:

$$x_t = ln(a_t) - ln(a_{t-1}),   \tag{27}$$

where a_t is the level at time t.

The multivariate regression was done using data from the period 1991-1992. The rough set analysis used data from 1991 only. The required data discretization was performed using arbitrarily selected thresholds, resulting in 5 values for all conditional variables. The resulting decision table was processed to produce the Pairwise Comparison Table used to generate classification rules. The multivariate analysis identified only 2 risk factors as significant, whereas the rough set analysis has identified only one redundant factor. The rough set model generated classification rules, which provided a correct prediction of the portfolio value. However, no further information about the accuracy of the experiment was provided as the

---

[26] OECD CPI and Business Confidence Index were provided on a monthly basis.

authors focused more on identifying the possible benefits of the method at that initial (as the authors said 'pioneristic') stage, i.e.:

1. Statistical multifactor models (MFM) are suitable for analysis of stable relationships, but may fail when applied to small data samples or during turbulent periods - an example of this phenomena is given by the subprime mortgage crisis, where wrong valuation and rating of US mortgages as well as disdain for liquidity risk caused multiple financial institutions to fail and the crisis spilled over to the Eurozone. In this respect, rough set analysis was said to perform well with small samples and react to all input signals without the need for any assumptions regarding their statistical properties, like in the MFM case.

2. From the portfolio and risk management perspective, risk model sensitivities and rough set decision rules are alternative representations of risk exposures, with rough sets being able to capture the inherent market behavior.

3. Ability to consider strictly qualitative information while managing portfolio is an increasingly attractive feature postulated by the behavioral finance, with the rough set model being well suited for this purpose.

Susmaga *et al.* [129] subsequently proposed the use of the rough set theory when selecting candidate stocks for an active portfolio management strategy. The ability of the rough set model to find the optimal (and minimal) set of attributes really driving the decision parameters out of the large input group of candidate attributes was given as the main advantage. The investors can reduce the amount of data needed to take the investment decision and are guided by decision rules described by a set of accuracy measures (coverage, accuracy) to classify candidate instruments.

The authors showed the applicability of the rough set theory, as implemented in the ProFIT system, to the stock classification problem on the example of a subset of stocks listed on the Toronto Stock Exchange between 1989 and 1993 (several volatile stock classes, like mining and financial stocks were excluded). For each year in the entry data sample, a decision table was constructed using the annual rate of return as the decision attribute and other quarterly financial data (earnings, earnings surprise, capitalization, price tendency ratios, etc.) as conditional attributes. Discretization of conditional attributes was performed using value thresholds recommended by financial experts. The decision attribute was discretized into four schemas consisting of 2 to 4 classes. The impact of the classification group schema (number of classes assigned to the decision attribute) on the strength and accuracy of the subsequently generated decision rules was shown as being significant, with the accuracy ratio of a 10-fold cross validation varying between 49.7% and 72.3%.

Reducts and resulting decision rules were generated separately for each year in the data sample, based on the decision attribute being an annual rate of return. This would potentially require the use of a different subset of entry conditional attributes for each year.

The authors proposed the notion of a Common Reduct applicable across all years in the data sample. This allowed a reduction of the number of initially used conditional attributes by 40%. The average accuracy of the model was 71.8%, as tested using 10-fold cross-validation, which indicated a large potential of the method when applied to the stock classification problem.

Shyng et al. [118] proposed a method to improve the quality and knowledge discovery ability of generated rules, as applied to the case of a personal investment portfolio analysis. The required behavioral analysis is well suited for rough set based models, as most of the information acquired from individuals (esp. their preferences and targets) is of a qualitative nature. The proposed Forward Search and Backward Trace (FSBT) algorithm seeks to extract more descriptive information from the analyzed data set than the one generated by classical rough set method. The considered information system contains sets of conditional and decision attributes. The method starts with the identification of decision sets and selection of one such set representing each value of the target classification concept, i.e. the case study was considered with the perceived risk exposure of the portfolio being conservative, moderate or aggressive. Therefore, three decision sets had to be selected, which would be representable for each type of risk appetite. The authors did not provide an automated algorithm for the selection of representative decision sets – these were selected using the rule of maximum support and expert judgment for final selection (if more than one set had the maximum support). The data objects supporting the selected decision sets (called *selected objects*) are subsequently identified and all objects matching their set of conditional attributes and the portfolio risk appetite (called *target objects*) are found. The proposed algorithm is therefore in fact looking for all objects belonging to the upper approximation of the decision class (defined by selection objects forming its lower approximation), which are matching the predefined target criteria (i.e. the actual decision attribute). The found target objects are then used for data analysis comparing their structure – the authors did not deliver more details on the last step but acknowledged that the proposed method serves the best as a supportive algorithm to prune and improve the set of generated rules in an expert based knowledge discovery process. The method was empirically tested on the sample of 200 questionnaires describing the individual profile, selected portfolio and risk appetite of the questioned persons. The conditional variables were qualitative or already divided into arbitrarily defined bins, e.g. monthly salary was divided into 5 bins. The method generated 14 rules and had 100% accuracy vs. rough sets generating more than 85 rules with lower than 100% accuracy (not provided)[27] .

The authors continued looking at the rough set based analysis of personal portfolio investment preferences in Shyng et al. [117], where a combination of rough sets and Formal Concept Analysis (FCA) was proposed. The rough set model was used in its classical capacity of generating decision rules, which were subsequently pruned using a simple support threshold parameter (all rules with support below or equal the threshold were discarded). The remaining rules

---

[27] It should however be noted, that no out-of-sample testing was described/used.

were analyzed using the FCA concept in order to enhance the amount of meta-information about the rules and used attributes (structural dependencies). The concept was tested using the same data set as the one utilized in Shyn *et al.* [118] with one difference, namely the 'intermediary' set of decision attributes was not used and replaced with the target decision attribute being the portfolio risk appetite. The pruning support threshold of one object was used. The authors selected 40 rules out of the 67 rules generated by rough sets to be analyzed using the FCA method. The insight in the attribute relationship gained using the FCA method allowed to simplify some rules beyond the form proposed by rough sets. It also provided an insight in the relative importance of the attributes using the FCA notion of sub- and super-concept.

D'Amato [30] showed the application of rough sets to the problem of real estate pricing classification in southern Italy. The applicability and advantages of the model was shown using a data sample of 30 properties, where the property price was used as the decision variable, and only 2 conditional variables, namely commercial area and parking availability were used. The ability to work with incomplete and scarce data but also the ability to apply the classification abilities of rough sets to the mass appraisal task were cited as the decisive advantage of the rough set approach.

The author subsequently presented in d'Amato [31] and d'Amato [32] the application of an extended rough set model, using the concept of Tolerance Rough Sets' *Value Tolerance Relation* (Stefanowski and Tsoukias [128]), to the mass appraisal problem and compared it with the traditional multiple regression analysis. Data was not discretized – instead, the flexibility in the definition of the equivalence relation proposed by the Tolerance Rough Sets was employed to introduce equivalence thresholds. The employed similarity measure was defined as follows:

$$R_j(x,y) = \frac{\left(max\left(0, min\left((a_j(x), a_j(y)\right) + \tau - max\left(a_j(x), a_j(y)\right)\right)\right)}{\tau}, \qquad (28)$$

where τ is the threshold parameter.

The matching between the given classified object and a candidate classification rules was also using the above similarity formula. The most 'similar' rule would be used to classify the given object. This replaces the standard rough set rule voting based on the coverage ratio.

In d'Amato [31] the data sample of 69 real estate properties described by 4 conditional variables was used to test the Tolerance Rough Sets model. A learning sample of 19 properties was used to generate classification rules tested on the remaining test sample of 50 properties. The equivalence thresholds were chosen arbitrarily and the predictive accuracy of the model compared with the predictive quality of the linear multiple regression. The forecast accuracy of the multiple regression analysis was deemed better than that of rough sets but the difficulties with the calibration of the τ threshold parameter were quoted as the possible main reason.

In d'Amato [32] the Tolerance Rough Sets model was improved by providing an objective definition of the equivalence threshold defined as a standard deviation of the respective attribute. The concept was tested using a sample of 600 real estate prices from Helsinki, Finland for January 2005. Data was divided into learning and testing samples with 390 and 210 entries, respectively. The prediction accuracy was comparable, with the rough set model being slightly better than the linear regression model when applied to the testing sample.

It is worth noting that all the rules generated and tested by the model had the in-sample support and accuracy of 1, i.e. were supported by only one data entry. This is surprising since, even though the data was not discretized, the usage of the tolerance relation should allow generating classification rules with higher support. Usage of many rules with weak support hints at overfitting and may diminish the predictive accuracy of the model. One of the possible reasons could be inappropriate discriminant thresholds, which need to be calibrated to the data. Direct discretization of input data seems to provide a better control over the data partitioning and is directly reflected in the generated rules unlike in the above cases, where the generated rules contained raw numeric values from the underlying data samples (it seems however to be a limitation of the used software rather than the model, as the tolerance relation can be considered when generating the rules).

Tolerance Rough Sets

Tolerance Rough Sets replace the equivalence relation used by the classical rough set theory by another eligible tolerance (Skowron and Stepaniuk [122,123]) or similarity relation (Yao and Wong [141]). The approach allows one to use any suitable similarity measure in the similarity relation definition for each attribute. A popular measure used in this approach is defined as follows (Kretowski and Stepaniuk [77], Stefanowski [127]):

$$s_a(x, y) = 1 - \frac{|(a(x) - a(y)|}{|a_{max} - a_{min}|} \qquad (29)$$

where:

- $a(x)$ and $a(y)$ are values of the attribute a used for the similarity verification between objects x and y, respectively.
- a_{max} and a_{min} are respective maximum and minimum values of the attribute a in the given set U.

The Tolerance Rough Sets model introduces a threshold parameter τ, which decides about the similarity of the compared objects, e.g. objects x and y are similar if $s_a(x, y) \geq \tau$.

Objects meeting the similarity condition construct so-called tolerance set $TS_{a,\tau}$:

$$TS_{a,\tau}(x) = \{y \in \mathbb{U} : s_a(x, y) \geq \tau\} \qquad (30)$$

If multiple attributes are used to assess the similarity of compared objects then an aggregated similarity measure $s_B(x, y)$ and a global similarity threshold τ have to be provided, which considers the attribute set $B \subseteq C$ to create the tolerance set $TS_{B,\tau}$.

The most popular approaches use additive or multiplicative aggregation of the similarity measures computed for individual attributes, for example:

$$(x, y) \in TS_{B,x} \text{ iff } \prod_{a \in B} s_a(x, y) \geq \tau \tag{31}$$

$$(x, y) \in TS_{B,\tau} \text{ iff } \frac{\sum_{a \in B} s_a(x, y)}{|B|} \geq \tau \tag{32}$$

The definitions of lower and upper set approximations follow then the classical RS definition:

$$\underline{B_\tau}(X) = \{x \in \mathbb{U} : TS_{B,\tau}(x) \subseteq X\}, \tag{33}$$

$$\overline{B_\tau}(X) = \{x \in \mathbb{U} : TS_{B,\tau}(x) \cap X \neq \emptyset\} \tag{34}$$

Hybridization

The combination of rough sets with other soft computing techniques was also applied to the problem of asset value forecasting.

Chen and Cheng [23] applied a fuzzy set model to discretize input data and the rough set based LEM2 algorithm to generate predictive decision rules used to the classify Initial Public Offering (IPO) returns. The authors proposed three variants using different methods of data selection and discretization. Unlike Teoh et al. [134], where the focus was on the analysis of time series identified by one conditional attribute being the closing price, the authors were looking at a classification problem described by multiple candidate condition variables. Furthermore, the initial data set and candidate attributes were manually selected and preprocessed by the authors, based on their expert knowledge of the market. The candidate set of attributes was then further reduced using several Feature Selection Methods (e.g. Chi-square, GainRatio, InfoGain, etc.). The authors limited the core set to three attributes, one of them being continuous (closing price). The continuous attribute value set was partitioned using the Minimize Entropy Principle Approach (MEPA). The partitioned set of values was then used to generate fuzzy membership functions, based on MEPA, and assigned the attribute data to one or more linguistic values (fuzzification). The maximum membership degree was then used to select the associated linguistic value, being the output of the discretization process.

An alternative decision table was generated using expert knowledge (selection of conditional attributes) and C4.5 decision tree algorithm.

The resulting decision table, where enumerated IPO returns served as the decision variable, was processed by the rough set based LEM2 algorithm in order to generate a set of decision rules. The generated decision rules were filtered by ignoring rules supported by less than 2 entries.

The hybrid models were tested using the IPO data set gathered by authors for Initial Public Offerings launched in the period 1985-2003 on the Taiwanese financial market. The selected data set contained 220 entries described by 11 attributes. The data set was randomly split into training and testing samples with the ratio of 67%/33%. The split procedure was repeated 10 times.

The model variant based on the discretization performed with the C4.5 decision tree had the best accuracy of 82%. This is not surprising as this model was able to consider more conditional attributes than the alternatives while generating the decision rules (10 vs. 3). The MEPA based model delivered a very good accuracy of 80% given the much smaller number of conditional attributes used (3).

The accuracy of the proposed hybrids was also compared to that of stand-alone classification algorithms, namely C4.5, Bayesian networks, neural network and traditional rough sets. The proposed combination of C4.5 and LEM2 algorithms proved again to deliver the best accuracy (82%).

While the empirical results show the usefulness and potential of hybrid methods, the authors acknowledged the potential of considering more conditional variables in the primary MEPA based solution, including macroeconomic (background or environmental) variables like an economic depression indicator. They also noted the necessity of applying rule filtering in order to prevent data overfitting and high dependency of the proposed models from the expert input in the data selection and preprocessing step.

Liu et al. [85] proposed a combination of rough sets, feature selection algorithms and ordered weighted average (OWA) operator applied in the context of an electronic industry revenue growth rate forecast, based on the related company data. The authors used the ability of rough sets to generate human readable classification rules and forecast the trend of the observed decision variable (growth rate) but did not employ the core ability of rough sets to generate reducts. Instead, the model employed multiple feature selection methods (Chi Squared, Gain Ratio, Information Gain, ReliefF, Symmetrical Uncertainty) to remove irrelevant information from the input data set. The resulting conditional variable data was then processed using the OWA operator (Yager [140]), which allowed aggregation of input variables into one representative value per data entry. Another quoted advantage of OWA was the ability to apply configurable weights to the variables being aggregated, as controlled by so called *orness* α parameter (another descriptive parameter used is the dispersion parameter). The decision variable, being the operating revenue growth trend, had 3 values, denoting the growth rate value buckets of less than 0%, 0%-100%, and more than 100%.The resulting decision table was then fed into the rough set model to generate classification rules used to forecast the revenue growth trend.

The proposed model was tested using quarterly data of publicly traded electronic companies listed in the Taiwan Economic Journal database for years 2004 and 2005. The data set consisted of 12 variables and contained 2413 records for 2004 and 2490 record for 2005. Data from 2004 and 2005 were processed separately. The rough set model was implemented using the LEM2 algorithm. The model performance was tested using 10-fold cross validation for multiple values of $\alpha = [0.0, 1.0]$, where the learning and testing samples contained 67% and 33% of data, respectively. The best achieved accuracy was 96.82% and out-performed classical rough set (74.89%), decision tree C4.5 (74.47%), Bayesian networks (71.01%), and multilayer perceptron neural network (73.28%). How-ever, the best performance was achieved for $\alpha = 1.0$, which means that the explanatory power was concentrated in one input variable (operational profit) only. This rather trivial correlation hints at bias in the input data set, which could be caused by focusing on only one industry. The authors acknowledged the need for further experiments on larger data sets to further improve and verify the accuracy of the proposed model.

Chen *et al.* [21] and Cheng and Chen [26] proposed a combination of mul-tiple techniques supporting a rough set based classifier applied to the problem of a profit growth prediction in financial industry. The ability of rough sets to rely only on the actual data and their ability to generate comprehensive rules based on the set of quantitative and qualitative variables was one of the rea-sons for their selection as the core classifier. The input data set consisted of financial variables describing the financial well-being of the analyzed companies and associated growth rate as the decision variable. The proposed processing steps generally consisted of an expert based initial data selection, preprocessing (outlier detection and data cleanup), data discretization, feature selection, and finally classification rule generation and pruning. The authors proposed three hybrid rough set based models differing in the selected data discretization and attribute selection methods, which produced the entry data set for the rough set based rule generation:

- Model A used the fuzzy set based Minimize Entropy Principle Approach (MEPA), applied in multiple other publications referenced in this paper, to partition the entry continuous variables. Subsequently, the similarity thresh-old is computed and the fuzzified values are mapped into the best matching linguistic values using the triangular fuzzy number method.
- Model B extends Model A by adding the feature selection step before ap-plying the MEPA based partitioning. The authors applied the correlation based feature selection, Chi-square, Consistency, Gain Ratio, and InfoGain methods to filter out insignificant entry attributes.
- Model C employs the C4.5 decision tree to discretize data, unlike models A and B employing the fuzzy sets theory. The input variables were discretized using the expert knowledge to calibrate the C4.5 decision tree partitioning generating the cutoff points for the discretization.

The so-prepared decision system was analyzed using the rough set LEM2 algo-rithm in order to generate decision rules.

The proposed hybrid models were tested using a data sample randomly split into the learning and testing sample containing 67% and 33% of data, respectively.

The data sample consisted of 636 entries containing 8 conditional variables and 1 decision variable (profit growth rate) describing quarterly data of 70 financial holding companies in the period 2004-2006. The conditional attributes were discretized using one of the discretization methods described above (following the model definition), whereas the PGR decision attribute was mapped into three classes using the expert knowledge of the authors (PGR< 0% ⇒poor; PGR = 0 − 50% ⇒good; PGR > 50% ⇒fair). The procedure was repeated 10 times and also used to train and test alternative classification models in order to generate performance benchmarks for the proposed hybrid models. The alternative models include decision tree, MLP neural network, Bayesian network, and rough sets. The proposed hybrid model C delivered the best average classification accuracy of 97.41%, with the model A being the second with 96.99% accuracy. Model B achieved 94.86% and was worse than the decision tree based model (95.05%). The remaining single model classifiers were worse than the hybrids with rough set model achieving 82.87% accuracy. The proposed hybrid model C also reduced the number of generated rules by almost 68.5% against the standalone rough set model. Models A and B were even better, reducing the number of rules vs. the standalone rough set model by 88% and 98.5%, respectively. The authors pointed out some weaknesses and room for future improvements including larger and more diversified data samples, including data from different industry sectors and geographic regions, a larger set of input variables, further combinations of multiple classifier models, and comparison with other techniques like genetic algorithms or fuzzy time series.

Chen and Cheng [20] continued the experiment described in Chen et al. [21] and Cheng and Chen [26], focusing on the combination of a decision tree (C4.5) and rough set classifier (i.e. the best performing model C) and using the same data set consisting of 636 entries containing 8 conditional variables and 1 decision variable (profit growth rate) describing quarterly data of 70 financial holding companies in the period 2004-2006. The decision variable (profit growth rate) was again partitioned into three classes but using different threshold, i.e. PGR< 0% ⇒poor; PGR= 0 − 100% ⇒ good; PGR> 100% ⇒fair). In addition, the data set was split into the testing and learning sample using a slightly different ratio of 66%/34% vs. the previously used 67%/33%. Interestingly, the tested hybrid model delivered the same accuracy (97.41%) but generated twice as many rules as reported before (50% vs. 68.5%). This may hint at high sensitivity of the model to the discretization of the decision variable.

The authors suggested a further improvement of the proposed method by considering the cost of false classification and incorporating macroeconomic and qualitative information, like governance quality into the forecasting process.

The problem of the profit rate forecast was also studied by Zhao et al. [156] who proposed a forecasting model being a combination of a neighborhood rough set and least square support vector machine (LS-SVM) models. The neighborhood rough set model was used as the pre-processing stage, responsible for data reduction and discretization, thanks to the concept of the neighborhood distance δ.

The prepared data was then analyzed using the LS-SVM model in order to generate the classification function for the company profit forecast. The proposed model was accompanied by the alternative analysis using the factor analysis and a combination of classical rough sets and SVM.

The classical rough set based model used the equal frequency binning for data discretization and Johnson's algorithm for generation of reducts. The models were tested using an initial data set described by 31 conditional variables describing the financial characteristics of the analyzed company, e.g. return on equity, total assets turnover, etc. The rough pre-processing stage reduced the number of variables to 5, whereas the alternative classical rough analysis generated a reduct with 12 variables. The proposed model of neighborhood rough sets and LS-SVM provided the best classification accuracy of 94% with the LS-SVM classification based on the rough preprocessing being second (90.6%). The standard SVM model performed worse regardless of the data preprocessing algorithm (neighborhood or classical rough sets).

Research on the application of rough sets to the portfolio modeling is relatively scarce, with the seminal work of Susmaga et al. [129] dated back to 1996. Portfolio modeling is inevitably dependent on the valuation of the underlying assets, and this is where most of the focus can be found, either directly within the context of portfolio stock selection, where interaction between selected assets and their weighing has to be considered (Greco et al. [47]) or on the individual asset level (Chen and Cheng [23], Liu et al. [85], Chen and Cheng [20]). The latter is also the actual underlying drive of the trading systems and time series forecasting research presented in the previous section. Generally, rough sets and their extensions, like dominance and tolerance relation based variants, were found to be a viable tool for stock selection, implicitly considering the relationships present in the analyzed portfolio data. Hybrid models were also proposed for valuation of assets with little or hard interpretable information achieving very promising results. The presented rough set based models regularly outperformed the alternative solutions employing neural networks, decision trees, and naive Bayesian classifiers. Consequently, it seems that rough set based models deliver interesting tools to analyze combined quantitative and qualitative input data, and generate interpretable and useful results. On the other side, the rough set based models are not as simple to use as traditional statistical tools, since especially the extended rough sets (variable precision, dominance or tolerance relation) require careful tuning of their parameters and recurring interpretation of output in order to achieve the best forecasting performance (d'Amato [32,31]). Hence the need for further research in optimal calibration and rule selection algorithms but this is true for all soft computing methods. A summary of the described rough set research focused on the portfolio management and asset valuation is given in Table 3.

Table 3. Summary of Rough Set based methods used for Portfolio Management and Asset Valuation

Area	Model	Discreti-sation Method	Sample Size	Condi-tional varia-bles	Accuracy	Accuracy vs. Others[28],[29]	Refer-ences
active portfolio man-age-ment	Rough Sets	Expert	30-405	2-10	71.8-100%[30]	>RS	[30], [117], [118], [129],
	Dominance Rough Sets	Expert	66	8	–	–	[47]
asset valua-	Tolerance Rough Sets	Expert	69-600	4	76.2-82.9%	<LRM	[31], [32]
	Hybrid models	Objec-tive[31]	220-4903	4-31	83.0-97.4%	>Bayes >DT >FA >Grey Model >RS >NN	[20], [21], [23], [26], [85], [156]

4   Conclusion and Future Research Areas

This paper provided a review of problems in the area of economy and finance prediction models, and the related empirical research and solutions using the classical rough set theory, its extensions and hybrid solutions including other knowledge discovery methods, delivered in the time period 1992-2012.

The research is very active and growing in sophistication with the recent trend going towards hybrid solutions combining rough sets with other knowledge discovery techniques, especially neural networks (being traditionally the most popular predictive model in finance), fuzzy sets, genetic algorithms and support vector machines (Zhou and Bai [157]). Another trend is an innovative combination of traditional time series modeling (e.g. ARIMA) with the rough set's ability to deliver reducts (Huang and Jane [62]). A very promising hybridization proposal attempts to combine rough sets with text mining techniques, which have been recently making significant inroads in finance, especially for stock market prediction (Yu et al. [151]), but also risk management (Beynon et al. [11]). The ability to incorporate qualitative signals from the text mining will make it possible to account for events not currently being captured by the traditional technical or fundamental analysis.

The research is however still suffering from a relatively low number of real-life studies on large data samples, accounting for noisy and imperfect data. The reported accuracy ratios are very sensitive to the selected conditional variables,

---

[28] Bayes=naive Bayes classifier; DT=decision tree; FA=factor analysis; LRM=linear regres-sion model; NN=neural network; RS=rough sets

[29] > is better; < is worse; = is comparable

[30] 100% on learning sample only reported by Shyng et al. [118]

[31] Chen et al. [21,26,20] used expert knowledge to discretize the decision variable and calibrate the discretization methods applied to conditional variables.

learning data sample, its discretization methods and definition of the decision variables. The performance of the proposed solutions is still one of open research challenges. There are however some attempts towards increasing the volumes of data being analyzed, going in the direction of real time analysis (Lee *et al.* [82,83]), which is an interesting area of research in the financial domain context, as real time streaming data is ubiquitous.

It is evident from the presented empirical research that the rough set model has multiple interesting applications in economy and finance, delivering successful solutions comparable to and often exceeding the performance of more established techniques, including statistical tools, neural networks or decision trees.

Consequently, more research is needed in high impact areas like risk management, (also outside of the credit risk domain), time series-aware stock market prediction and incorporation of qualitative events coming from text data mining in the prediction process used to time the market and value and manage portfolio assets.

References

1. Ahn, B., Cho, S., Kim, C.: The integrated methodology of rough set theory and artificial neural network for business failure prediction. Expert Systems with Applications 18(2), 65–74 (2000)
2. Al-Qaheri, H., Hassanien, A., Abraham, A.: Discovering stock price prediction rules using rough sets. Neural Network World Journal (2008)
3. Ang, K., Quek, C.: RSPOP: Rough set-based pseudo outer-product fuzzy rule identification algorithm. Neural Computation 17(1), 205–243 (2005)
4. Ang, K., Quek, C.: Stock trading using RSPOP: A novel rough set-based neuro-fuzzy approach. IEEE Transactions on Neural Networks 17(5), 1301–1315 (2006)
5. Atsalakis, G., Valavanis, K.: Surveying stock market forecasting techniques–Part II: Soft computing methods. Expert Systems with Applications 36(3), 5932–5941 (2009)
6. Bahrammirzaee, A.: A comparative survey of artificial intelligence applications in finance: artificial neural networks, expert system and hybrid intelligent systems. Neural Computing and Applications 19(8), 1165–1195 (2010)
7. Baltzersen, J.: An attempt to predict stock market data: a rough sets approach. Ph.D. Dissertation (1996)
8. Bazan, J., Skowron, A., Synak, P.: Market data analysis: A rough set approach. ICS Research Reports 6, 94 (1994)
9. Beynon, M.: Reducts within the variable precision rough sets model: a further investigation. European Journal of Operational Research 134(3), 592–605 (2001)
10. Beynon, M., Peel, M.: Variable precision rough set theory and data discretisation: an application to corporate failure prediction. Omega 29(6), 561–576 (2001)
11. Beynon, M., Clatworthy, M., Jones, M.: The prediction of profitability using accounting narratives: a variable-precision rough set approach. Intelligent Systems in Accounting, Finance and Management 12(4), 227–242 (2004)
12. Bezdek, J.: Pattern recognition with fuzzy objective function algorithms. Kluwer Academic Publishers (1981)

13. Bhardwaj, G., Swanson, N.: An empirical investigation of the usefulness of ARFIMA models for predicting macroeconomic and financial time series. Journal of Econometrics 131(1), 539–578 (2006)

14. Bioch, J., Popova, V.: Bankruptcy prediction with rough sets. ERIM Report Series Reference No. ERS-2001-11-LIS (2003)

15. Bose, I.: Deciding the financial health of dot-coms using rough sets. Information & Management 43(7), 835–846 (2006)

16. Boudreau-Trudel, B., Zaras, K.: Comparison of Analytic Hierarchy Process and Dominance-Based Rough Set Approach as Multi-Criteria Decision Aid Methods for the Selection of Investment Projects. American Journal of Industrial and Business Management 2(1), 7–12 (2012)

17. Capotorti, A., Barbanera, E.: Credit scoring analysis using a fuzzy probabilistic rough set model. Computational Statistics & Data Analysis 56(4), 981–994 (2012)

18. Chen, S.-M.: Forecasting enrollments based on fuzzy time series. Fuzzy Sets and Systems 81(3), 311–319 (1996)

19. Chen, S.-H., Wang, P.: Computational intelligence in economics and finance. Springer (2004)

20. Chen, Y.-S., Cheng, C.-H.: Forecasting PGR of the financial industry using a rough sets classifier based on attribute-granularity. Knowledge and Information Systems 25(1), 57–79 (2010)

21. Chen, Y.-S., Cheng, C.-H., Chen, D.-R.: A fuzzy-based rough sets classifier for forecasting quarterly PGR in the stock market (Part I). International Journal of Innovative Computing, Information and Control 7(2), 555–569 (2011)

22. Chen, Y.-S.: Classifying credit ratings for Asian banks using integrating feature selection and the CPDA-based rough sets approach. Knowledge-Based Systems 26, 259–270 (2012)

23. Chen, Y.-S., Cheng, C.-H.: A soft-computing based rough sets classifier for classifying IPO returns in the financial markets. Applied Soft Computing 12(1), 462–475 (2012)

24. Cheng, C.-H., Chen, T.-L., Wei, L.-Y.: A hybrid model based on rough sets theory and genetic algorithms for stock price forecasting. Information Sciences 180(9), 1610–1629 (2010)

25. Cheng, J.-H., Chen, H.-P., Lin, Y.-M.: A hybrid forecast marketing timing model based on probabilistic neural network, rough set and C4. 5. Expert systems with Applications 37(3), 1814–1820 (2010)

26. Cheng, C.-H., Chen, Y.-S.: A fuzzy-based rough sets classifier for forecasting quarterly PGR in the stock market (part II). International Journal of Innovative Computing, Information and Control 7(3), 1209–1228 (2011)

27. Cornelis, C., De Cock, M., Radzikowska, A.: Fuzzy rough sets: from theory into practice. Handbook of Granular Computing. Wiley, Chichester (2008)

28. Cornelis, C., Jensen, R., Hurtado, G.D.S.: Attribute selection with fuzzy decision reducts. Information Sciences 180(2), 209–224 (2010)

29. Cox, E.: Fuzzy modeling and genetic algorithms for data mining and exploration. Morgan Kaufmann (2005)

30. d'Amato, M.: Appraising property with rough set theory. Journal of Property Investment & Finance 20(4), 406–418 (2002)

31. d'Amato, M.: A comparison between MRA and rough set theory for mass appraisal. A case in Bar. International Journal of Strategic Property Management 8(4), 205–217 (2004)

32. d'Amato, M.: Comparing rough set theory with multiple regression analysis as automated valuation methodologies. International Real Estate Review 10(2), 42–65 (2007)

33. Daubechies, I., et al.: Ten lectures on wavelets 61. SIAM (1992)

34. Lin, T., Tremba, J.: Attribute Transformations on Numerical Databases. Applications to Stock Market and Economic Data. In: Terano, T., Liu, H., Chen, A.L.P. (eds.) PAKDD 2000. LNCS, vol. 1805, pp. 181–192. Springer, Heidelberg (2000)

35. Dembczyński, K., Greco, S., Kotłowski, W., Słowiński, R.: Statistical model for rough set approach to multicriteria classification. In: Kok, J.N., Koronacki, J., Lopez de Mantaras, R., Matwin, S., Mladenič, D., Skowron, A. (eds.) PKDD 2007. LNCS (LNAI), vol. 4702, pp. 164–175. Springer, Heidelberg (2007)

36. Demyanyk, Y., Hasan, I.: Financial crises and bank failures: a review of prediction methods. Omega 38(5), 315–324 (2010)

37. Deng, J.-L.: Introduction to grey system theory. The Journal of Grey System 1(1), 1–24 (1989)

38. Dimitras, A., Zanakis, S., Zopounidis, C.: A survey of business failures with an emphasis on prediction methods and industrial applications. European Journal of Operational Research 90(3), 487–513 (1996)

39. Dimitras, A., Slowinski, R., Susmaga, R., Zopounidis, C.: Business failure prediction using rough sets. European Journal of Operational Research 114(2), 263–280 (1999)

40. Doumpos, M., Zopounidis, C.: Multi–Criteria Classification Methods in Financial and Banking Decisions. International Transactions in Operational Research 9(5), 567–581 (2002)

41. Dubois, D., Prade, H.: Rough fuzzy sets and fuzzy rough sets*. International Journal of General System 17(2-3), 191–209 (1990)

42. Dymowa, L.: Soft computing in economics and finance, vol. 6. Springer (2011)

43. Fayyad, U., Irani, K.: Multi-interval discretization of continuous-valued attributes for classification learning (1993)

44. Flesch, R.: A new readability yardstick. The Journal of Applied Psychology 32(3), 221 (1948)

45. Golan, R., Edwards, D.: Temporal rules discovery using datalogic/R+ with stock market data. In: Rough Sets, Fuzzy Sets and Knowledge Discovery, pp. 74–81. Springer (1994)

46. Golan, R., Ziarko, W.: A methodology for stock market analysis utilizing rough set theory. In: Proceedings of the IEEE/IAFE 1995 Computational Intelligence for Financial Engineering, pp. 32–40 (1995)

47. Greco, S., Cascio, S., Matarazzo, B.: Rough set approach to stock selection: An application to the Italian Market. In: Modelling Techniques for Financial Markets and Bank Management, pp. 192–211. Springer (1996)

48. Greco, S., Matarazzo, B., Slowinski, R.: Rough set approach to multi-attribute choice and ranking problems. In: Multiple Criteria Decision Making, pp. 318–329. Springer (1997)

49. Greco, S., Matarazzo, B., Slowinski, R.: A new rough set approach to evaluation of bankruptcy risk. In: Operational Tools in the Management of Financial Risks, pp. 121–136. Springer (1998)

50. Greco, S., Matarazzo, B., Slowinski, R.: Rough approximation of a preference relation by dominance relations. European Journal of Operational Research 117(1), 63–83 (1999)

51. Greco, S., Matarazzo, B., Słowiński, R., Stefanowski, J.: Variable consistency model of dominance-based rough sets approach. In: Ziarko, W.P., Yao, Y. (eds.) RSCTC 2000. LNCS (LNAI), vol. 2005, pp. 170–181. Springer, Heidelberg (2001)
52. Greco, S., Matarazzo, B., Slowinski, R.: Rough approximation by dominance relations. International Journal of Intelligent Systems 17(2), 153–171 (2002)
53. Greco, S., Matarazzo, B., Slowinski, R., Zanakis, S.: Global investing risk: a case study of knowledge assessment via rough sets. Annals of Operations Research 185(1), 105–138 (2011)
54. Griffiths, B., Beynon, M.: Expositing stages of VPRS analysis in an expert system: Application with bank credit ratings. Expert Systems with Applications 29(4), 879–888 (2005)
55. Grzymala-Busse, J.: LERS-a system for learning from examples based on rough sets. In: Intelligent Decision Support, pp. 3–18. Springer (1992)
56. Grzymala-Busse, J.: A new version of the rule induction system LERS. Fundamenta Informaticae 31(1), 27–39 (1997)
57. Grzymala-Busse, J., Ziarko, W.: Data mining and rough set theory. Communications of the ACM 43(4), 108–109 (2000)
58. Hashemi, R., Le Blanc, L., Rucks, C., Rajaratnam, A.: A hybrid intelligent system for predicting bank holding structures. European Journal of Operational Research 109(2), 390–402 (1998)
59. Herbert, J., Yao, J.: Time-series data analysis with rough sets. CIEF 4, 908–911 (2005)
60. Hu, Q., Yu, D., Liu, J., Wu, C.: Neighborhood rough set based heterogeneous feature subset selection. Information Sciences 178(18), 3577–3594 (2008)
61. Huang, K.: Application of VPRS model with enhanced threshold parameter selection mechanism to automatic stock market forecasting and portfolio selection. Expert Systems with Applications 36(9), 11652–11661 (2009)
62. Huang, K., Jane, C.-J.: A hybrid model for stock market forecasting and portfolio selection based on ARX, grey system and RS theories. Expert Systems with Applications 36(3), 5387–5392 (2009)
63. Jankowski, A., Skowron, A.: Practical Issues of Complex Systems Engineering: Wisdom Technology Approach. Springer, Heidelberg (in preparation, 2014)
64. Jensen, R., Shen, Q.: Finding rough set reducts with ant colony optimization. In: Proceedings of the 2003 UK Workshop on Computational Intelligence, vol. 1 (2003)
65. Jie, Z., Yan, L., Xin, L.: Research on financial crisis prediction model based on Rough Sets and Neural Network. In: 2011 International Conference on E-Business and E-Government (ICEE), pp. 1–4 (2011)
66. Jorion, P.: Value at risk: the new benchmark for managing financial risk 2. McGraw-Hill, New York (2007)
67. Kasabov, N.: Evolving fuzzy neural networks for supervised/unsupervised online knowledge-based learning. IEEE Transactions on Systems, Man, and Cybernetics, Part B: Cybernetics 31(6), 902–918 (2001)
68. Kasabov, N., Song, Q.: DENFIS: dynamic evolving neural-fuzzy inference system and its application for time-series prediction. IEEE Transactions on Fuzzy Systems 10(2), 144–154 (2002)
69. Kawasaki, S., Binh, N., Bao, T.: Hierarchical document clustering based on tolerance rough set model. In: Zighed, D.A., Komorowski, J., Żytkow, J.M. (eds.) PKDD 2000. LNCS (LNAI), vol. 1910, pp. 458–463. Springer, Heidelberg (2000)

70. Khoza, M., Marwala, T.: A rough set theory based predictive model for stock prices. In: 2011 IEEE 12th International Symposium on Computational Intelligence and Informatics (CINTI), pp. 57–62 (2011)

71. Kim, K.-J., Han, I.: The extraction of trading rules from stock market data using rough sets. Expert Systems 18(4), 194–202 (2001)

72. Koczkodaj, W., Orlowski, M., Marek, V.: Myths about rough set theory. Communications of the ACM 41(11), 102–103 (1998)

73. Kohonen, T.: Self-organizing maps. Springer Series in Information Sciences, vol. 30. Springer, Berlin (2001)

74. Komorowski, J., Pawlak, Z., Polkowski, L., Skowron, A.: Rough sets: A tutorial. Rough fuzzy hybridization: A new trend in decision-making, 3–98 (1999)

75. Kot, D., Greco, S.S.: Stochastic dominance-based rough set model for ordinal classification. Information Sciences 178(21), 4019–4037 (2008)

76. Kovalerchuk, B., Vityaev, E.: Data mining in finance: advances in relational and hybrid methods. Springer (2000)

77. Kretowski, M., Stepaniuk, J.: Selection of objects and attributes a tolerance rough set approach. In: Proceedings of the Ninth International Symposium on Methodologies for Intelligent Systems, Zakopane, Poland (1996)

78. Kryszkiewicz, M.: Maintenance of reducts in the variable precision rough set model. In: Rough Sets and Data Mining, pp. 355–372. Springer (1996)

79. Kryszkiewicz, M.: Rough set approach to incomplete information systems. Information Sciences 112(1), 39–49 (1998)

80. Kumar, A.: New techniques for data reduction in a database system for knowledge discovery applications. Journal of Intelligent Information Systems 10(1), 31–48 (1998)

81. Kumar, A., Agrawal, D., Joshi, S.: Multiscale rough set data analysis with application to stock performance modeling. Intelligent Data Analysis 8(2), 197–209 (2004)

82. Lee, S., Ann, J., Oh, K., Kim, T., Lee, H., Song, C.: Using Rough Set to Support Investment Strategies of Rule-Based Trading with Real-Time Data in Futures Market. In: 42nd Hawaii International Conference on System Sciences, HICSS 2009, pp. 1–10 (2009)

83. Lee, S., Ahn, J., Oh, K., Kim, T.: Using rough set to support investment strategies of real-time trading in futures market. Applied Intelligence 32(3), 364–377 (2010)

84. Liu, G., Zhu, Y.: Credit assessment of contractors: a rough set method. Tsinghua Science & Technology 11(3), 357–362 (2006)

85. Liu, J.-W., Cheng, C.-H., Chen, Y.-H., Chen, T.-L.: OWA rough set model for forecasting the revenues growth rate of the electronic industry. Expert Systems with Applications 37(1), 610–617 (2010)

86. McKee, T.: Predicting bankruptcy via induction. Journal of Information Technology 10(1), 26–36 (1995)

87. McKee, T.: Developing a bankruptcy prediction model via rough sets theory. International Journal of Intelligent Systems in Accounting, Finance & Management 9(3), 159–173 (2000)

88. McKee, T., Lensberg, T.: Genetic programming and rough sets: A hybrid approach to bankruptcy classification. European Journal of Operational Research 138(2), 436–451 (2002)

89. McKee, T.: Rough sets bankruptcy prediction models versus auditor signalling rates. Journal of Forecasting 22(8), 569–586 (2003)

90. Mienko, R., Slowinski, R., Stefanowski, J.: Rule Classifier Based on Valued Closeness Relation: ROUGHCLASS version 2.0. Poznan University of Technology Research Report RA-95/002, Pozan, Poland (1995)
91. Mrózek, A., Skabek, K.: Rough sets in economic applications. In: Rough Sets in Knowledge Discovery 2, pp. 238–271. Springer (1998)
92. Murphy, J.: Technical Analysis of the Financial Markets: A Comprehensive Guide to Trading Methods and Applications. New York Institute of Finance (1999)
93. Nair, B., Mohandas, V., Sakthivel, N.: A Decision tree- Rough set Hybrid System for Stock Market Trend Prediction. International Journal of Computer Applications 6(9) (2010)
94. Øhrn, A.: Rosetta technical reference manual. Department of Computer and Information Science, Norwegian University of Science and Technology (NTNU), Trondheim, Norway, 1-66 (2000)
95. Pai, P.-F., Chen, S.-Y., Huang, C.-W., Chang, Y.-H.: Analyzing foreign exchange rates by rough set theory and directed acyclic graph support vector machines. Expert Systems with Applications 37(8), 5993–5998 (2010)
96. Pawlak, Z.: Information systems theoretical foundations. Information Systems 6(3), 205–218 (1981)
97. Pawlak, Z.: Rough sets. International Journal of Computer & Information Sciences 11(5), 341–356 (1982)
98. Pawlak, Z.: Rough Sets Theoretical Aspects of Reasoning about Data. Kluwer Academic Publishers, London (1991)
99. Pawlak, Z., Skowron, A.: Rough sets: some extensions. Information Sciences 177(1), 28–40 (2007)
100. Pawlak, Z., Skowron, A.: Rudiments of rough sets. Information Sciences 177(1), 3–27 (2007)
101. Qizhong, Z.: An approach to rough set decomposition of incomplete information systems. In: 2nd IEEE Conference on Industrial Electronics and Applications, ICIEA 2007, pp. 2455–2460 (2007)
102. Radzikowska, A., Kerre, E.: A comparative study of fuzzy rough sets. Fuzzy Sets and Systems 126(2), 137–155 (2002)
103. Ravi Kumar, P., Ravi, V.: Bankruptcy prediction in banks and firms via statistical and intelligent techniques–A review. European Journal of Operational Research 180(1), 1–28 (2007)
104. Ruggiero, M.: How to build a system framework. Futures 23(12), 50–56 (1994)
105. Ruggiero, M.: Rules are made to be traded. AI in Finance Fall, 35-40 (1994)
106. Ruggiero, M.: Turning the key. Futures 23(14), 38–40 (1994)
107. Ruggiero, M.: Cybernetic Trading Strategies: developing a profitable trading system with state-of-the-art technologies 68. Wiley (1997)
108. Ruizhong, W.: Analyses the Financial Data of Stocks Based Rough Set Theory. In: 2012 Eighth International Conference on Computational Intelligence and Security (CIS), pp. 387–390 (2012)
109. Ruzgar, N.S., Unsal, F., Ruzgar, B.: Predicting business failures using the rough set theory approach: The case of the Turkish banks. International Journal of Mathematical models and Methods in Applied Sciences 2, 57–64 (2008)
110. Sanchis, A., Segovia, M., Gil, J., Heras, A., Vilar, J.: Rough sets and the role of the monetary policy in financial stability (macroeconomic problem) and the prediction of insolvency in insurance sector (microeconomic problem). European Journal of Operational Research 181(3), 1554–1573 (2007)
111. Seiford, L., Zhu, J.: An acceptance system decision rule with data envelopment analysis. Computers & Operations Research 25(4), 329–332 (1998)

112. Shen, L., Tay, F.E.H.: Classifying market states with WARS. In: Leung, K.-S., Chan, L., Meng, H. (eds.) IDEAL 2000. LNCS, vol. 1983, pp. 280–285. Springer, Heidelberg (2000)

113. Shen, L.: Data mining techniques based on rough sets theory. Ph.D. Dissertation (2003)

114. Shen, L., Loh, H.: Applying rough sets to market timing decisions. Decision Support Systems 37(4), 583–597 (2004)

115. Shen, Q., Jensen, R.: Rough sets, their extensions and applications. International Journal of Automation and Computing 4(3), 217–228 (2007)

116. Shuai, J.-J., Li, H.-L.: Using rough set and worst practice DEA in business failure prediction. In: Ślęzak, D., Yao, J., Peters, J.F., Ziarko, W.P., Hu, X. (eds.) RSFD-GrC 2005. LNCS (LNAI), vol. 3642, pp. 503–510. Springer, Heidelberg (2005)

117. Shyng, J.-Y., Shieh, H.-M., Tzeng, G.-H.: An integration method combining Rough Set Theory with formal concept analysis for personal investment portfolios. Knowledge-Based Systems 23(6), 586–597 (2010)

118. Shyng, J.-Y., Shieh, H.-M., Tzeng, G.-H., Hsieh, S.-H.: Using FSBT technique with Rough Set Theory for personal investment portfolio analysis. European Journal of Operational Research 201(2), 601–607 (2010)

119. Skalko, C.: Rough sets help time the OEX. Journal of Computational Intelligence in Finance 4(6), 20–27 (1996)

120. Skowron, A., Stepaniuk, J., Jankowski, A., Bazan, J., Swiniarski, R.: Rough Set Based Reasoning About Changes. Fundamenta Informaticae 119(3-4), 421–437 (2012)

121. Skowron, A., Rauszer, C.: The discernibility matrices and functions in information systems. In: Intelligent Decision Support, pp. 331–362. Springer (1992)

122. Skowron, A., Stepaniuk, J.: Generalized approximation spaces. In: Soft Computing, Simulation Councils, San Diego, pp. 18–21 (1995)

123. Skowron, A., Stepaniuk, J.: Tolerance approximation spaces. Fundamenta Informaticae 27(2), 245–253 (1996)

124. Slowinski, R., Zapounidis, C.: Application of the rough set approach to evaluation of bankruptcy risk. International J. of Intelligent Systems in Accounting, Finance & Management 4, 27–41 (1995)

125. Slowinski, R., Zopounidis, C.: Rough-set sorting of firms according to bankruptcy risk. In: Applying Multiple Criteria Aid for Decision to Environmental Management, pp. 339–357. Springer (1994)

126. Slowinski, R., Zopounidis, C., Dimitras, A.: Prediction of company acquisition in Greece by means of the rough set approach. European Journal of Operational Research 100(1), 1–15 (1997)

127. Stefanowski, J.: On rough set based approaches to induction of decision rules. Rough Sets in Knowledge Discovery 1(1), 500–529 (1998)

128. Stefanowski, J., Tsoukiàs, A.: Valued tolerance and decision rules. In: Ziarko, W.P., Yao, Y. (eds.) RSCTC 2000. LNCS (LNAI), vol. 2005, pp. 212–219. Springer, Heidelberg (2001)

129. Susmaga, R., Michalowski, W., Slowinski, R.: Identifying regularities in stock portfolio tilting. Tech. rep. (1997)

130. Triana, P.: VaR: The number that killed us. Futures Magazine (2010)

131. Tan, A., Quek, C., Yow, K.: Maximizing winning trades using a novel RSPOP fuzzy neural network intelligent stock trading system. Applied Intelligence 29(2), 116–128 (2008)

132. Tay, F., Shen, L.: Economic and financial prediction using rough sets model. European Journal of Operational Research 141(3), 641–659 (2002)

133. Tay, F., Shen, L.: A modified Chi2 algorithm for discretization. IEEE Transactions on Knowledge and Data Engineering 14(3), 666–670 (2002)
134. Teoh, H., Cheng, C.-H., Chu, H.-H., Chen, J.-S.: Fuzzy time series model based on probabilistic approach and rough set rule induction for empirical research in stock markets. Data & Knowledge Engineering 67(1), 103–117 (2008)
135. Tremba, J., Lin, T.: Attribute transformations for data mining II: Applications to economic and stock market data. International Journal of Intelligent Systems 17(2), 223–233 (2002)
136. Wang, X.-Y., Wang, Z.-O.: Stock market time series data mining based on regularized neural network and rough set. In: Proceedings of 2002 International Conference on Machine Learning and Cybernetics, vol. 1, pp. 315–318 (2002)
137. Wang, Y.-F.: Mining stock price using fuzzy rough set system. Expert Systems with Applications 24(1), 13–23 (2003)
138. Wroblewski, J.: Finding minimal reducts using genetic algorithms. In: Proceedings of Second International Joint Conference on Information Science, pp. 186–189 (1995)
139. Xiao, Z., Yang, X., Pang, Y., Dang, X.: The prediction for listed companies' financial distress by using multiple prediction methods with rough set and Dempster–Shafer evidence theory. Knowledge-Based Systems 26, 196–206 (2012)
140. Yager, R.: On ordered weighted averaging aggregation operators in multicriteria decisionmaking. IEEE Transactions on Systems, Man and Cybernetics 18(1), 183–190 (1988)
141. Yao, Y., Wong, S.: Generalization of rough sets using relationships between attribute values. In: Proceedings of the 2nd Annual Joint Conference on Information Sciences, pp. 30–33 (1995)
142. Yao, J., Teng, N., Poh, H.-L., Tan, C.: Forecasting and analysis of marketing data using neural networks. J. Inf. Sci. Eng. 14(4), 843–862 (1998)
143. Yao, J., Li, Y., Tan, C.: Option price forecasting using neural networks. Omega 28(4), 455–466 (2000)
144. Yao, J., Tan, C.: Time dependent directional profit model for financial time series forecasting. In: Proceedings of the IEEE-INNS-ENNS International Joint Conference on Neural Networks, IJCNN 2000, vol. 5, pp. 291–296 (2000)
145. Yao, Y., Zhao, Y.: Attribute reduction in decision-theoretic rough set models. Information Sciences 178(17), 3356–3373 (2008)
146. Yao, J., Herbert, J.: Financial time-series analysis with rough sets. Applied Soft Computing 9(3), 1000–1007 (2009)
147. Yao, P.: Hybrid classifier using neighborhood rough set and SVM for credit scoring. In: International Conference on Business Intelligence and Financial Engineering, BIFE 2009, pp. 138–142 (2009)
148. Yeh, C.-C., Chi, D.-J., Hsu, M.-F.: A hybrid approach of DEA, rough set and support vector machines for business failure prediction. Expert Systems with Applications 37(2), 1535–1541 (2010)
149. Yeh, C.-C., Lin, F., Hsu, C.-Y.: A hybrid KMV model, random forests and rough set theory approach for credit rating. Knowledge-Based Systems (2012)
150. Yeung, D., Chen, D., Tsang, E., Lee, J., Xizhao, W.: On the generalization of fuzzy rough sets. IEEE Transactions on Fuzzy Systems 13(3), 343–361 (2005)
151. Yu, L., Wang, S., Lai, K.: A rough-set-refined text mining approach for crude oil market tendency forecasting. International Journal of Knowledge and Systems Sciences 2(1), 33–46 (2005)
152. Yu, H.-K.: Weighted fuzzy time series models for TAIEX forecasting. Physica A: Statistical Mechanics and its Applications 349(3), 609–624 (2005)

153. Zadeh, L.: Fuzzy sets. Information and Control 8(3), 338–353 (1965)
154. Zhang, Y.-Q., Wan, X.: Statistical fuzzy interval neural networks for currency exchange rate time series prediction. Applied Soft Computing 7(4), 1149–1156 (2007)
155. Zhang, Q.-F., Zhao, S.-Y., Bai, Y.-C.: On the application of rough sets to data mining in economic practice. In: 2009 International Conference on Machine Learning and Cybernetics, vol. 1, pp. 272–276 (2009)
156. Zhao, G., Yan, W., Li, Y.: LS-SVM Financial Achievement Prediction Based on Targets Optimization of Neighborhood Rough Sets. In: Shen, G., Huang, X. (eds.) CSIE 2011, Part I. CCIS, vol. 152, pp. 171–178. Springer, Heidelberg (2011)
157. Zhou, J., Bai, T.: Credit risk assessment using rough set theory and GA-based SVM. In: The 3rd International Conference on Grid and Pervasive Computing Workshops, GPC Workshops 2008, pp. 320–325 (2008)
158. Ziarko, W., Golan, R., Edwards, D.: An application of datalogic/R knowledge discovery tool to identify strong predictive rules in stock market data. In: Proceedings of AAAI Workshop on Knowledge Discovery in Databases, Washington, DC, pp. 89–101 (1993)
159. Ziarko, W.: Variable precision rough set model. Journal of Computer and System Sciences 46(1), 39–59 (1993)
160. Zighed, D., Rabaseda, S., Rakotomalala, R.: FUSINTER: a method for discretization of continuous attributes. International Journal of Uncertainty, Fuzziness and Knowledge-Based Systems 6(03), 307–326 (1998)
161. Zimmermann, H.: Fuzzy set theory-and its applications. Springer (2001)

Algorithms for Similarity Relation Learning from High Dimensional Data[*]

Andrzej Janusz

Faculty of Mathematics, Informatics, and Mechanics, The University of Warsaw,
Banacha 2, 02-097 Warszawa, Poland
janusza@mimuw.edu.pl

Abstract. The notion of similarity plays an important role in machine learning and artificial intelligence. It is widely used in tasks related to a supervised classification, clustering, an outlier detection and planning. Moreover, in domains such as information retrieval or case-based reasoning, the concept of similarity is essential as it is used at every phase of the reasoning cycle. The similarity itself, however, is a very complex concept that slips out from formal definitions. A similarity of two objects can be different depending on a considered context. In many practical situations it is difficult even to evaluate the quality of similarity assessments without considering the task for which they were performed. Due to this fact the similarity should be learnt from data, specifically for the task at hand. This paper presents a research on the problem of similarity learning, which is a part of author's PHD dissertation. It describes a similarity model, called Rule-Based Similarity, and shows algorithms for constructing this model from available data. The model utilizes notions from the rough set theory to derive a similarity function that allows to approximate the similarity relation in a given context. It is largely inspired by the idea of Tversky's feature contrast model and it has several analogical properties. In the paper, those theoretical properties are described and discussed. Moreover, the paper presents results of experiments on real-life data sets, in which a quality of the proposed model is thoroughly evaluated and compared with the state-of-the-art algorithms.

Keywords: Rule-Based Similarity, Similarity Learning, Rough Set Theory, Tversky's Model, Case-Based Reasoning, Feature Extraction.

1 Introduction

For many centuries the idea of similarity has inspired researchers from different fields, in particular philosophers, psychologists and mathematicians. Since

[*] The research was supported by the grants DEC-2011/01/B/ST6/03867 and DEC-2012/05/B/ST6/03215 from the National Research Centre, and the National Centre for Research and Development (NCBiR) under the grant SP/I/1/77065/10 by the strategic scientific research and experimental development program: "Interdisciplinary System for Interactive Scientific and Scientific-Technical Information".

J.F. Peters and A. Skowron (Eds.): Transactions on Rough Sets XVII, LNCS 8375, pp. 174–292, 2014.
© Springer-Verlag Berlin Heidelberg 2014

Plato and his student, Aristotle, people have been trying to systematize the world around them by creating ontologies and grouping similar objects, living organisms or natural phenomena based on their characteristics. Over the years, many of the great discoveries have been made by scientists and inventors who noticed some resemblance between processes or objects, and on that basis formed a theory describing them.

Although human mind is capable of effortlessly assessing similarities between objects, there is no single methodology of selecting or building similarity models appropriate for a wide range of complex object classes and domains. This dissertation deals with a problem of learning a similarity relation or constructing a similarity function from data with a particular focus on high dimensional object domains. Apart from an overview of several well-known similarity learning methods, a rule-based model of similarity is proposed, whose flexibility allows to overcome many practical issues related with the commonly used approaches. This model and its two extensions, which are designed specifically to facilitate dealing with extremely high dimensional objects, are tested in extensive experiments in order to show their practical usefulness.

1.1   Motivation and Aims

The ability to identify similar objects is believed to play a fundamental role in the process of human decision making and learning [1–3]. Stefan Banach was known to say that:

> "Good mathematicians see analogies. Great mathematicians see analogies between analogies."

The notion of similarity itself, however, slips out from the formal scientific definitions [4,5]. Despite this fact, similarity or reasoning by analogy is being utilized by numerous machine learning algorithms in applications ranging from a supervised classification to unsupervised clustering and an outlier detection [6–8]. Knowing how to discriminate similar cases (or objects) from those which are dissimilar in a desired context would enable a more accurate classification and detection of unusual or dangerous situations or behaviours. Unfortunately, due to difficulties related to an a priori selection of a similarity model, which are particularly apparent when a metric space representation of objects is high dimensional, the performance of similarity-based machine learning algorithms may be limited [9].

A scope of this dissertation is a problem of learning how to recognize whether two objects are similar in a pre-specified context. A variety of methods have been used in order to construct similarity models and define a relation which would combine intuitive properties postulated by psychologists with a good performance in real-life applications. Among those a huge share was based on distance measures. In that approach, objects are treated as points in a metric space of their attributes and the similarity is a non-increasing function of the distance between them. Objects are regarded as similar if they are close enough in this space [9–11]. Such models may often be improved by assigning weights to attributes which express their importance to the model. Tuning those weights results in better fitting the relation to a data set and can be regarded as an example of similarity learning. Algorithms for a computationally efficient optimization of parameters for common similarity measures were investigated by numerous researchers, e.g. [12–19].

One may argue that the relation of this kind is very intuitive because objects which have many similar values of attributes are likely to be similar. However, researchers like Amos Tversky [5, 10, 20] empirically showed that in some contexts, similarity does not necessarily have properties like symmetry or subadditivity which are implied by distance measures. This situation occurs particularly frequent when we compare objects of great complexity, often described by a large number of attributes. The explanation for this may lie in the fact that complex objects can be similar in some aspects and dissimilar in others. Hence, some additional knowledge about the context is needed to decide which of the similarity aspects are more important [5, 21].

Moreover, the dependencies between local and global similarities may be highly non-linear and in order to capture them it is necessary to extract some higher-level features of objects. Since there usually are numerous possible features to consider, this task can rarely be performed by human experts. Instead, the higher-level characteristics of objects and methods for their aggregation need to be derived from available data. Of course, as in all types of machine learning tasks, a similarity learning algorithm needs to balance complexity and efficiency [7, 8]. The construction of an overly complex similarity model will take too much time and resources to be applicable to real-life problems. Such a model may also be over-fitted to available data and yield poor performance in assessing the similarity of new objects.

The aim of this dissertation is to address those issues by proposing a similarity learning model called Rule-Based Similarity. The main motivation for that model comes from Tversky's works on the feature contrast model of similarity [5]. Instead of embedding objects into a metric space of their attributes, in the proposed approach the objects are represented by sets of higher-level features which can be more semantically meaningful than the low-level attribute values. In the model, such new features are defined by rules extracted from data, analogically to a rule-based object representation discussed in [22]. Unlike in that approach, however, in Rule-Based Similarity the new features are not treated as regular attributes but rather, they are regarded as arguments *for* or *against* the similarity of the compared objects. By combining the set representation with techniques developed within the theory of rough sets, the model tries to aggregate those arguments and to express the similarity in a context dictated by a given task (e.g. supervised classification or semantic clustering), and by other objects present in the data. In this way, the resulting similarity function is more likely to reflect natural properties of similarity without loosing its practical usefulness and reliability.

Due to the subjectivity and complexity of the similarity notion, those appealing qualities can not be justified based only on theoretical properties and intuitions. The second goal of this dissertation is to provide results of thorough experiments in which the performance of Rule-Based Similarity was evaluated on many different data sets. Usefulness of this model in practical tasks, such as a supervised classification and an unsupervised cluster analysis, was compared with other similarity models as well as to the state-of-the-art in a given domain. The results of those tests may be used as arguments confirming the validity of the proposed model design.

1.2   Main Contributions

In the dissertation the problem of learning a similarity relation for a predefined data analysis task is discussed. Expectations regarding the construction and general properties of similarity models are formulated. Major challenges related to this problem are characterised and some practical solutions are proposed. Finally, the validity of the proposed methods is shown through extensive experiments on real-life data. Hence the main contributions of this dissertation are threefold:

1. A discussion on properties of the similarity relation from the point of view of data analysis and artificial intelligence.
2. A proposition of a similarity model and some construction algorithms that combine intuitive expectations with efficiency in practical applications.
3. An implementation and an experimental evaluation of the proposed similarity model on a wide range of data sets and in different use scenarios.

In particular, after reviewing observations of psychologists regarding the nature of the similarity, definition of a *proper similarity function* is proposed in Subsection 3.3. It aims at providing a more formal description of an abstract *similarity function* concept. Intuitively, pairs of objects for which a proper similarity function takes high values are more likely to be in the real similarity relation, relative to a predefined context. An example of such a context can be a classification of objects from the investigated domain. In that case, a similarity learning process can be guided by the fundamental properties of the similarity for classification, which are stated in Subsection 3.2.

The context of a similarity assessment is imposed by a purpose for which the evaluation is performed. It is also influenced by a presence of other objects. Those general observations together with the computational effectiveness constitute a basis for the desirable properties of similarity learning models which are given in Subsection 4.1. They are treated as requirements and a motivation for designing the similarity model which is the main scope of this dissertation.

The proposed Rule-Based Similarity (RBS) model and its two extensions are described in Section 5. Subsection 5.2 shows the construction of the basic version of RBS, designed for learning the similarity in a classification context from regular data tables. Additionally, this subsection offers an intuitive interpretation of the model and explains its relations with the rough set theory. An important aspect of the construction of RBS is the computation of a decision reduct for each of the decision classes occurring in the data. This often needs to be done for data sets containing numerical attributes. Algorithm 2 shows how to compute a reduct in such a case. Some of the basic mathematical properties of the RBS similarity function are discussed in Subsection 5.3. In this subsection it is also shown that under certain conditions the proposed function is a proper similarity function for a similarity relation in the context of a classification.

The first extension of RBS, which is designed to efficiently handle extremely high dimensional data sets, is presented in Subsection 5.4. Its core is an algorithm for the computation of a diverse set of dynamic decision reducts (Algorithm 3). By combining randomization with the greedy heuristic for the computation of reducts this algorithm enables an efficient construction of robust sets of higher-level features. Due to the diversity of the sets, those features correspond to different similarity aspects. The similarity function which is proposed for this model, aggregates the local similarities analogically to aggregations of classifier ensembles.

The second of the proposed extensions is described in Subsection 5.5. The purpose of this model is to facilitate the similarity learning from textual corpora. Unlike the previous models, unsupervised RBS does not require information regarding decision classes and can be used for cluster analysis. To extract higher-level features it uses a combination of Explicit Semantic Analysis with a novel algorithm for the computation of information bireducts (Algorithm 4).

All the models proposed in this dissertation were thoroughly evaluated in experiments described in Section 6. RBS was compared to several other similarity learning techniques in the classification context on a variety of data tables. The tests were performed on benchmark tables (Subsection 6.1) as well as on real-life microarray data sets containing tens of thousands attributes (Subsection 6.2). Finally, tests with the unsupervised RBS were conducted and their results were described in Subsection 6.3.

Most of the partial results of this dissertation were presented at international conferences and workshops. They were published in conference proceedings and respectable journals. For example, the publications related to the construction and the applications of Rule-Based Similarity include [23–29]. There are also several other research directions of the author that had a significant influence on the design of the proposed similarity learning models. Among them, the most important considered the problem of feature selection and learning with ensembles of single and multi-label classifiers [30–37]. Moreover, the research on unsupervised version of Rule-Based Similarity was largely influenced by the author's previous work on the semantic information retrieval and Explicit Semantic Analysis, which was conducted within the SYNAT project [38–40].

1.3   Plan of the Dissertation

The dissertation is divided into seven main sections. This introductory section aims to provide a brief description of the considered problem and to help a reader with navigation through the remaining part of the text.

Section 2 is devoted to the theory of rough sets. Its main role is to introduce the basic concepts and notations used in the subsequent sections. It is divided into three subsections. Subsection 2.1 introduces the notions of information and decision systems. It also discusses fundamental building blocks of the rough set theory such as the indiscernibility relation and the notions of a concept, decision logic language and rules. Subsection 2.2 explains the rough set view on the approximation of vague or imprecise concepts. It gives the definition of a rough set and shows elementary properties of lower and upper approximations. Further in this subsection there is a discussion on finding appropriate approximation spaces for constructing approximations of concepts and relations. The last subsection of the second section (Subsection 2.3) focuses on rough set methods for selecting informative sets of attributes. It gives definitions of the classical information and decision reducts, and then it reviews several extensions of this important notion, such as approximate reducts, dynamic reducts and a novel concept of decision bireducts.

Section 3 introduces similarity as a relation between objects and discusses its main properties. It also provides an overview of the most well-known similarity models and gives examples of their practical applications. The section is divided into five subsections. The first one (Subsection 3.1) starts with a discussion on psychological properties of similarity as a semantic relation. After this introduction, the importance of setting a similarity evaluation in a context which is appropriate for a task is highlighted in Subsection 3.2. This discussion is followed by definitions of a proper similarity function and similarity-based classification rules in Subsection 3.3 and then, an overview of similarity model evaluation methods is given. The next subsection (Subsection 3.4) summarises the most commonly used similarity models. The distance metric-based similarity modelling is characterized and then, the subsection explains Tversky's feature contrast model as an alternative to the distance-based approach. The subsection ends with a brief description of hierarchical similarity modelling methods. The section concludes with Subsection 3.5, which is a survey on applications of similarity models in machine learning. It shows how the similarity can be employed for a predictive data analysis and visualization and briefly discusses the Case-Based Reasoning framework. It ends with a usage example of similarity functions for unsupervised learning in cluster analysis.

Section 4 focuses on similarity learning methods. Its first subsection (Subsection 4.1) defines the problem of similarity learning and lists some desirable properties of a good similarity learning model. Subsection 4.2 presents examples of four popular approaches to the similarity learning task. It summarises methods that use feature extraction techniques in order to improve a similarity model by selecting attributes which are relevant in a given context or by constructing new ones. Next, there is an overview of a very popular approach that utilizes a genetic algorithm to tune parameters of a predefined similarity function. Then, it shows how a similarity relation can be induced and optimized in a tolerance approximation space. The last example, concerns a specific task of using Explicit

Semantic Analysis for learning a semantic representation of texts which can be used to better evaluate their similarity.

Section 5 describes the idea of Rule-Based Similarity which is the main contribution of this dissertation. Subsection 5.1 discusses intuitions and motivations for this model. The following subsection (Subsection 5.2) reveals construction details of the model and Subsection 5.3 discusses some of its mathematical properties. The next two subsections show how Rule-Based Similarity can be adjusted to efficiently learn the similarity in contexts defined by two different tasks related to analysis of high dimensional data. Namely, Subsection 5.4 focuses on similarity learning from high dimensional data for a classification purpose and Subsection 5.5 deals with the problem of unsupervised similarity learning for clustering of textual documents. The last subsection of the section (Subsection 5.6) summarises the properties of Rule-Based Similarity.

Section 6 provides results of experiments in which the proposed model was tested on benchmark and real-life data sets. Each of its subsections is devoted to a series of experiments on different types of data. Subsection 6.1 investigates the performance of Rule-Based Similarity in the context of classification on standard and high dimensional data tables. First, it describes the data sets used in this series of tests. Then, it briefly characterises the competing similarity models and discusses the results of the comparisons between them. Subsection 6.2 presents the evaluation of the dynamic extension to Rule-Based Similarity on microarray data. This subsection starts with a discussion of general properties of microarrays as an example of extremely high dimensional data. It shows how efficient Dynamic Rule-Based Similarity can be for coping with the few-objects-many-attributes problem, in comparison to the state-of-the-art in the microarray data classification. The last subsection (Subsection 6.3) presents an example of an application of the unsupervised extension to Rule-Based Similarity. At the beginning it explains the methodology of the experiment and clarifies how the compared similarity models were evaluated. Next, it characterizes the models which were used in the comparison and summarises the results.

Finally, the last Section 7 concludes the dissertation. Subsection 7.1 draws a summary of the discussed problems and Subsection 7.2 proposes some directions for future development of the rule-based models of similarity.

2   Theory of Rough Sets

The theory of rough sets, proposed by Zdzisław Pawlak in 1981 [41], provides a mathematical formalism for reasoning about imperfect data and knowledge [42–44]. Since their introduction, rough sets have been widely used in numerous real-life applications related to intelligent knowledge discovery, such as classification, clustering, approximation of concepts, discovering of patterns and dependencies in data [34, 42, 43, 45–49]. They were also used for hierarchical modelling of complex objects [45, 50], as well as approximation of relations and functions [47, 51, 52].

The notion of similarity has always been important for researchers in the field of rough sets. Several extensions of the classical discernibility-based rough sets were proposed, in which a similarity relation was used to generalized rough approximations [43, 53–58]. Similarity was also utilized in order to explain relations between rough sets and fuzzy sets and interpret fuzziness in the rough set setting [59]. On the other hand, some similarity measures were motivated by the rough set theory [60].

In this dissertation similarity is viewed as a relation whose properties may vary depending on a specific context. Since without any additional knowledge the similarity can be regarded as an arbitrary relation, it needs to be learnt from available data. The similarity relation is vague in nature [5,21,61]. For this reason the rough set theory seems suitable for this purpose. It does not only offer intuitive foundations for modelling complex relations, but also provides practical tools for extracting meaningful features and defining important aspects of similarity between considered objects [48,62]. Those aspects often correspond to higher-level characteristics or concepts which can also be vague. To better cope with such a multi-level vagueness there were proposed models that combine the rough set and fuzzy set theories into rough-fuzzy or fuzzy-rough models [63–65].

The similarity learning model described in this dissertation (Section 5) derives from the theory of rough sets. To better explain their construction, the following subsections briefly overview selected aspects of the rough sets and introduce some basic notation used in the remaining parts of this thesis. Subsection 2.1 gives definitions of fundamental concepts, such as an *information system* or a *decision rule*. Subsection 2.2 provides an insight on approximation spaces and explains the basic principles of a rough set approximation. Subsection 2.3 describes a rough set approach to the problem of data dimensionality reduction. In its last part, the notion of reducts is extended to bireducts and some interesting properties of decision bireducts are discussed.

2.1   Introduction to Rough Sets

The theory of rough sets deals with problems related to reasoning about vagueness in data [41]. Its main assumption is that with every object of the considered universe Ω there is some associated information which can be represented in a tabular form as attribute-value entries. Available objects which are characterized by the same information are indiscernible - it is not possible to make any distinction between them. Those elementary sets of indiscernible objects are used to model uncertainty of vague concepts.

In this dissertation, every concept is associated with a set of objects $X \subset \Omega$. It is usually assumed that information regarding belongingness of objects to X is available for at least a finite subset of objects $U \subset \Omega$. This subset is called a *training set*. When solving practical problems we are often interested in finding an accurate but comprehensible description of a concept X in terms of features of objects from the training set U. Ideally, this description should fit to all objects from Ω. In the rough set terminology, the process of finding an appropriate description of a concept is referred to as an approximation of X. In a more general context of machine learning, this task is often called a *classification problem*. The main part of this dissertation is focusing on similarity models which can be used to facilitate the classification.

Within the rough set approach, vagueness or vague concepts correspond to sets of objects which can not be precisely described using available information. To enable reasoning about such concepts, they are associated with two crisp sets which can be unambiguously defined [42–44]. The first set is the largest possible subset of available data that contains only objects which surely belong to the concept. The second set is the smallest possible set which surely contains all objects belonging to the concept in the available data. Together, those two set allow to handle vagueness without a need for introducing artificial functions, as it is done in the fuzzy set theory [66]. This subsection overviews the basic notions of the rough set theory which are used in further parts of this dissertation.

Information and Decision Systems. In the rough set theory, available knowledge about object $u \in U$ is represented as a vector of information about values of its *attributes*. An attribute can be treated as a function $a : U \to V_a$ that assigns values from a set V_a to objects from U. In a vast majority of cases, those functions are not explicitly given. However, we can still assume their existence if for any object from U we are able to measure, compute or obtain in other way the corresponding values of its attributes.

All available information about objects from U can be stored in a structure called an *information system*. Formally, an information system $\mathbb{S}$ can be defined as a tuple:

$$\mathbb{S} = (U, A) \tag{1}$$

where U is a finite non-empty set of objects and A is a finite non-empty set of attributes. The most common representation of the information system is a table whose rows correspond to objects from U and columns are associated with attributes from A. There are however some other information system representation forms [67]. A simple example of an information system represented in the tabular form is given in Table 1.a (on the left).

It is usually assumed that information about values of all the attributes from A can be obtained for any object, including those which are not present in U. In such a case, those attributes are often called *conditional attributes*. However, there might also exist some special characteristic of objects from U, which can be used to define a partitioning of U into disjoint sets. Such a characteristic may correspond to, e.g. belongingness of the objects to some concept. In this case, it is possible to define an attribute, called a *decision* or *class attribute*, that reflects this characteristic. In order to deliberately emphasize its presence, an information system with a defined decision attribute is called a *decision system* and is denoted by $\mathbb{S}_d = (U, A \cup \{d\})$, where $A \cap \{d\} = \emptyset$. A tabular representation of a decision system is sometimes called a *decision table* and the disjoint sets of objects with different values of the decision attribute are called *categories* or *decision classes*. Table 1.b shows an exemplary decision system $\mathbb{S}_d$ with a binary decision attribute d (on the right).

Table 1. An exemplary information system $\mathbb{S}$ (Table (a)) and a decision system $\mathbb{S}_d$ with a binary decision attribute (Table (b))

	a_1	a_2	a_3	a_4	a_5	a_6	a_7	a_8
u_1	1	2	2	0	0	1	0	1
u_2	0	1	1	1	1	0	1	0
u_3	1	2	0	1	0	2	1	0
u_4	0	1	0	0	1	0	0	1
u_5	2	0	1	0	2	1	0	0
u_6	1	0	2	0	2	0	0	2
u_7	0	1	1	2	0	2	1	0
u_8	0	0	0	2	1	1	1	1
u_9	2	1	0	0	1	1	0	0

(a)

	a_1	a_2	a_3	a_4	a_5	a_6	a_7	a_8	d
u_1	1	2	2	0	0	1	0	1	1
u_2	0	1	1	1	1	0	1	0	1
u_3	1	2	0	1	0	2	1	0	1
u_4	0	1	0	0	1	0	0	1	0
u_5	2	0	1	0	2	1	0	0	1
u_6	1	0	2	0	2	0	0	2	0
u_7	0	1	1	2	0	2	1	0	1
u_8	0	0	0	2	1	1	1	1	0
u_9	2	1	0	0	1	1	0	0	0

(b)

Unlike in the case of conditional attributes, a value of a decision attribute may be unknown for objects from $\Omega \setminus U$. Therefore, the approximation of concepts (a classification problem) can sometimes be restated as a prediction of decision attribute values for objects which are not included in the training set. In many practical applications, such as the topical classification of textual documents [32, 35], it might be convenient to define more than one decision attribute. In such a case, a decision system will be denoted by $\mathbb{S}_D = (U, A \cup D)$, where D is a set of decision attributes and $A \cap D = \emptyset$, and the prediction of the decision values will be called a *multi-label classification* problem.

In many practical applications the assumption regarding availability of information concerning values of conditional attributes in decision systems is not true. Real-life decision systems often have *missing* attribute values and some dedicated techniques for analysing this kind of data have been developed within the theory of rough sets [68–70]. The reasons for lack of partial information about particular objects might be diverse. The semantics of different kinds of missing values have also been studied [69, 71, 72]. Although this problem remains a vital research direction, handling data with missing or vague information lies outside the scope of this dissertation.

Indiscernibility Relation. In the rough set theory objects from U are seen through the information that can be used to describe them. This fact implies that in a case when information available for two different objects does not differ (i.e. values on all attributes are the same), those objects are regarded *indiscernible*.

Definition 1 (Indiscernibility relation)
Let $\mathbb{S} = (U, A)$ be an information system and let $B \subseteq A$. We will say that $u_1, u_2 \in U$ are satisfying the indiscernibility relation IND_B with regard to the attribute set B iff they have equal attribute values for every $a \in B$:

$$(u_1, u_2) \in IND_B \Leftrightarrow \forall_{a \in B} a(u_1) = a(u_2).$$

Otherwise u_1 and u_2 will be regarded discernible.

It is easy to observe that the indiscernibility is in fact an equivalence relation in U (i.e. it is reflexive, symmetric and transitive). An indiscernibility class of an object u with regard to an attribute set B will be denoted by $[u]_B$:

$$[u]_B = \{u' \in U : \forall_{a \in B} \ a(u') = a(u)\} . \tag{2}$$

Therefore, using the indiscernibility relation it is possible to define a *granulation* of objects described by an information system $\mathbb{S}$ into disjoint subsets. For any $B \subseteq A$ it will be denoted by $U/B = \{[u]_B : u \in U\}$. For example, the indiscernibility class of an object u_1 with regard to $\{a_1, a_3\}$ in the information system from Table 1.a (on the left) is $[u_1]_{\{a_1, a_3\}} = \{u_1, u_6\}$ and $U/\{a_1, a_3\} = \{\{u_1, u_6\}, \{u_2, u_7\}, \{u_3\}, \{u_4, u_8\}, \{u_5\}, \{u_9\}\}$.

Many different equivalence relations in U can be defined using different attribute subsets. The indiscernibility relations with regard to single attributes can serve as a basis for the construction of equivalence relations defined by any subset of attributes. For any two subsets of attributes $B, B' \subseteq A$ and any $u \in U$, the following equations hold:

$$[u]_B = \bigcap_{a \in B} [u]_{\{a\}} , \tag{3}$$

$$[u]_{B \cup B'} = [u]_B \cap [u]_{B'} , \tag{4}$$

$$B \subseteq B' \Rightarrow [u]_{B'} \subseteq [u]_B . \tag{5}$$

When constructing an approximation of a concept it is important to investigate a relation between indiscernibility classes with regard to conditional attributes and with regard to decision attributes.

Definition 2 (Consistent decision system)
A decision system $\mathbb{S}_d = (U, A \cup D)$ will be called consistent iff

$$\forall_{u \in U} [u]_A \subseteq [u]_D. \tag{6}$$

Otherwise $\mathbb{S}_d$ will be called inconsistent.

Several extensions of the indiscernibility notion can be found in the rough set literature. For example, generalizations based on a tolerance relation [56, 73] or a predefined similarity relation [54, 57, 58] have been proposed in order to define better approximations of concepts. In other approaches the definition of indiscernibility has been modified to facilitate generation of decision rules from incomplete data [69, 72].

Descriptions and Rules. The rough set theory is often utilized to provide description of *concepts* from the considered universe. Any concept can generally be associated with a subset of objects from U which belong or match to it. In general, decision attributes in a decision system can usually be interpreted as expressing the property of belongingness to some concept. Given some information (e.g. in the form of a decision system) about characteristics (values of attributes) of objects corresponding to the considered concept one may try to describe it using a *decision logic language* [74].

Decision logic language L_A is defined over an alphabet consisting of a set of attribute constants (i.e. names of attributes from A) and a set of attribute value constants (i.e. symbols representing possible attribute values). The attribute and attribute value constants can be connected using the equity symbol $=$ to form attribute-value pairs ($a = v$, where $a \in A$ and $v \in V_a$), which are regarded as atomic formulas of the language L_A. The atomic formulas can be combined into compound formulas of L_A using connectives from a set $\{\neg, \wedge, \vee, \rightarrow, \equiv\}$ called negation, conjunction, alternative, implication and equivalence, respectively. If ϕ and ψ are in L_A, then $\neg(\phi)$, $(\phi \wedge \psi)$, $(\phi \vee \psi)$, $(\phi \rightarrow \psi)$ and $(\phi \equiv \psi)$ are in L_A. The atomic formulas of a compound formula (the attribute-value pairs) are often called *descriptors* and the formula itself is sometimes called a *description* of some concept.

The satisfiability of a formula ϕ from L_A by an object from an information system $\mathbb{S} = (U, A)$, which is denoted by $u \models_{\mathbb{S}} \phi$ or by $u \models \phi$ if $\mathbb{S}$ is understood, can be defined recursively:

1. $u \models (a = v) \Leftrightarrow a(u) = v$.
2. $u \models \neg\phi \Leftrightarrow$ not $u \models \phi$.
3. $u \models (\phi \wedge \psi) \Leftrightarrow u \models \phi$ and $u \models \psi$.
4. $u \models (\phi \vee \psi) \Leftrightarrow u \models \phi$ or $u \models \psi$.
5. $u \models (\phi \rightarrow \psi) \Leftrightarrow u \models (\neg\phi \vee \psi)$.
6. $u \models (\phi \equiv \psi) \Leftrightarrow u \models (\phi \rightarrow \psi)$ and $u \models (\psi \rightarrow \phi)$.

Each description (a formula) ϕ in a decision logic language L_A can be associated with a set of objects from U that satisfy it. This set is called a *meaning* of the formula in an information system $\mathbb{S} = (U, A)$ and is denoted by $\phi(U) = \{u \in U : u \models \phi\}$. Moreover, we will say that a formula ϕ is *true* or *consistent* in $\mathbb{S}$ if and only if its meaning is equal to the whole set U (i.e. $\phi(U) = U$). Otherwise a formula is *inconsistent* in $\mathbb{S}$.

It is worth noticing that an indiscernibility class of any object u described in $\mathbb{S} = (U, A)$ can be expressed as a meaning of a formula in the language L_A as $[u]_A = \phi(U)$, where $\phi = \big(a_1 = a_1(u) \wedge \ldots \wedge a_i = a_i(u) \wedge \ldots \wedge a_m = a_m(u)\big)$, and $m = |A|$. Based on equations 3, 4 and 5 this can be generalized to indiscernibility classes with regard to any subset of attributes. For example, in the information system $\mathbb{S}$ from Table 1.a the meaning of $\phi = (a_1 = 1 \wedge a_3 = 2)$ is $\phi(U) = \{u_1, u_6\} = [u_1]_{\{a_1, a_3\}}$. One example of a formula that is consistent in $\mathbb{S}$ is $(a_7 = 0 \vee a_7 = 1)$.

In the rough set data analysis, knowledge about dependencies between conditional attributes and decision attributes of a decision system are often represented using special formulas called *decision rules*.

Definition 3 (Decision rules)

Let A and D be conditional and decision attribute sets of some decision system. Moreover, let $L_{A \cup D}$ be a decision logic language and π be a formula of $L_{A \cup D}$. We will say that π is a decision rule iff the following conditions are met:

1. $\pi = (\phi \to \psi)$,
2. ϕ and ψ are conjunctions of descriptors,
3. ϕ is a formula of L_A and ψ is a formula of L_D.

The right hand side of a decision rule $\pi = (\phi \to \psi)$ (i.e. ψ) will be called a *consequent* or a *successor* of a rule and the left hand side will be called an *antecedent* or a *predecessor* (i.e. ϕ). The antecedent of π will be denoted by $lh(\pi)$ and the consequent of π will be marked by $rh(\pi)$. It is important to note that the above definition of a decision rule is more specific than the original definition from [74]. In fact the definition used in this dissertation corresponds to *P-basic decision rules* from Pawlak's original paper.

Decision rules aim at providing partial descriptions of concepts indicated by the decision attributes. They can be learnt from a decision system and then used to predict decision classes of new objects, provided that values of conditional attributes of those objects are known. For example, from the decision system $\mathbb{S}_d$ shown in Table 1.b we can induce decision rules:

$$\pi_1 = \big((a_4 = 0 \wedge a_6 = 1) \to (d = 1)\big)$$

and

$$\pi_2 = \big((a_2 = 1 \wedge a_3 = 1) \to (d = 1)\big).$$

The meaning of π_1 in $\mathbb{S}_d$ is the set $\pi_1(U) = \{u_1, u_2, u_3, u_4, u_5, u_6, u_7, u_8\} = U \setminus \{u_9\}$, whereas the meaning of π_2 in $\mathbb{S}_d$ is $\pi_2(U) = U$. The first rule is inconsistent in $\mathbb{S}_d$, whereas the second rule is true in $\mathbb{S}_d$. However, the second rule is more general than the first one, since meanings of the antecedents of those rules have different cardinalities: $|lh(\pi_1)(U)| = |\{u_1, u_5, u_9\}| = 3$, and $|lh(\pi_2)(U)| = |\{u_2, u_7\}| = 2$. We may say that those rules are true with different degrees in $\mathbb{S}_d$, thus their predictive power is different.

There is also a different type of rules within the rough set theory, which can be particularly useful for analysing dependencies in data with multiple decision values, namely, *inhibitory rules* [75].

Definition 4 (Inhibitory rules)
Let A and D be conditional and decision attribute sets of a decision system. Moreover, let $L_{A \cup D}$ be a decision logic language and π be a formula of $L_{A \cup D}$. We will say that π is an inhibitory rule iff the following conditions are met:

1. $\pi = (\phi \to \neg\psi)$,
2. ϕ and ψ are conjunctions of descriptors,
3. ϕ is a formula of L_A and ψ is a formula of L_D.

An inhibitory rule tells us that an object which satisfies the predecessor of this rule[1] cannot belong to a pointed decision class. The inhibitory rules can be seen as a complement to decision rules as they often provide means to classify objects

---

[1] In the remaining parts of this dissertation such objects will also be regarded to as *matching* the rule.

which are difficult to cover by the traditional rules [75]. They are particularly useful for constructing classifiers in a presence of a highly imbalanced distribution of decision values. It needs to be noted, however, that a cardinality of a set of all possible inhibitory rules for a given data is usually much greater than that of all decision rules.

Usefulness of a rule for prediction of decision classes of new objects (or just *classification*, in short) can be quantitatively assessed using rule quality measures. There exist many measures that aim at evaluating the strength of dependency between the antecedent and the consequent of rules [76–78]. However, the bigger part of them is based on the notions of rule's *support* and *confidence*. The support of a rule π is defined as:

$$supp(\pi) = \frac{|lh(\pi)(U)|}{|U|}$$

and the confidence of π is:

$$conf(\pi) = \frac{|lh(\pi)(U) \cap rh(\pi)(U)|}{|lh(\pi)(U)|} = 1 - \frac{|U \setminus \pi(U)|}{|lh(\pi)(U)|} .$$

From the second equation it follows that the confidence factor of a rule π equals 1 iff the rule is consistent in $\mathbb{S}_d$. To prove it, it is sufficient to show that $U \setminus \pi(U) = lh(\pi)(U) \setminus rh(\pi)(U)$. This equity, however, is a straight consequence of a definition of the meaning of an implication:

$$u \in \pi(U) \Leftrightarrow u \vDash \big(lh(\pi) \to rh(\pi)\big) \Leftrightarrow u \vDash \big(\neg lh(\pi) \vee rh(\pi)\big)$$
$$\Leftrightarrow \big(u \in U \setminus lh(\pi)(U)\big) \vee \big(u \in rh(\pi)(U)\big).$$

If so, then:

$$u \in \big(U \setminus \pi(U)\big) \Leftrightarrow u \in \Big(U \setminus \big(U \setminus lh(\pi)(U)\big)\Big) \cap \Big(U \setminus rh(\pi)(U)\Big)$$
$$\Leftrightarrow u \in \Big(lh(\pi)(U) \setminus rh(\pi)(U)\Big).$$

The confidence of a rule is often interpreted as an indicator whether the rule is true. We may say that a rule is true in a degree corresponding to its confidence. An example of a rule quality measure that, in a sense, combines the desirable properties of the support and confidence coefficients is *Laplace m-estimate* defined as $laplace_m(\pi) = \frac{|lh(\pi)(U) \cap rh(\pi)(U)| + m \cdot p}{(|lh(\pi)(U)| + m)}$, where m and p are positive parameters. Values of m and p usually correspond to a number of decision classes, and prior probability of the $rh(\pi)$, respectively [79]. Unlike the confidence, this measure favours rules with a higher support.

Intuitively, the support of a rule expresses how large data fragment the rule describes, i.e. measures its generality, whereas the confidence says how often the rule truly indicates consequent for objects belonging to the meaning of its antecedent. For instance, the support of the rule π_1 from the previous example is $3/9 = 1/3$ and its confidence is $2/3$. At the same time the support and the confidence of π_2 are $2/9$ and 1, respectively. In order to compare those rules we may also use the Laplace m-estimate for $m = 2$ and $p = 0.5$: $laplace_2(\pi_1) = 3/5$ whereas $laplace_2(\pi_2) = 3/4$. Rough set methods usually derive rules using descriptions of indiscernibility classes in $\mathbb{S}_d$.

Each formula in the language L_A corresponds to a unique set of objects but there is no guarantee that for a given subset of objects $X \subset U$ there exists a formula ϕ whose meaning equals X. Moreover, several different formulas may have exactly the same meaning in $\mathbb{S}$. A set of objects represented in an information system $\mathbb{S}$ that can be exactly described by some formula in a language L_A is called a *definable set* in $\mathbb{S}$. More formally, the set X will be called definable in $\mathbb{S} = (U, A)$ iff there exists a formula ϕ of the language L_A, such that $\phi(U) = X$. Subsets of U that are not definable will be called *undefinable*. The family of all definable sets in $\mathbb{S}$ will be denoted by $DEF(\mathbb{S})$.

Concepts corresponding to undefinable sets can be approximated using definable sets. A typical task in the rough set data analysis is to find an optimal approximation of a predefined concept using knowledge represented by a decision system and describe it using formulas, such as decision and inhibitory rules. Such an approximation is usually expected to be accurate not only for known objects from U, but also for the new ones which were not available when the approximation was learnt. For this purpose many rough set techniques employ the Minimal Description Length (MDL) principle and constrain the language used to describe and reason about the data. This approach to the problem of approximating the undefinable sets is the most characteristic feature of the rough set theory [44, 80].

2.2   Rough Set Approximations

In the rough set theory any arbitrary set of objects X can be approximated within an information system $\mathbb{S} = (U, A)$ by a pair of definable sets $App(X) = (\underline{X}, \overline{X})$, called a *rough set* of X in $\mathbb{S}$. The set $\underline{X}$ is the largest definable set which is contained in X. Analogically, the set $\overline{X}$ is the smallest definable set which contains X. The sets $\underline{X}$ and $\overline{X}$ are called a *lower* and *upper approximation* of X in $\mathbb{S}$, respectively.

Lower and Upper Approximations. The lower and upper approximations can also be constructively defined using the notion of indiscernibility classes. Let $X \subseteq \Omega$ represent an arbitrary concept. The rough set of X in $\mathbb{S} = (U, A)$ with regard to a set of attributes $B \subseteq A$ is a pair $App_B(X) = (\underline{X}, \overline{X})$, where

$$\underline{X} = \{u \in U : [u]_B \subseteq X\},$$
$$\overline{X} = \{u \in U : [u]_B \cap X \neq \emptyset\}.$$

The sets $\underline{X}$ and $\overline{X}$ constructed for an attribute set $B \subseteq A$ are called B-lower and B-upper approximations and the pair $App_B(X) = (\underline{X}, \overline{X})$ is sometimes called a B-*rough set* of X in $\mathbb{S}$. However, when the set B is fixed (or irrelevant) we will call the sets $\underline{X}$ and $\overline{X}$ simply the lower and upper approximations of X.

Of course, since an indiscernibility class of any object in U is a definable set in $\mathbb{S}$, the definitions of a rough set by definable sets and indiscernibility classes are equivalent. The lower and upper approximations can also be defined in several other equivalent ways, which might be convenient when dealing with specific problems [48, 56]. The above definition makes it obvious that the lower approximation of a concept can be described using predecessors of consistent rules, whereas the description of the upper approximation may require some rules with the confidence factor lower than 1. This fact will be used during the construction of a similarity model proposed in Subsection 5.2.

For the classical definition of rough set and for any $B \subseteq A$, the lower and upper approximations of $X \subseteq U$ have several interesting properties:

(L1) $\underline{X} \in DEF(\mathbb{S})$	(U1) $\overline{X} \in DEF(\mathbb{S})$
(L2) $X \in DEF(\mathbb{S}) \Rightarrow \underline{X} = X$	(U2) $X \in DEF(\mathbb{S}) \Rightarrow \overline{X} = X$
(L3) $\underline{X} \subseteq X$	(U3) $X \subseteq \overline{X}$
(L4) $\underline{X} = U \setminus \overline{(U \setminus X)}$	(U4) $\overline{X} = U \setminus \underline{(U \setminus X)}$
(L5) $\underline{(X \cap Y)} = \underline{X} \cap \underline{Y}$	(U5) $\overline{(X \cup Y)} = \overline{X} \cup \overline{Y}$
(L6) $\underline{(X \cup Y)} \supseteq \underline{X} \cup \underline{Y}$	(U6) $\overline{(X \cap Y)} \subseteq \overline{X} \cap \overline{Y}$
(L7) $X \subseteq Y \Rightarrow \underline{X} \subseteq \underline{Y}$	(U7) $X \subseteq Y \Rightarrow \overline{X} \subseteq \overline{Y}$
(L8) $\underline{X} = \underline{(\underline{X})}$	(U8) $\overline{X} = \overline{(\overline{X})}$
(L9) $\underline{X} = \overline{(\underline{X})}$	(U9) $\overline{X} = \underline{(\overline{X})}$

where $App_B(X) = (\underline{X}, \overline{X})$. Proofs of those properties are omitted since they are quite obvious and have already been presented in rough set literature (e.g. [80]). The properties (L4) and (U4) show that the lower and upper approximations are, in a sense, dual operations. In general, the other properties with the same number may be regarded as dual. The properties (L1-2) and (U1-2) say that the two approximations are definable set (also called *crisp sets*). The properties (L3) and (U3) imply that for any set X, $\underline{X} \subseteq X \subseteq \overline{X}$. By the properties (L5-7) and (U5-7) it is shown that the operations of the lower and upper approximation are monotonic with regard to set inclusion, and the properties (L8-9), (U8-9) state that chains of rough set approximations are stable.

A B-rough set of a given set X defines a partitioning of objects from an information system into three disjoint sets called a B-*positive region*, B-*boundary region* and B-*negative region*. The positive region corresponds to the lower approximation of X - it contains objects that surely belong to the considered concept. It is usually denoted by $POS_B(X)$. The boundary region $BND_B(X)$ consists of objects whose belongingness is unclear (relative to a given set of attributes). It can be expressed as a difference between the upper and lower approximations: $BND_B(X) = \overline{X} - \underline{X}$. Finally, the negative region $NEG_B(X)$ contains objects that definitely do not belong to X, since they are outside its upper approximation: $NEG_B(X) = U \setminus \overline{X}$. Figure 1 shows rough set regions of an exemplary concept.

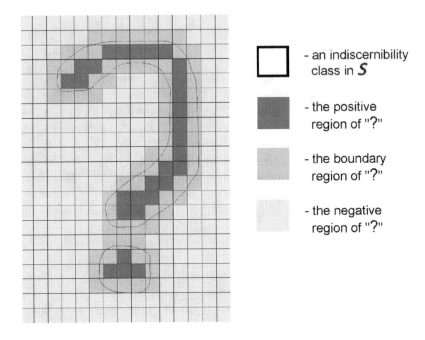

Fig. 1. An exemplary rough set approximation of a concept

Zdzisław Pawlak in his early works on rough sets suggested an intuitive measure of rough approximation accuracy:

$$\alpha(App_B(X)) = \frac{|\underline{X}|}{|\overline{X}|} \quad . \tag{7}$$

The accuracy measure α expresses how well a given concept is modelled by its rough set. This measure is closely related to *roughness* of a set:

$$\rho_B(X) = \frac{|BND_B(X)|}{|\overline{X}|} = 1 - \alpha\big(App_B(X)\big) \quad . \tag{8}$$

It is important to realize that the accuracy and roughness evaluate the rough approximations only on the available objects from an information system. Unfortunately, a close approximation on known data does not necessarily lead to a reliable assessment of new cases due to the over-fitting problem [7,8]. However, those measures are still useful for tasks such as the feature selection, where they can help evaluating the impact of including or excluding an attribute from a given attribute set [81–83].

Approximation Spaces. Although the rough set approximation of a concept is defined only for known objects from S it can be easily extended to all objects from Ω by considering descriptions of the lower and upper approximations. If the aim of the rough set analysis is to create a predictive model, then the quality of approximation on previously unseen cases is much more important than for the objects described in the decision table. To ensure this property, it is often necessary to modify representation of objects in the decision system by reducing unimportant or misleading attributes or by constructing new ones which are more informative. Such an operation influences the shape of the family of definable sets in $\mathbb{S}$, i.e. it changes the *approximation space* [56, 73] constructed for $\mathbb{S}$.

More formally, an approximation space is a tuple $\mathbb{A} = (U, IND)$, where U is a subset of known objects from Ω and $IND \subset U \times U$ is an indiscernibility relation [56]. This notion can be generalized by introducing two important concepts, namely an *uncertainty function* and a *f-membership function*.

Definition 5 (Uncertainty function)
Let $U \subseteq \Omega$. A function $I : U \to \mathbb{P}(U)$ will be called an uncertainty function iff the following conditions are met:

1. $\forall_{u \in U} u \in I(u)$.
2. $u_1 \in I(u_2) \Leftrightarrow u_2 \in I(u_1)$.

The uncertainty function assigns neighbourhoods to objects from the set U. The conditions from Definition 5 imply that the uncertainty function defines a *tolerance relation*, i.e. a relation that is reflexive and symmetric [56]. However, in rough set literature this condition is sometimes weakened to consider any reflexive relation [15].

The sets defined by the uncertainty function may be utilized to measure a degree in which an object belongs to a given concept. It is usually done using an *f*-membership function.

Definition 6 (*f*-membership function)
Let $U \subseteq \Omega$, $I : U \to \mathbb{P}(U)$ be an uncertainty function, $f : [0,1] \to [0,1]$ be a non-decreasing function and $\eta : U \times \mathbb{P}(U) \to \mathbb{R}$ be a function defined as:

$$\eta_I(u, X) = \frac{|I(u) \cap X|}{|I(u)|} \ . \tag{9}$$

A function $\mu = f(\eta)$ will be called an f-membership function.

If f is an identity function, then the f-membership function will be called simply a *membership function*. This type of an f-membership function coupled with a data driven uncertainty function will be explicitly used in the construction of the similarity model described in Section 5.

Having defined the uncertainty and the membership functions, a generalized approximation space can be defined as a tuple $\mathbb{A} = (U, I, \mu)$, where U is a subset of known objects from Ω, $I : U \to \mathbb{P}(U)$ is an uncertainty function and μ is an f-membership function.

In the classical rough set theory, the uncertainty function I often associates objects with their indiscernibility classes (i.e. $I(u) = I_B(u) = [u]_B$ for $B \subseteq A$) and the f-membership function has a form of $\mu(u, X) = \mu_B(u, X) = \frac{|[u]_B \cap X|}{|[u]_B|}$. For example, if we consider the information system from Table 1.a and the uncertainty function $I(u) = [u]_{\{a_1, a_2\}}$, the neighbourhood of u_2 would be $I(u_2) = \{u_2, u_4, u_7\}$. Furthermore, a degree to which u_2 belongs to the decision class with a label 1 with regard to I is equal to $\mu(u_2, \{d = 1\}(U)) = 2/3$.

In this way, the function I can be used to generalize the indiscernibility relation and define a new family of sets that can serve as building-blocks for constructing approximations. Coupled with the rough membership function, it leads to a more flexible definition of the lower and upper approximations:

$$\underline{X} = \{u \in U : \mu_I(u, X) = 1\} \ , \tag{10}$$

$$\overline{X} = \{u \in U : \mu_I(u, X) > 0\} \ . \tag{11}$$

Of course, if I is a description identity function, those definitions are equivalent to the classical ones. There also exist further generalizations of rough approximations, such as the variable precision rough set model [84,85] which introduces an additional parameter allowing to weaken the zero-one bounds in the above definitions.

The uncertainty function can be defined, for example, by combining transformations of object representation space (the set of attributes) with the classical indiscernibility. Such a transformation may include reduction of the information describing objects to attributes which are truly related to the considered problem, as well as an extension of the attribute set by new, often higher-level features.

Approximation of Relations. The rough approximations allow not only to express the uncertainty about concepts but also to model arbitrary relations between objects from Ω [51]. In fact, the notion of approximation spaces was generalized in [51] to allow defining approximations of sets in $\mathbb{U} = U_1 \times \ldots \times U_k$, where $U_i \subset \Omega$ are arbitrary sets of objects. Since the scope of this dissertation is on a similarity which can be seen as a binary relation (see Section 3), only this type of relations will be considered in this subsection.

A binary relation r between objects from a given set U is a subset of a Cartesian product of this set ($r \subseteq U \times U$). Having a subset of objects from Ω we may try to approximate an arbitrary binary relation $r \subseteq \Omega \times \Omega$ within the set $U \times U$ by considering a generalized approximation space, defined as a tuple $\mathbb{A}_2 = (U \times U, I_2, \mu_2)$, where $U \subset \Omega$, $I_2 : U \times U \to \mathbb{P}(U \times U)$ is a generalized uncertainty function and $\mu_2 : (U \times U) \times (\mathbb{P}(U \times U)) \to \mathbb{R}$ is a generalized rough membership function.

The functions I_2 and μ_2 can be easily defined by an analogy with the case of a regular approximation space. Their simplified graphical interpretation is shown in Figure 2. However, the meaning of an indiscernibility class of a pair of objects needs to be adjusted. In general, a pair (u_1, u_2) can be characterised by three possibly different sets of features – features specific to u_1, features specific to u_2 and those which describe u_1 and u_2 as a pair. This fact is utilized in a construction of the Rule-Based Similarity (RBS) model proposed in Section 5. In this model, objects are represented in a new feature space that allows for a robust approximation of a similarity relation. Such approximation is likely to be

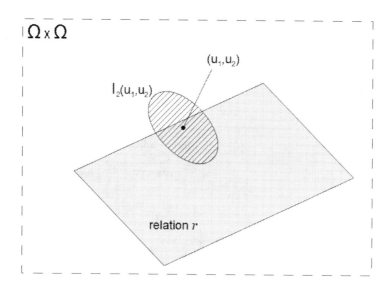

Fig. 2. A graphical interpretation of as uncertainty function for approximation of a binary relation. In this case a membership function value $\mu_2\big((u_1, u_2), r\big)$ could be defined as a ratio between a size of the intersection of $I_2(u_1, u_2)$ and r, and the size of whole $I_2(u_1, u_2)$.

precise not only on training data but also in a situation when the model is used for assessment of resemblance of the training cases to completely new objects.

Approximations of a binary relation may have two desirable properties that indicate their quality, namely the *consistence* and *covering* properties defined below:

Definition 7 (Consistence property)
Let $U \subseteq \Omega$ and r be a binary relation in Ω. We will say that a binary relation r' is consistent with r in U iff the implication

$$(u_1, u_2) \in r' \Rightarrow (u_1, u_2) \in r$$

holds for every $u_1, u_2 \in U$.

Definition 8 (Covering property)
Let $U \subseteq \Omega$ and r be a binary relation in Ω. We will say that a binary relation r' covers r in U iff the implication

$$(u_1, u_2) \in r \Rightarrow (u_1, u_2) \in r'$$

holds for every $u_1, u_2 \in U$.

A fact that a relation r' is consistent with r in U will be denoted by $r' \subseteq_U r$. Analogically, a fact that r' covers r in U will be denoted by $r' \supseteq_U r$.

An approximation of a relation that has the consistence property can be seen as a kind of a rough set lower approximation, whereas an approximation that covers a binary relation can be treated as its upper approximation. Those two notions will be used in Section 3 to characterize a class of similarity functions that is the main scope of this dissertation.

2.3   Attribute Reduction

The problem of finding a representation of objects, which is appropriate in a given task, can be seen as a process of adaptation of an approximation space, therefore it is closely related to the rough sets in general. Zdzisław Pawlak wrote in [86] that discovering redundancy and dependencies between attributes is one of the fundamental and the most challenging problems of the rough set philosophy. The rough set theory provides intuitive tools for selecting informative features and constructing new ones. The most important of such tools are the notions of *information* and *decision reducts*.

Rough Set Information Reduction. In many applications information about objects from a considered universe has to be reduced. This reduction is necessary in order to limit resources that are needed by algorithms analysing the data or to prevent crippling their performance by noisy or irrelevant attributes [7,87,88]. This vital problem has been in the scope of the rough set theory since its beginnings [42,44,80] and has been investigated by numerous researchers [33,34,81,82, 89,90].

Typically, in the rough set theory selecting compact yet informative sets of attributes is conducted using the notion of indiscernibility, by computing so called reducts [80,91].

Definition 9 (Information reduct)
Let $\mathbb{S} = (U, A)$ be an information system. A subset of attributes $IR \subseteq A$ will be called an information reduct iff the following two conditions are met:

1. *For any $u \in U$ the indiscernibility classes of u with regard to IR and A are equal, i.e. $[u]_A = [u]_{IR}$.*
2. *There is no proper subset $IR' \subsetneq IR$ for which the first condition holds.*

An information reduct IR can be interpreted as a set of attributes that are sufficient to discern among as many objects described in $\mathbb{S}$ as the whole attribute set A. At the same time the reduct is minimal, in a sense that no further attributes can be removed from IR without losing the full discernibility property. Analogically, it is possible to define a decision reduct DR for a decision system $\mathbb{S}_d$:

Definition 10 (Decision reduct)
Let $\mathbb{S}_d = (U, A \cup \{d\})$ be a decision system with a decision attribute d that indicates belongingness of objects to an investigated concept. A subset of attributes $DR \subseteq A$ will be called a decision reduct iff the following two conditions are met:

1. *For any $u \in U$ if the indiscernibility class of u relative to A is a subset of some decision class, its indiscernibility class relative to DR should also be a subset of that decision class, i.e. $[u]_A \subseteq [u]_d \Rightarrow [u]_{DR} \subseteq [u]_d$.*
2. *There is no proper subset $DR' \subsetneq DR$ for which the first condition holds.*

Unlike in the definition of information reducts, a decision reduct needs only to sustain the ability to discriminate objects from different decision classes. For example, $\{a_1, a_3, a_6\}$ and $\{a_3, a_5, a_6, a_7\}$ are information reducts of the information system from Table 1.a while $\{a_3, a_5\}$ and $\{a_3, a_6\}$ are decision reducts of the corresponding decision system.

The minimality of reducts stays in accordance with the Minimum Description Length (MDL) rule. Depending on an application, however, the minimality requirement for the reducts may sometimes be relaxed in order to ensure inclusion of the key attributes to the constructed model. In some cases keeping relevant but highly interdependent attributes may have a positive impact on model's performance [87, 88]. For this reason within the theory of rough sets a notion of decision superreduct is considered which is a set of attributes that discerns all objects from different decision classes but does not need to be minimal.

Usually for any information system there are numerous reducts. In the rough set literature there are described many algorithms for attribute reduction. The most commonly used are the methods utilizing discernibility matrices and the boolean reasoning [42, 82, 91, 92], and those which use a greedy or randomized search in the attribute space [26, 34, 93, 94].

In [80] it is shown that a decision reduct can consist only of strongly and weakly relevant attributes (it cannot contain any irrelevant attribute)[2] if the available data is sufficiently representative for the universe at scope. However, in real-life situations this requirement is rarely met. Very often, especially when analysing high dimensional data, some dependencies between attribute values and decisions are not general – they are specific only to a given data set. In such a case attributes which are in fact irrelevant might still be present in some decision reducts.

Generalizations of Reducts. Many researchers have made attempts to tackle the problem of attribute relevance in decision reducts. Apart from devising heuristic algorithms for performing the attribute reduction that are more likely to select relevant features, a significant effort has been made in order to come up with some more general definitions of the reducts.

It has been noticed that subsets of attributes which preserve discernibility of a slightly lower number of objects from different decision classes than the whole attribute set tend to be much smaller than the regular reducts. Usually objects that are described with fewer attributes have larger discernibility classes which correspond to more general decision rules. This observation motivated introduction of the notion of an approximate decision reduct [94, 96–99].

Definition 11 (Approximate decision reduct)
Let $\mathbb{S}_d = \left(U, A \cup \{d\}\right)$ be a decision system with a decision attribute d and let ϵ be a real non-negative number, $\epsilon \in [0, 1)$. Additionally, let $|POS_B(d)|$ denote the number of objects whose indiscernibility classes with regard to an attribute set $B \subseteq A$ are subsets of a single decision class, i.e. $|POS_B(d)| = |\{u \in U : [u]_B \subseteq$

---

[2] The strong and weak relevance of attributes is understood as in [95].

$[u]_d\}|$. A subset of attributes $ADR \subseteq A$ will be called an ϵ-approximate decision reduct iff the following two conditions are met:

1. ADR preserves discernibility in $\mathbb{S}_d$ with a degree of $1 - \epsilon$, i.e.
   $|POS_{ADR}(d)| \geq (1 - \epsilon) \cdot |POS_A(d)|$.
2. There is no proper subset $ADR' \subsetneq ADR$ for which the first condition holds.

Of course, for $\epsilon = 0$ this definition is equivalent to the definition of regular decision reducts. For the decision system from Table 1.b the attribute subsets $\{a_1, a_3\}$ and $\{a_5, a_6\}$ are examples of the 0.3-approximate decision reducts.

The ϵ-approximate reducts can also be defined using differently formulated conditions. For example, instead of relying on the sizes of positive regions of decision classes, the approximate decision reducts can be defined based on the conditional entropy [99] of an attribute set or the number of discerned object pairs [96]. In fact, any measure of dependence between a conditional attribute subset and the decision, which is monotonic with regard to inclusion of new attributes, can be used [94].

A different generalization of the decision reducts, called dynamic decision reducts, has been proposed in [100]. In this approach a stability of a selected attribute set is additionally verified by checking if all the attributes are still necessary when only some smaller random subsets of objects are considered.

Definition 12 (Dynamic decision reduct)

Let $\mathbb{S}_d = (U, A \cup \{d\})$ be a decision system with a decision attribute d and let $RED(\mathbb{S}_d)$ be a family of all decision reducts of $\mathbb{S}_d$. Moreover, let ϵ and δ be real numbers such that $\epsilon, \delta \in [0, 1)$. A subset of attributes $DDR \subseteq A$ will be called an (ϵ, δ)-dynamic decision reduct iff for a finite set of all subsystems of $\mathbb{S}_d$, denoted by $SUB(\mathbb{S}_d, \epsilon)$, such that for each $\mathbb{S}'_d = (U', A, d) \in SUB(\mathbb{S}_d, \epsilon)$, $U' \subset U$ and $|U'| \leq (1 - \epsilon) \cdot |U|$, the following two conditions are met:

1. DDR is a decision reduct of $\mathbb{S}_d$ ($DDR \in RED(\mathbb{S}_d)$).
2. DDR is a decision reduct of sufficiently many $\mathbb{S}'_d \in SUB(\mathbb{S}_d, \epsilon)$, i.e. $|\{\mathbb{S}'_d \in SUB(\mathbb{S}_d, \epsilon) : DDR \in RED(\mathbb{S}'_d)\}| \geq (1 - \delta) \cdot |SUB(\mathbb{S}_d, \epsilon)|$.

Intuitively, if none of the attributes selected as belonging to a decision reduct is redundant when considering only subsets of objects, then the reduct can be seen as insensitive to data disturbances. Due to this characteristic the dynamic decision reducts are more likely to define robust decision rules [100, 101]. Additionally, the dynamic decision reducts tend to be more compact than the regular reducts. For example, from two decision reducts $DR_1 = \{a_3, a_5\}$ and $DR_2 = \{a_1, a_2, a_8\}$ of the decision system $(U, A \cup \{d\})$ from Table 1.b, only the first one is a $(0.1, 0)$-dynamic decision reduct, since DR_2 is not a reduct of a decision system $\mathbb{S}'_d = (U \setminus \{u_4\}, A, d)$.

Both of those generalizations of the decision reducts have been successfully used in applications, such as constructing ensembles of predictive models [102], discovering of approximate dependencies between attributes [94, 97] and attribute ranking [34]. In this dissertation it is also showed how the dynamic reducts can be utilized for learning of a similarity function [27, 28] (see also Section 4). The definitions of approximate and dynamic reducts for information systems can be given analogously to those for the decision systems, thus they are omitted.

Notion of Bireducts. The original definition of a decision reduct is quite restrictive, requiring that it should provide the same level of information about decisions as the complete set of available attributes. On the other hand, the approximate reducts, which are usually smaller and provide a more reliable basis for constructing classifiers [94, 103], can be defined in so many ways that selecting the optimal one for a given task is very difficult. The choice of the method may depend on a nature of particular data sets and on a purpose for the attribute reduction. Moreover, computation of the approximate decision reducts may require tuning of some unintuitive parameters, such as the threshold for a stopping criterion (ϵ).

Another issue with the approximate reducts is related to the problem of building classifier ensembles [102–106]. Combining multiple classifiers is efficient only if particular models tend to make errors on different areas of the universe at scope. Although, in general, there is no computationally feasible solution that can guarantee such a diversity, several heuristic approaches exist. For instance, one may focus on the classifier ensembles learnt from reducts that include as different attributes as possible. In this way one may increase stability of the classification and improve the ability to represent data dependencies to the users. Unfortunately, the common approximate reduct computation methods do not provide any means for controlling which parts of data are problematic for particular reducts. As a result, when building an ensemble where individual reducts are supposed to correctly classify at least 90% of the training objects, we may fail to anticipate that each of the resulting classifiers will have problems with the same 10% of instances.

To tackle the above challenges, a new extension of the original notion of a reduct was proposed [29, 36], called a *decision bireduct*. In this approach the emphasis is on both, a subset of attributes that describes the decision classes and a subset of objects for which such a description is possible.

Definition 13 (Decision bireduct)
Let $\mathbb{S}_d = (U, A \cup \{d\})$ be a decision system. A pair (B, X), where $B \subseteq A$ and $X \subseteq U$, is called a decision bireduct, iff B is a decision reduct of a subsystem (X, A, d) and the following properties hold:

1. *B discerns all pairs of objects from different decision classes in X and there is no proper subset $C \subsetneq B$ for which such a condition is met.*
2. *There is no $Y \supsetneq X$ such that B discerns all pairs of objects from different decision classes in (Y, B, d).*

It is important to realize that a decision subsystem (X, B, d) is always consistent (all indiscernibility classes in (X, B, d) are subsets of the decision classes), regardless of the consistency of the original system. However, a decision bireduct

(B, X) can be regarded as an inexact functional dependence in $\mathbb{S}_d$ linking the subset of attributes B with the decision d, just as in a case of approximate reducts. The objects in X can be used to construct a classifier based on B and the objects from $U \setminus X$ can be treated as outliers. The computation of bireducts can be seen as searching for an approximation space that allows to generate meaningful decision rules. Such rules are local, since they are defined only for objects from X. However, by neglecting the potentially noisy outliers, the rules induced from the decision bireducts (e.g. by considering the indiscernibility classes of objects from X) are more likely to be robust [36]. It has been noted that bireduct-based ensembles tend to cover much broader areas of data than the regular reducts, which leads to better performance in classification problems [36].

3   Notion of Similarity

The notion of similarity has been in a scope of interest for many decades [20,21, 107]. Knowing how to discriminate similar cases (or objects) from those which are dissimilar in a context of a decision class would enable us to conduct an accurate classification and to detect unusual situations or behaviours. Although human mind is capable of effortless assessing the resemblance of even very complex objects [108,109], mathematicians, computer scientists, philosophers and psychologist have not come up with a single methodology of building similarity models appropriate for a wide range of complex object classes or domains.

A variety of methods were used in order to construct such models and define a relation which would combine an intuitive structure with a good predictive power. Among those a huge share was based on some distance measures. In that approach objects are treated as points in a metric space of their attributes and the similarity is a decreasing function of the distance between them. Objects are regarded as similar if they are close enough in this space. Such models may be generalized by introducing a list of parameters to the similarity function, e.g. weights of attributes. Tuning them results in the relation better fitting to a dataset. Algorithms for computationally efficient optimization of parameters for common similarity measures in the context of information systems were studied in, for instance, [15,16,19].

One may argue that the relation of this kind is very intuitive because objects which have many similar values of attributes are likely to be similar. However, Amos Tversky [5,20] showed in empirical studies that in some contexts similarity does not necessarily have features like symmetry or subadditivity which are implied by distance measures. This situation occurs particularly often when we compare objects of great complexity. The explanation for this may lie in the fact that complex objects can be similar in some aspects and dissimilar in others. A dependency between local and global similarities may be highly non-linear and in order to model it we need to learn this dependency from the data, often relying on the domain knowledge provided by an expert.

This section discusses general properties of the similarity understood as a binary relation between objects from a considered universe. The following Subsection 3.1 introduces the notion of a similarity relation and explains some difficulties related to the formal definition of this idea. In its last part it describes how a performance of a similarity model can be quantitatively evaluated. Next, Subsection 3.4 briefly overviews the most commonly used approaches to the problem of modelling the similarity relation. Its main focus is on showing the differences between the distance-based model and the approach proposed by Amos Tversky [5, 11]. Finally, the last subsection (Subsection 3.5) shows exemplary applications of the similarity in fields such as Case-Based Reasoning and Cluster Analysis.

3.1   Similarity as a Relation

The similarity can be treated as a binary relation τ between objects from a universe Ω. Importance of this relation is unquestionable. In fact, many philosophers and cognitivists believe that the similarity plays a fundamental role in a process of learning from examples as well as acquiring new knowledge in general [3, 4, 108, 109]. Unfortunately, even though a human mind is capable of assessing similarity of even complex objects with a little effort, the existing computational models of this relation have troubles with accurate measuring of the resemblance between objects.

Vagueness of a Similarity Relation. Numerous empirical studies of psychologists and cognitivists showed that human perception of similar objects depends heavily on external factors, such as available information, personal experience and a context [5, 20, 21]. As a consequence, properties of a similarity relation may vary depending on both the universe and the context in which it is considered (see, e.g. [5, 21]). The similarity relation can be characterized only for a specific task or a problem. For instance, when comparing a general appearance of people in the same age, the similarity relation is likely to have a property of the symmetry. However, in a case when we compare people of a different age this property would not necessarily hold (e.g. a son is more similar to his father than the opposite). In a general case even the most basic properties, such as the reflexivity, can be questioned [5]. Figure 3 shows a drawing from two different perspectives. It can either be similar to itself or dissimilar, depending on whether we decide to consider its perspective.

The subjective nature of a similarity assessment makes it impossible to perfectly reflect the similarity using a single model. Capturing personal preferences would require tailoring the model to individual users. This could be hypothetically possible only if some personalized data was available and it would require some form of an automatic learning method. Even though in many applications it is sufficient to model similarity assessments of an "average user", a model which is designed for a given task and which takes into account the considered context, have much better chances to accurately measure the resemblance than an a priori selected general-purpose model.

Fig. 3. A single drawing from two different perspectives

Additionally, due to the fact that it is impossible to determine any specific features of the similarity without fixing its context, if no domain knowledge is available, it may be treated as a vague concept. In order to model it, all properties of this relation have to be derived from information at hand. Such information can usually be represented in an information system. That is another argument motivating the need for development of algorithms for learning domain-specific similarity relations from data. One possible approach to this task is to utilize the theory of rough sets (see Section 2) to construct an approximation of τ, which will be denoted by τ^*.

Within the rough set theory, relations can be approximated just as any other concept (see Subsection 2.2). The problem of approximation of binary relations was investigated by researchers since the beginnings of the rough sets [47,51,57]. If no additional knowledge is available this task is much more difficult than, for instance, a classification. It may be regarded as a problem of assigning binary decision labels to pairs of instances from the universe Ω in an unsupervised manner, using information about a limited number of objects described in available information system. It is important to realize that the resulting approximation τ^* has to be reliable not only for objects at hand but also for new ones. For this reason, in practical situations, in order to properly approximate the similarity it is necessary to utilize some domain knowledge and to specify a context for the relation.

3.2   Similarity in a Context

Several independent studies showed how important is to consider an appropriate context while judging a similarity between objects [4, 5, 21, 107]. Two stimuli presented to a representative group of people can be assessed as similar or dissimilar, depending on whether some additional information is given about their classification or whether they are shown along with some other characteristic objects.

For instance, if we consider cars in a context of their class[3], then Chevrolet Camaro will be more similar to Ford Mustang than Ford Tempo. However, if we change the context to a make of a car the assessment would be completely different.

A selection of the context for the similarity has a great impact on features or in other words factors, that influence the judgements [5, 21]. In the previous example, a feature such as a colour of a car would be irrelevant in the context of car's class. Nevertheless, it might be important in the context of a make of a car, since some car paints could be exclusively used by specific car producers.

When constructing a similarity model for a given data, the context for the relation can usually be inferred from a purpose which motivates performing the analysis. If a task is to cluster the given data into subsets of closely related objects in an unsupervised way and without any additional knowledge, then the context will probably be a general appearance of objects. However, if we know that, for example, the data describe textual documents, it is possible to consider them in a context of their semantics (their meaning – see Section 6.3). Furthermore, if the similarity model is created for a task such as a diagnosis of a specific condition based on a genetic profile of tissue samples, then a classification into severity stages of the condition will probably be the best context to choose (see experiments in Subsections 6.1 and 6.2). In the last case the information specifying the context will usually correspond to a decision attribute in the data table.

It is also reasonable to consider similarity of two objects in a context of other objects in the data. For instance, a banana will be more similar to a cherry when considered in a data set describing dairy and meat products, vegetables and fruits, than in a case when the data is related only to different kinds of fruits. In those two cases, different aspects of the similarity would have to be taken into account, and as a consequence, the same attributes of the fruits would have different importance.

In general, similar objects are expected to have similar properties with regard to the considered context. Since in this dissertation the main focus is on the similarity in the context of a classification, the above principle can be reformulated in terms of the consistency of the similarity with the decision classes of objects[4]. More formally:

Definition 14 (Consistency with a classification)
Let Ω be a universe of considered objects and let d be a decision attribute which can be used to divide objects from Ω into indiscernibility classes $\Omega/\{d\}$. Moreover, let τ denote a binary relation in Ω. We will say that τ is consistent with the classification indicated by d iff the following implication holds for every $u_1, u_2 \in \Omega$:

$$(u_1, u_2) \in \tau \Rightarrow d(u_1) = d(u_2) \,. \tag{12}$$

The above property will be referred to as the main feature of the similarity for the classification. It is also often assumed that a similarity relation in the context

---

[3] The official classification of cars is discussed, e.g., in a Wikipedia article *Car classification* (http://en.wikipedia.org/wiki/Car_classification).
[4] The consistency of two relations within a given set is defined in Subsection 2.2

of a classification needs to be reflexive, namely $\forall_{u \in \Omega}(u, u) \in \tau$. Additionally, for objects which are described by a set of conditional attributes A, the reflexivity is understood in terms of indescernibility. In particular, we will say that τ is reflexive if and only if $\forall_{(u,u') \in IND_A}(u, u') \in \tau$. This assumption, however, can be true only if there are no two objects in Ω which are identical in all aspects but belong to different decision classes. Binary relations in Ω that have the above two properties will be regarded as *possible similarity relations* in the context of the classification. The set of all such relations will be denoted by:

$$R = \{\tau : \tau \subseteq_\Omega IND_{\{d\}} \wedge \tau \supseteq_\Omega IND_A\}.$$

In the remaining parts of the dissertation it is assumed that one of such relations $\tau \in R$ is fixed and considered as the reference similarity relation in the specified classification context. It should be noted, however, that different scenarios for inducing this relation from data may or may not assume the availability of knowledge regarding τ for the training data. In the second case, which applies to the similarity model proposed in Section 5, only information about the properties of τ is utilized in the learning process.

The property from Definition 14 can be used to guide the process of constructing approximations of the relation τ. It infers that a desirable approximation τ^* should also be consistent with the decision classes indicated by d. In practice, however, this condition can be verified only for the known objects described in a decision system. Moreover, in real life applications it may sometimes be slightly relaxed in order to increase the recall of the approximation. Nevertheless, the knowledge that any set of objects that are similar to a given one must have the same decision can be used to limit a search space for features that can conveniently represent pairs of objects in an approximation space, as discussed in Subsection 2.2. It is also the fundamental assumption used in the construction of the Rule-Based Similarity in Section 4.

Although an approximation of a similarity in a context of classification can be made only using known objects from a given decision system, it has to allow an assessment of whether an arbitrary object from $\Omega \setminus U$ is similar to an object from U. To make this possible, an assumption is made that for objects from $\Omega \setminus U$ we can retrieve values of their conditional attributes (without the need for referring to their decision class, which may remain unknown).

There can be many approximations of a similarity relation for a given decision table $\mathbb{S}_d = (U, A \cup \{d\})$. For example, one can always define a trivial approximation for which no pair of objects is similar or a naive one, for which only objects from U that are known to belong to the same decision class can be similar. Therefore, in practical applications it is crucial to have means to evaluate quality of an approximation and estimate how close it is to the real similarity for the objects that are not described in $\mathbb{S}_d$, i.e. $\{u' \in \Omega : u' \notin U\}$. Since there is no available information regarding those objects, the Minimum Description Length rule (MDL) is often used to select the approximation which can be simply characterized but is sufficiently precise.

3.3   Similarity Function and Classification Rules

In a variety of practical situations it is convenient to express a degree in which one object is similar to another. For instance, in machine learning many classification methods construct rankings of training objects based on their similarity to a considered test case (e.g. the k *nearest neighbours* algorithm [7, 9, 15, 17]).

To assess the level of similarity between a pair of objects, a special kind of function is used, called a *similarity function*. Usually, such a function for a considered data set is given a priori by an expert for a whole $\Omega \times \Omega$ set, independently of the available data and the context. Intuitively, however, a similarity function for an information system $\mathbb{S} = (U, A)$ should be a function $Sim : U \times \Omega \to \mathbb{R}$, whose values are "high" for objects in a true similarity relation and becomes "low" for objects not in this relation. Such a function could be used to define a family of approximations of the similarity relation τ by considering the sets $\tau_{(\lambda)}^{Sim} = \{(u_1, u_2) \in U \times U : Sim(u_1, u_2) \geq \lambda\}$ for any $\lambda \in \mathbb{R}$. If a function Sim is appropriate for a given relation, then at least some of the approximations $\tau_{(\lambda)}^{Sim}$ should be consistent with τ (see Definition 7) for available data. To further formalize this notion for the purpose of this dissertation a concept of a *proper similarity function* is proposed:

Definition 15 (Proper similarity function)
Let τ be a similarity relation between objects from Ω, $U \subseteq \Omega$ be a subset of known reference objects and $Sim : U \times \Omega \to \mathbb{R}$ be a function. We will say that Sim is a proper similarity function for the relation τ within the set U iff there exist $\epsilon_1, \epsilon_2 \in \mathbb{R}$, $\epsilon_1 > \epsilon_2$, such that the following conditions hold:

1. $\left| \tau_{(\epsilon_1)}^{Sim} \right| > 0$ *and* $\tau_{(\epsilon_1)}^{Sim} \subseteq_U \tau$ *(see Def. 7)*,
2. $\left| (U \times U) \setminus \tau_{(\epsilon_2)}^{Sim} \right| > 0$ *and* $\tau_{(\epsilon_2)}^{Sim} \supseteq_U \tau$ *(see Def. 8)*.

A value of a similarity function for a pair (u_1, u_2) will be called *a similarity degree* of u_1 relative to u_2. Each of the sets $\tau_{(\lambda)}^{Sim}$ can be regarded as an approximation of the similarity relation τ. The first condition from Definition 15 requires that, starting from some ϵ_1, all the approximations $\tau_{(\lambda)}^{Sim}$ defined by a proper similarity function were subsets of the true similarity relation. It means that the precision of the approximation defined as $prec_{\tau}(\tau_{(\lambda)}^{Sim}) = \frac{|\tau_{(\lambda)}^{Sim} \cap \tau|}{|\tau_{(\lambda)}^{Sim}|}$ equals 1 for all $\lambda \geq \epsilon_1$ such that $|\tau_{(\lambda)}^{Sim}| > 0$. One practical implication of this fact is that in the context of classification, for sufficiently large λ, objects in each pair from $\tau_{(\lambda)}^{Sim}$ must belong to the same decision class (see Definition 14).

The second condition in the definition of a proper similarity function requires that there exists a border value ϵ_2 such that all $\tau_{(\lambda)}^{Sim}$ for $\lambda \leq \epsilon_2$ were supersets of the true similarity relation τ. By an analogy to the rough sets, $\tau_{(\epsilon_1)}^{Sim}$ and $\tau_{(\epsilon_2)}^{Sim}$ can be treated as a lower and upper approximation of the similarity, respectively (see Subsection 2.2).

Of course, one function can be a proper similarity function in one context but not in the other. If a function Sim is a proper similarity function for a given similarity relation, we will say that this relation is approximable by Sim.

Table 2. A similarity relation in a context of classification for the objects from the decision system depicted in Table 1.b (on the left) and a table with the corresponding values of a similarity function (on the right).

$\tau = \big\{ (u_1, u_1), (u_1, u_7),$
$\quad (u_2, u_2), (u_2, u_5),$
$\quad (u_2, u_7), (u_3, u_3),$
$\quad (u_3, u_7), (u_4, u_4),$
$\quad (u_4, u_8), (u_5, u_2),$
$\quad (u_5, u_5), (u_5, u_7),$
$\quad (u_6, u_6), (u_6, u_8),$
$\quad (u_7, u_1), (u_7, u_2),$
$\quad (u_7, u_5), (u_7, u_7),$
$\quad (u_8, u_4), (u_8, u_8),$
$\quad (u_9, u_4), (u_9, u_9) \big\}$

A similarity matrix:

	u_1	u_2	u_3	u_4	u_5	u_6	u_7	u_8	u_9
u_1	1.00	0.50	0.42	0.09	0.51	0.23	0.66	0.09	0.09
u_2	0.50	1.00	0.42	0.09	1.00	0.20	0.67	0.08	0.43
u_3	0.42	0.42	1.00	0.00	0.42	0.00	0.62	0.00	0.32
u_4	0.07	0.09	0.00	1.00	0.08	0.50	0.00	1.00	0.48
u_5	0.50	1.00	0.42	0.09	1.00	0.20	0.66	0.08	0.12
u_6	0.20	0.25	0.00	0.50	0.20	1.00	0.09	0.50	0.20
u_7	0.68	0.66	0.60	0.00	0.66	0.07	1.00	0.00	0.00
u_8	0.09	0.09	0.00	1.00	0.12	0.50	0.00	1.00	0.33
u_9	0.09	0.50	0.32	0.50	0.11	0.20	0.00	0.33	1.00

For example, Table 2 shows a similarity relation between objects described in the decision system from Table 1.b and a similarity matrix displaying values of some similarity function for all the pairs of the objects. As one can easily notice, the relation τ is consistent with the decision classes. Moreover, the similarity function used to generate the matrix is a proper similarity function for τ within the considered set of objects because for $\lambda = 0.66$ the corresponding approximation $\tau_{(0.66)}^{Sim} = \big\{ (u_1, u_1), (u_1, u_7), (u_2, u_2), (u_2, u_5), (u_2, u_7), (u_3, u_3), (u_4, u_4),$ $(u_4, u_8), (u_5, u_2), (u_5, u_5), (u_5, u_7), (u_6, u_6), (u_7, u_1), (u_7, u_2), (u_7, u_5), (u_7, u_7),$ $(u_8, u_4), (u_8, u_8), (u_9, u_9) \big\}$ is consistent with τ and for $\lambda = 0.50$ the approximation $\tau_{(0.50)}^{Sim} = \big\{ (u_1, u_1), (u_1, u_2), (u_1, u_5), (u_1, u_7), (u_2, u_1), (u_2, u_2), (u_2, u_5),$ $(u_2, u_7), (u_3, u_3), (u_3, u_7), (u_4, u_4), (u_4, u_6), (u_4, u_8), (u_5, u_1), (u_5, u_2), (u_5, u_5),$ $(u_5, u_7), (u_6, u_4), (u_6, u_6), (u_6, u_8), (u_7, u_1), (u_7, u_2), (u_7, u_3), (u_7, u_5), (u_7, u_7),$ $(u_8, u_4), (u_8, u_6), (u_8, u_8), (u_9, u_2), (u_9, u_4), (u_9, u_9) \big\}$ covers the relation τ.

Is is worth to notice that a proper similarity function does not need to be symmetric. For instance, in the previous example $Sim(u_1, u_4) \neq Sim(u_4, u_1)$. It is also important to realize that a similarity function does not need to be non-negative. The negative values of a similarity function are usually interpreted as an indication that the compared objects are more dissimilar than they are similar. However, the majority of commonly used similarity functions are non-negative.

A similarity function allows to order objects from U according to their degree of similarity to any given object from the considered universe. It is important to notice, that the similarity function allows to compute the similarity coefficient of u from the set of known objects U to any object from the universe Ω, given that it is possible to determine its attribute values. In particular, information about a decision class of the second object does not need to be available. That property may be used to define several simple, case-based classification methods. For instance, if the available training objects are described in a decision system $\mathbb{S}_d = (U, A \cup \{d\})$, an object $y \in \Omega$ can be assigned to a decision class of the most similar object from U:

$$1\text{-NN}_{Sim}(y) = d\left(\underset{u \in U}{\arg\max}\ Sim(u, y) \right). \tag{13}$$

This formula can be easily generalized to a k-nearest neighbours rule by introducing a voting scheme for deciding a class of the investigated case [7,8,17,110,111]. A voting scheme can also be applied in a case when in U there are several objects which are equally similar to y and belong to different decision classes. There are numerous voting schemes that aim at optimizing the classification performance. A basic heuristic is a majority voting by the k most similar objects from the system $\mathbb{S}_d$. Some more complex voting schemes may additionally take into account the actual similarity function values to weight the votes of the neighbours. The relative "importance" of a vote may also be adjusted by considering empirical probabilities of the decision classes.

A similarity function may also be used to define a slightly different kind of a classification rule. In the λ-majority classification, an object $y \in \Omega$ is assigned to a decision class which is the most frequent within the set of objects regarded as similar to y. Particularly, if we denote $C_\lambda(y) = \{u \in U : Sim(u, y) \geq \lambda\}$, then y can be classified as belonging to one of the l decision classes $d_1, ..., d_l$ of d using the formula:

$$\lambda\text{-majority}_{Sim}(y) = \underset{d_j \in \{d_1,...,d_l\}}{\arg\max}\ \left|\{u \in U : u \in C_\lambda(y) \wedge d(u) = d_j\}\right|. \tag{14}$$

The λ-majority classification assigns objects to a class with the highest number of similar examples, according to an approximation of the similarity relation by the set $\tau_{(\lambda)}^{Sim}$. It also can make use of different voting schemes, such as object weighting by a similarity degree or considering sizes of the decision classes. Some exemplary similarity functions and their applications in the context of the classification task are discussed in Subsection 3.5.

The ability to assess a similarity degree is also useful in an unsupervised data analysis (see Subsection 3.5). For instance, various similarity functions are commonly used by clustering algorithms to form homogeneous groups of objects. Moreover, similarity functions may be more convenient to use for an evaluation of a similarity model, since the implicit verification of a similarity relation approximation may require checking all pairs of objects. More application examples of similarity functions for supervised and unsupervised learning are discussed in Subsection 3.5.

Evaluation of Similarity Models. A similarity relation in a given context can be approximated using many different methods. However, a quality of two different approximations will rarely be the same. In order to be able to select the one which is appropriate for a considered problem there have to be defined some means of measuring a compliance of the approximation with the real similarity relation.

An objective evaluation of similarity assessment is a problem that has always accompanied research on similarity models. Although there have been developed many methods for measuring the quality of a similarity model, the most of them can be grouped into three categories. The main criteria for this division is a required involvement of human experts.

In the first category there are methods which measure compliance of the assessment returned by a model with human-made similarity ratings. Such an approach includes researches in which human subjects are asked to assess the similarity between pairs of objects (called stimuli). Next, those assessments are compared with an output of the tested model and some statistics measuring their correspondence are computed. For instance, Tversky in [5] describes a study in which people were asked about a similarity between particular pairs of countries. As a part of this study, two independent groups of participants had to assess the similarity degrees between the same pairs of countries, but with an inverse ordering (i.e. one group assessed how similar is country A to B, whereas the second judged the similarity of B to A). Based on those ratings, Tversky showed that there is a statistically significant asymmetry in the average similarity judgements within those two groups and used this finding as an argument for viability of his feature contrast model (see Subsection 3.4). In a different study on the similarity of vehicles [5], Tversky measured the correlation between the average assessments made by human subjects and the results of his model. In this way he was able to show that taking into account both common and distinctive features of objects, his model can better fit the data than in a case when those sets of characteristics are considered separately.

The main advantage of this approach is that it allows to directly assess the viability of the tested model to a given problem. Average assessments made by human subjects define the ground truth similarity relation which the model tries to approximate. By using well-defined statistical measures of compliance between two sets of judgements it is possible not only to objectively evaluate the model but also to quantitatively compare it to different models and decide which one is better.

However, such a direct approach has some serious disadvantages. It usually requires a lot of time and resources to gather a meaningful amount of data from human participants. This does not only increase the overall cost of the model but also limits the possible test applications to relatively small data sets. Additionally, it is sometimes difficult to design an environment for manual assessment of the similarity in a desired context. Since there are many factors that can influence human judgement, the similarity ratings obtained in this way can be biased. Due to those practical reasons, usage of this evaluation method is very rare for data sets with more than a few hundreds of stimuli.

The second category of similarity model evaluation methods consists of measures that verify compliance of the tested model with constraints imposed by domain experts. Usually, even when a data set is too large to evaluate similarity degrees between every pair of objects, experts are able to define some rules that must be satisfied by a good similarity model. Such rules may be either very general (e.g. *less complex objects should be more similar to the more complex ones than the opposite*) or very specific (e.g. *object u_1 and u_2 must not be indicated as similar*). The quality of a model is then expressed as a function of a cardinality of a set of violated rules.

Table 3. Summary of typical similarity model evaluation methods

Correlation with average similarity ratings	
Advantages:	Disadvantages:
– direct assessment of a model – simple and intuitive evaluation	– requires human-made ratings – deficiencies in data availability – possibility of a context bias
Measures of compliance with constraints	
Advantages:	Disadvantages:
– semi-direct model assessment – simpler for experts	– requires experts to impose constraints by labelling or grouping – possible inconsistencies
Measures of classification accuracy	
Advantages:	Disadvantages:
– no human involvement required – no limitations on data availability or quality – applicable for large data sets	– indirect model assessment – can be used only in the context of classification

Experts may also provide some feedback regarding truly relevant characteristics of some objects in the considered context. This information can be utilized to heuristically assess the similarity degree of the preselected objects and those values may be used as a reference during the evaluation of similarity models. In a more general setting, this type of quality assessment can be used to measure quality in a semantic clustering task [29] and motivates the semi-supervised clustering algorithms [112]. This approach is used in experiments described in Subsection 6.3 to evaluate the similarity models for scientific articles, constructed in the context of their semantic similarity.

The main advantage of this approach is that it is usually much more convenient for experts to specify constraints rather than indicate exact similarity values. Since such rules may be local and do not need to cover all pairs of objects, they might be applied to evaluate a similarity model on a much larger data. One major drawback is the possible inconsistency within the constraints defined by different experts. Also the evaluation cost which is related to the employment of human experts cannot be neglected.

Finally, the last category consists of methods that can only be applied in the context of classification. Similarity models are often built in order to support decision making or to facilitate a prediction of classes of new objects. If a model is designed specifically for this purpose, it is reasonable to evaluate its performance by measuring the quality of predictions made with a use of similarity-based decision rules (see Definitions 13 and 14). Since the main feature of similarity in a context of classification (Definition 14) imposes a kind of a constraint on desired assessments of similarity, this approach can be seen as a special case of the methods from the second category. However it differs in that, it does not need the involvement of human experts.

The biggest advantage of this approach is the lack of restrictions on evaluation data availability. It makes it possible to automatically test a similarity model even on huge data sets, which makes the evaluation more reliable. Due to those practical reasons this particular method was used in many studies, including [14–16, 23, 26–28]. It was also used in experiments conducted for the purpose of this dissertation which are described in Subsections 6.1 and 6.2. Table 3 summarizes the above discussion on the methods for evaluation of similarity models.

3.4   Commonly Used Similarity Models

This subsection overviews the most commonly used similarity models. The presented approaches differ in the constrains on the way they approximate the similarity relation. For instance, the distance-based models restrict the approximations to relations which are reflexive and symmetric. However, all the models discussed in this subsection have one property in common. They can be used to approximate the similarity in a way that is independent of a particular data domain or a context. For this reason the resulting approximations are often not optimal and expert knowledge is needed to decide whether it is worth to apply a selected model to a given problem.

Distance-Based Similarity Modelling. The most commonly used in practical applications are the distance-based similarity models. A basic intuition behind this approach is that each object from a universe Ω can be mapped to some point in an attribute value vector space. It is assumed that in this space there is a metric defined which allows to assess a distance between any two points. Such a metric will be called a *distance function* or a *distance measure*.

Definition 16 (Distance measure)
Let Ω be a universe of objects and let $Dist : \Omega \times \Omega \to \mathbb{R}^+ \cup \{0\}$ be a non-negative real function. We will say that $Dist$ is a distance measure if the following conditions are met for all $u_1, u_2, u_3 \in \Omega$:

1. $Dist(u_1, u_2) = 0 \Leftrightarrow u_1 = u_2$ *(identity of indiscernibles),*
2. $Dist(u_1, u_2) = Dist(u_2, u_1)$ *(symmetry),*
3. $Dist(u_1, u_2) + Dist(u_2, u_3) \geq Dist(u_1, u_3)$ *(triangle inequality).*

If the objects from Ω are described by attributes from a set A, then the first condition can be generalized by considering the indiscernibility classes of u_1 and u_2: $Dist(u_1, u_2) = 0 \Leftrightarrow (u_1, u_2) \in IND_A$. This particular variation of the distance measure definition will be used in the later sections. Moreover, if a given function does not fulfill the third condition (the triangle inequality) but meets the other two it is called a *semidistance* or a *semimetric*.

A typical example of a distance measure is the Euclidean distance, which is a standard metric in Euclidean spaces:

$$Dist_E(u_1, u_2) = \sqrt{\sum_{a \in A} \big(a(u_1) - a(u_2)\big)^2}. \tag{15}$$

Another example of a useful distance measure is the Manhattan distance:

$$Dist_M(u_1, u_2) = \sum_{a \in A} |a(u_1) - a(u_2)|. \tag{16}$$

Both of the above metrics are generalized by the Minkowski distances, which can be regarded as a parametrized family of distance measures:

$$Dist_p(u_1, u_2) = \left(\sum_{a \in A} |a(u_1) - a(u_2)|^p \right)^{1/p}. \tag{17}$$

Figure 4 presents shapes of circles in spaces with Minkowski metric for different values of the parameter p.

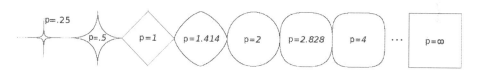

Fig. 4. Shapes of circles in spaces with different Minkowski distances

A different example of an interesting distance function in a $\mathbb{R}^{|A|}$ space is the Canberra distance:

$$Dist_C(u_1, u_2) = \sum_{a \in A} \frac{|a(u_1) - a(u_2)|}{|a(u_1)| + |a(u_2)|}. \tag{18}$$

It is mostly used for data with non-negative attribute values scattered around the centre since it has a property that its value becomes unity when the attributes are of opposite sign.

All the above metrics work only for objects described by numeric attributes. There are however numerous metrics which can be applied to cases with symbolic attributes. The most basic of those is the Hamming distance:

$$Dist_H(u_1, u_2) = |\{a \in A : a(u_1) \neq a(u_2)\}|. \tag{19}$$

Typically, the Hamming distance is used for the assessment of a proximity between binary strings. It can also be utilized for comparison of any equally sized strings, but in such a case the edit distance[5] is more commonly employed, since it allows to compare strings of different length.

Another example of a distance defined for objects with binary attributes is the binary distance:

$$Dist_b(u_1, u_2) = \frac{|\{a \in A : a(u_1) \neq a(u_2)\}|}{|\{a \in A : a(u_1) \neq 0 \vee a(u_2) \neq 0\}|}. \tag{20}$$

---

[5] A value of the edit distance is equal to the minimal number of edit operations needed to transform one string into another. It is often called the Levenshtein distance.

The binary distance can be applied to any type of symbolic data after transformation of each symbolic attribute to its binary representation.

A common choice of a measure in high dimensional numeric spaces is the cosine distance. It measures the angle between two vectors:

$$Dist_{arcc}(u_1, u_2) = \arccos(\cos(u_1, u_2)) \tag{21}$$

$$= \arccos\left(\frac{\sum\limits_{a \in A|} a(u_1) \cdot a(u_2)}{\sqrt{\sum\limits_{a \in A} \left(a(u_1)\right)^2} \cdot \sqrt{\sum\limits_{a \in A} \left(a(u_2)\right)^2}}\right). \tag{22}$$

The cosine between two vectors is equivalent to their scalar product divided by a product of their norms. A distance defined in this way is a proper metric only for points from $(\mathbb{R}^+)^{|A|}$ and lying on a sphere. To avoid computation of the arc-cosine, in applications this distance function is simplified to a form:

$$Dist_c(u_1, u_2) = 1 - \frac{\sum\limits_{a \in A|} a(u_1) \cdot a(u_2)}{\sqrt{\sum\limits_{a \in A} \left(a(u_1)\right)^2} \cdot \sqrt{\sum\limits_{a \in A} \left(a(u_2)\right)^2}}. \tag{23}$$

It needs to be noted, however, that $Dist_c$ is only a semimetric.

The main advantage of the cosine distance is that it can be efficiently computed even for extremely high dimensional but sparse data[6]. In such a case, representations of all objects can be normalized by dividing all attribute values by a norm of the corresponding vectors in Euclidean metric space. After this transformation, the distance can be computed by multiplying only attributes with non-zero values for both points and summing the results. For this reason the cosine distance is commonly used in information retrieval [113, 114] and textual data mining [46, 115, 116].

It can be easily noted that the most of the above distance measures can be seen as a composition of two functions. The first one is applied independently for each of the attributes to measure how different their values are in the compared objects. The second one aggregates those measurements and expresses the final distance. For example, in a case of the Minkowski distance the first function is $f(x, y) = |x - y|$ and the second is $F_p(x_1, \ldots, x_{|A|}) = \left(\sum\limits_{i=1,\ldots,|A|} x_i^p\right)^{1/p}$. Such functions are called a *local distance* and a *global distance*, respectively. It can be shown that a large share of distance measures can be constructed by composing a distance in a one-dimensional space and a norm in a $|A|$-dimensional vector space (a proof of this fact can be found e.g. in [117]). This fact is often called *a local-global principle*.

By applying different local distance types to attributes, it is possible to measure distances between objects described by a mixture of numerical and nominal

---

[6] Sparse data is data with little non-zero attribute values.

features. One example of such a measure is the Gower distance. It uses the absolute value of difference and the equivalence test for numerical and nominal attributes, respectively, and then it aggregates the local distances using the standard Euclidean norm.

In the distance-based approach, a similarity is a non-increasing function of a distance between representations of two objects. The transformation from a distance to a similarity values is usually done using some simple monotonic function such as the linear transform (Equation 24) or the inverse transform (Equation 25). Many other functions, such as common kernels, can also be used.

$$Sim_{lin}(u_1, u_2) = C - Dist(u_1, u_2) \tag{24}$$

$$Sim_{inv}(u_1, u_2) = \frac{1}{Dist(u_1, u_2) + C} \tag{25}$$

In the above equations C is a constant, which is used to place the similarity values into appropriate interval. Some other scaling methods can sometimes be additionally applied to secure that the similarity values stay in a desired range for pairs of objects from a given information system.

The usage of distance measures for computation of a similarity makes the resulting model inherit some of the properties of metrics. For instance, any distance-based approximation of the similarity will always have the property of reflexivity and symmetry, which might be undesirable. Moreover, if a similarity function is based on a globally predefined distance measure, it does not take into account the influence of particular characteristics of objects in a given context and treats all the attributes alike. The distinction between the local and global distances makes it possible to partially overcome this issue by introducing additional parameters which express the importance of the local factors to the global similarity. One example of such similarity measure is based on the generalized Minkowski distance:

$$Dist_{w,p}(u_1, u_2) = \left(\sum_{a \in A} w_a \cdot \left| a(u_1) - a(u_2) \right|^p \right)^{1/p}. \tag{26}$$

In this model, the vector of parameters $w = (w_{a_1}, \ldots, w_{a_{|A|}})$ can be set by domain experts or can be tuned using one of the similarity function learning techniques discussed in Subsection 4.2.

From the fact that any distance-based similarity approximation has to be reflexive, it follows that a distance-based similarity function can be a proper similarity function only in a case when the true similarity is also reflexive. In practical situations the similarity may not have this property. For instance, when it is considered in the context of classification and there are some inconsistencies in the data.

Feature Contrast Model. Although the distance-based similarity models were successfully applied in many domains to support a decision making or to discover groups of related objects (examples of such applications are given in Subsection 3.5), it has been noted that such models are rarely optimal, even if they were chosen by experts. For instance, in [9] the usefulness of classical distance-based measures for a classification task is being questioned for data sets with a high number of attributes. Additionally, a priori given distance-based similarity functions neglect a context for comparison of the objects.

Those observations were confirmed by psychologists studying properties of human perception of similar objects [4,5,20,21]. One of the first researchers who investigated this problem was Amos Tversky. In 1977, influenced by results of his experiments on properties of similarity, he came up with *a contrast model* [5]. He argued that the distance-based approaches are not appropriate for modelling similarity relations due to constraints imposed by the mathematical features of the distance metrics such as the symmetry or subadditivity [5, 11]. Even the assumption about the representation in a multidimensional metric space was contradicted [10, 11, 118].

For instance, the lack of symmetry of a similarity relation is apparent when we consider examples of statements about similarity judgements such as "a son resembles his father" or "an ellipse is similar to a circle". Indeed, the experimental studies conducted by Tversky revealed that people tend to assign a significantly lower similarity scores when the comparison is made the other way around [5]. Moreover, even the reflexivity of the similarity is problematic, since in many situations a probability that an object will be judged by people as similar to itself is different for different objects [5].

In his model of a similarity Tversky proposed that the evaluation of a similarity degree was conducted as a result of a binary features matching process. In this approach, the objects are represented not as points in some metric space but as sets of their meaningful characteristics. Those characteristics should be qualitative rather than quantitative and their selection should take into consideration the context in which the similarity is judged. For example, when comparing cars in a context of their class (see discussion in Subsection 3.2) a relevant feature of a car could be that *its size is moderate* but a feature *its colour is red* probably does not need to be considered.

Tversky also noticed that the similarity between objects depends not only on their common features but also on the features that are considered distinct. Such features may be interpreted as arguments for or against the similarity. He proposed the following formula to evaluate the similarity degree of compared stimuli:

$$Sim_T(x,y) = \theta f(X \cap Y) - \Big(\alpha f(Y \setminus X) + \beta f(X \setminus Y)\Big), \qquad (27)$$

where X and Y are sets of binary characteristics of the instances x, y, f is an interval scale function and the non-negative constants θ, α, β are the parameters. In Tversky's experiments f usually corresponded to the cardinality of a set.

Tversky argued that if the *ideal* similarity function for a given domain *sim* meets certain assumptions[7], there exist values of the parameters θ, α, β and an interval scale f that for any objects a, b, c, d, $Sim_T(a,b) > Sim_T(c,d) \Leftrightarrow sim(a,b) > sim(c,d)$.

Tversky's contrast model is sometimes expressed using a slightly different formula, known as the Tversky index:

$$Sim_T(x,y) = \frac{\theta f(X \cap Y)}{\theta f(X \cap Y) + \alpha f(Y \setminus X) + \beta f(X \setminus Y)}. \tag{28}$$

In this form values of the similarity function are bounded to the interval $[0,1]$. For appropriate values of the θ, α, β parameters and a selection of the interval scale function, this formula generalizes many common similarity functions. For example, if $\theta = \alpha = \beta = 1$ and f corresponds to the cardinality, Tversky index is equivalent to Jaccard similarity coefficient or Jaccard index[8]. When $\theta = 1$ and $\alpha = \beta = 0.5$ the Tversky's formula becomes equivalent to the Dice similarity coefficient.

Depending on the values of θ, α, β the contrast model may have different characteristics, e.g., for $\alpha \neq \beta$ the model is not symmetric. In Formula (27) $\theta f(X \cap Y)$ can be interpreted as corresponding to the strength of arguments for the similarity of x to y, whereas $\alpha f(Y \setminus X) + \beta f(X \setminus Y)$ may be regarded as a strength of arguments against the similarity. Using that model Tversky was able to create similarity rankings of simple objects, such as geometrical figures, which were more consistent with evaluations made by humans than the rankings constructed using standard distance-based similarity functions. Still, it needs to be noted that in those experiments, features to characterise the objects as well as the parameter settings were either chosen manually or they were extracted from results of a survey among volunteers who participated in the study. Although such an approach is suitable to explore small data, it would not be practical to use it for defining relevant features of objects described in large real-life data sets.

It is important to realize that in practical application, the features which can be used to characterize objects in the contrast model are usually on a much higher abstraction level than attributes from typical data sets. This fact makes it difficult to apply Tversky's model for predictive analysis of data represented in information systems. The problem is particularly evident when the analysed data are high dimensional. In such a case, manual construction of the important features is infeasible, even for domain experts.

For instance, microarray data sets contain numerical information about expression levels of tens of thousands genes. Within an information system, each

---

[7] Tversky made assumptions regarding viability of *the feature matching* approach, about *the monotonicity* of *sim* with regard to the common and distinct feature sets, *the independence* of the evaluation with regard to the common and distinct feature sets, *the solvability* of similarity equations and *the invariance* of the impact of particular feature sets on the similarity evaluation [5].

[8] It is easy to notice, that a function $1 - $Jaccard index corresponds to the binary distance discussed in Subsection 3.4.

gene corresponds to a different attribute. For such data, the appropriate features to use for Tversky's model may be interpreted as questions about activity of a particular gene or a group of genes, e.g.: *Are the Cytochrome C related genes overexpressed?* Since there is a huge number of genes and a function of many of them still remains unknown, experts are unable to manually select all the potentially important features of a given data sample. Additionally, there can be exponentially many binary characteristics for a data set and checking which of them can be used to characterize an object would be inefficient computationally. Those are the main motivations for a development of automated feature extraction methods and the similarity learning model which is proposed in Section 5.

Hierarchical and Ontology-Based Similarity Models. Similarity models are often built for very complex objects or processes with a predefined structure [45,119,120]. In such a case, a direct assessment of a similarity can be problematic, because two complex objects are likely to be similar in some aspects but dissimilar in other. Tversky's contrast model tries to overcome this issue by considering higher-level characteristics of objects and separately handling their common and distinctive features.

However, as it was pointed out in Subsection 3.4, typical data stored in information systems contain information only about relatively low-level, mostly numeric attributes. In order to define the higher-level features either domain knowledge or some learning techniques need to be employed. If the first eventuality is possible (i.e., an analyst has access to expert knowledge about the domain of interest), experts can provide description how to transform the attribute values into some more abstract but at the same time more informative characteristics.

For very complex objects a one aggregation step in construction of new features might be insufficient. Different features constructed from basic attributes might be correlated or might still require some generalization before they are able to express some relevant aspect of the similarity in a considered context. In this way, a whole hierarchy of features can be built. Such a structure is sometimes called *a similarity ontology* for a given domain.

Figure 5 shows a similarity ontology constructed for the car example discussed in previous sections. It was constructed for one of the data sets used as a benchmark in experiments described in Section 6 (i.e., the *Cars93* data). In this particular context (a class of a car) the similarity between two cars can be considered in aspects such as capacity, driving parameters, economy, size and value. Those local similarities can be aggregated to neatly express the global similarity, however the aggregation needs to be different for objects from different decision classes. For instance, the size aspect may be more important when assessing the similarity to a car from the *Full-size* class than in a case when the comparison is made to a *Sporty* car.

For this reason, in the hierarchical approach to approximating the similarity relation experts are required to provide local similarity functions and class-dependent aggregation rules. In this way the experts can give the model desirable

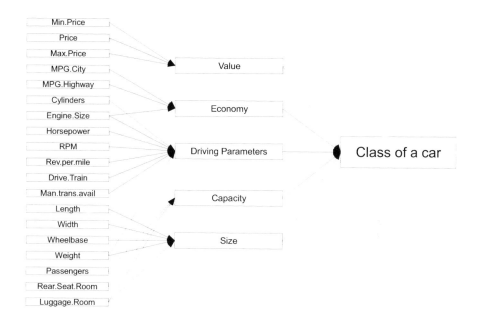

Fig. 5. An exemplary similarity ontology for comparison of cars in the context of their type

properties. For example, even if only very simple distance-base local similarities are used for computation of the similarity in each of the aspects, the resulting model can still be not symmetric.

Some experiments with a hierarchical similarity models are described in [23, 117]. This approach was also successfully used for case-based prediction of a treatment plan for infants with respiratory failure [45,120]. In that study, expert knowledge was combined with supervised learning techniques to assess the similarity of new cases in different aspects or abstraction levels. The incorporation of medical doctors into the model building process helped to handle the temporal aspect of data and made the results more intuitive for potential users.

One major drawback of the hierarchical similarity modelling is that it is extensively dependent on availability of domain knowledge. In the most of complex problems such knowledge is not easily obtainable. Additionally, the construction of a similarity ontology requires a significant effort from domain experts, which makes the model expensive. On the other hand, due to a vague and often abstract nature of the higher-level features which can influence human judgements of similarity, some expert guidance seems inevitable. Due to this fact, in practical applications the expert involvement needs to be balanced with automatic methods for learning the similarity from data.

3.5   Similarity in Machine Learning

Similarity models play an important role among the machine learning techniques. Their application ranges from supervised classification and regression problems to automatic planning and an unsupervised cluster analysis. In this subsection, three major application areas of similarity models are discussed. They correspond to similarity-based classification models, case-base reasoning framework and clustering algorithms, respectively. Although the presented list of examples is by no means complete, it shows how useful in practice is the ability to reliably assess the similarity between objects.

Similarity in Predictive Data Analysis and Visualization. One of the most common application areas of the similarity modelling is the classification task. Models of similarity in this context can actually be constructed for two reasons. The first and obvious one is to facilitate classification of new, previously unseen objects, based on available data stored in an information system.

The most recognized similarity-based classification algorithm is the k-nearest neighbours [7, 14, 17, 110, 111]. It is an example of a lazy classification method which does not have a learning phase. Instead, for a given test case, it uses a predefined similarity measure to construct a ranking of the k most similar objects from a training data base (the neighbours). In the classical approach the measure is based on the Euclidean distance. The decision class of the tested object is chosen based on classes of the neighbours using some voting scheme [13, 110, 111]. This approach can be seen as an extension of the simplest similarity-based classification rule (Definition 13). It can be generalized even further by, for example, considering the exact similarity function values during the voting or assigning weights to training objects that express their representativeness for the decision class. The k-nearest neighbours algorithm can also be used to predict values of a numeric decision attribute (regression) or to perform a multi-label classification [32]. However, in all those applications the correct selection of a similarity model is the factor that has the biggest influence on the quality of predictions.

The models of a similarity in the classification context may also be constructed for a different purpose. The information about relations between objects from an investigated universe is sometimes as important as the ability to classify new cases. It can be used, for instance, to construct meaningful visualizations of various types of data [121, 122]. Such visualizations can be obtained by changing the representation of objects from original attributes to similarities. It allows to display the objects in a graph structure or a low-dimensional metric space. Such a technique is called *multidimensional scaling* (MDS) [118, 123].

Changing the representation of objects may also be regarded as a preprocessing step in a more complex data analysis process. For example, similarity degrees to some preselected cases can serve as new features. Such a feature extraction method (see [88]) can significantly improve classification results of common machine learning algorithms [124]. To make it possible, a proper selection of the reference objects is essential. One way of doing this requires a selection of a single object from each decision class, such that its average similarity to other objects from its class is the highest. Another possibility is the random embedding technique [125] in which the reference objects are chosen randomly and the quality of the selection is often verified on separate validation data.

Case-Based Reasoning Framework. The similarity-based classification can be discussed in a more general framework of algorithmic problem solving. Case-based reasoning is an example of a computational model which can be used to support complex decision making. It evolved from a model of dynamic memory proposed by Roger Schank [2] and is related to the prototype theory in cognitive science [1, 3].

A case-based reasoning model relies on an assumption that *similar problems*, also called cases, should have *similar solutions*. It is an analogy to the everyday human problem solving process. For example, students who prepare for a math exam usually solve exercises and learn proofs of important theorems, which helps them in solving new exercises during the test. The reasoning based on previous experience is also noticeable in work of skilled professionals. For instance, medical doctors diagnose a condition of a patient based on their experience with other patients with similar symptoms. When they propose a treatment, they need to be aware of any past cases in which such a therapy had an undesired effect.

In a typical case-based reasoning approach, each decision making or a problem solving process can be seen as a cycle consisting of four phases [6] (see Figure 6). In the first phase, called *retrieve*, a description of a new problem (case) is compared with descriptions stored in an available knowledge base and the matching cases are retrieved. In the second phase, called *reuse*, solutions (or decisions) associated with the retrieved cases are combined to create a solution for the new problem. Then, in the *revise* phase, the solution is confronted with the real-life and some feedback on its quality is gathered. Lastly, in the *retain* phase, a decision is made whether the new case together with the revised solution are worth to be remembered in the knowledge base. If so, the update is made and the new example extends the system.

The notion of similarity is crucial in every phase of the CBR cycle. The cases which are to be retrieved are selected based on their similarity degree to the new case. Often, it is required that those cases were not only highly similar to the reference object but that they were also maximally dissimilar to each other [6, 107]. In the reuse phase, the similarity degrees may be incorporated into the construction of the new solution, for example, as weights during a voting. Additionally, information about similarities between solutions associated with the selected cases may be taken into account during the construction of new ones. Next, during the revision of the proposed solution, its similarity to the truly optimal one needs to be measured, in order to assess an overall quality of the given CBR system and to find out what needs to be improved in the future. Finally, when the corrected solution to the tested case is ready, its similarity degrees to the cases from the knowledge base can be utilized again to decide whether to save the new case or not.

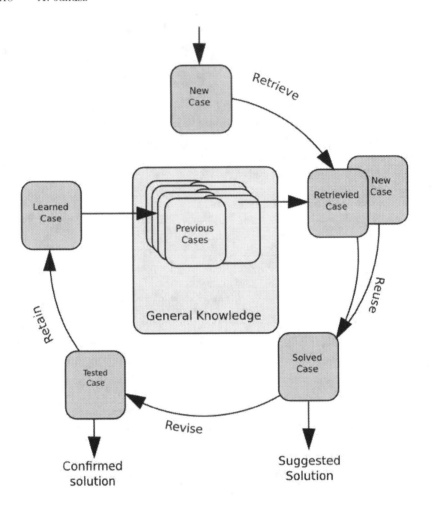

Fig. 6. A full Case-Based Reasoning cycle (based on a schema from [6])

It is worth mentioning that the classical k-NN algorithm can be seen as a very basic CBR model [16], hence the similarity in a context of classification plays a special role in the case-based modelling. However, case-based reasoning may be used for solving much more complex problems than a simple classification, such as treatment planning or recognition of behavioural patterns [45, 119, 120]. The rough set theory has proven to be very useful for construction of CBR systems dedicated to complex problem solving [57, 60, 126].

Similarity in Cluster Analysis. The concept of similarity is also used for solving problems related with unsupervised learning. One example of such a task is clustering of objects into homogeneous groups [7, 8, 127].

In the clustering task the similarity can be used for two reasons. Since homogeneity of a cluster corresponds to the similarity between its members, similarity

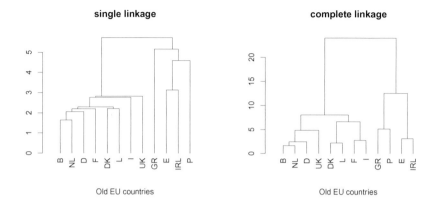

Fig. 7. An example of two clustering trees computed for the *agriculture* data set using the *agnes* algorithm [127] with the *single* (on the left) and *complete* (on the right) linkage functions

measures are used by clustering algorithms to partition objects into groups. The most representative example of such an algorithm is *k-means* [112, 127].

In the classical version of *k-means* objects are treated as points in the Euclidean space and the similarity between points is identified with their proximity. However, the algorithm can be easily modified to use any distance-based similarity function. A pseudo code of such a modification of *k-means*, called *k-centroids*, is given below (Algorithm 1).

In typical implementations of *k-means*, when the similarity function is a linear function of Euclidean distance between points, the selection of new cluster centres is trivial. Coordinates of the new centres are equal to mean coordinates of the corresponding cluster members. However, if some non-standard similarity functions are used, the computation of the new centres requires solving an optimization problem and may become much more complex. Therefore, in many cases it is more convenient to use the *k-medoids* [127] algorithm which restricts the set of possible cluster centres to actual members of the group. This algorithm is also known to be more robust than *k-means* since it is not biased by outliers in the data [127, 128].

In the context of the clustering, it is also possible to consider a similarity between groups of objects (clusters). This notion is especially important for algorithms that construct a hierarchy of clusters. So-called *hierarchical clustering* methods, instead of dividing the objects into a fixed number of groups, compute a series of nested partitions with a number of groups ranging from 1 (all objects are in the same group) to the total number of objects in a data set (every object is a separate group). Figure 7 shows an example of two clustering trees computed for the *agriculture*[9] data set.

---

[9] This data set describes a relation between a percentage of the population working in agriculture to Gross National Product (GNP) per capita in the old EU countries (in 1993).

Algorithm 1. The *k-centroids* algorithm.

Input: an information system $\mathbb{S} = (U, A)$;
  a desired number of clusters k;
  a similarity function $Sim : U \times U \to \mathbb{R}$;

Output: a grouping vector $g = (g_1, \ldots, g_{|U|})$, where $g_i \in \{1, \ldots, k\}$;

1 **begin**
2 | $endFlag = F$;
3 | Randomly select k initial cluster centres c_j, $j = 1, \ldots k$;
4 | **while** $endFlag == F$ **do**
5 | | Assign each $u \in U$ to the nearest (most similar) cluster centre:
6 | | **for** $i \in \{1, \ldots, |U|\}$ **do**
7 | | | $g_i = \arg\max_j \big(Sim(u_i, c_j)\big)$;
8 | | **end**
9 | | Compute new cluster centres $\bar{c}_j$, $j = 1, \ldots k$, such that $\sum_{i:g_i=j} \big(Sim(u_i, \bar{c}_j)\big)$ is minimal;
10 | | **if** $\forall_{j \in \{1, \ldots, k\}}(\bar{c}_j == c_j)$ **then**
11 | | | $endFlag = T$;
12 | | **end**
13 | | **else**
14 | | | **for** $j \in \{1, \ldots, k\}$ **do**
15 | | | | $c_j = \bar{c}_j$;
16 | | | **end**
17 | | **end**
18 | **end**
19 | **return** $g = (g_1, \ldots, g_{|U|})$;
20 **end**

In the agglomerative approach to hierarchical clustering, at each iteration of the algorithm two most similar groups are merged into a larger one (the bottom-up approach). There can be many ways to estimate the similarity between two clusters. Typically, it is done using some *linkage function*. The most commonly used linkage functions are *single linkage*, *average linkage* and *complete linkage* [8, 127]. They estimate the similarity between two groups by, respectively, the maximum, average and minimum from similarities between pairs of objects, such that one object is in the first group and the other is in the second.

The second reason for using the similarity in the clustering task is related to the problem of evaluation of a clustering quality. This issue can be seen as a complement to the evaluation of similarity measures, which was discussed in Subsection 3.3. Given reference values of similarity degrees between pairs of considered objects (for instance by domain experts) it is possible to assess the semantic homogeneity of a grouping. It can be done, for example, by using a function that is normally employed as an *internal* clustering quality measure[10]

---

[10] A clustering quality measure is called internal if its value is based solely on the data that were used by the clustering algorithm.

but with the reference similarities as an input. Such an approach is utilized in experiments described in [29], as well as those presented in Subsection 6.3 of this dissertation.

4   Similarity Relation Learning Methods

The notion of similarity, discussed in the previous section, is a complex concept whose properties are subjective in nature and strongly depend on a context in which they are considered [5,10,11,20,118]. Due to this complexity, it is extremely difficult to model the similarity based only on an intuition and general knowledge about a domain of interest (see the discussion in Subsection 3.1). For decades this fact has motivated research on methods which would allow to approximate a similarity relation or to estimate values of a similarity function, using additional samples of data.

Many of the similarity learning methods concentrate on tuning parameters of some a priori given (e.g. by an expert) similarity functions. This approach is most noticeably present in the distance-based similarity modelling where the similarity function is monotonically dependent on a distance between representations of objects in an information system (for more details see Subsection 3.4). Distance measures can usually be constructed using the *local-global principle* [15, 16, 117] which divides the calculation of the distances into two phases – local, in which objects are compared separately on each of their attributes, and global, in which the results of comparisons are aggregated. This separation of the local and the global distance computation allows to conveniently parametrize the function with weights assigned to the local distances. Using available data and reference similarity values, those weights can be tuned in order to better fit the resulting similarity model to the given task.

Although the distance-based models for learning the similarity relation are predominant, they are not free from shortcomings. These defects are in a large part due to the usage of distance-based similarity function which can be inappropriate for modelling the similarity in a given context (see discussion in Subsections 3.1, 3.2 and 3.4). Additionally, such an approach usually fails to capture higher-level characteristics of objects and their impact on the similarity relation. These limitations often lead to approximations of the similarity which are not consistent with human perception [5, 21].

To construct an approximation of the similarity which would truly mimic judgements of human beings it is necessary to go a step further than just relying on lower-level sensory data. The similarity learning process needs to support extraction of new higher-level characteristics of objects that might be important in the considered context. Since such abstract features are likely to correspond to vague concepts, some approximate reasoning methods need to be used in order to identify their occurrence in the objects. Additionally, the aggregation of local similarities also needs to be dependent on data and should not enforce any specific algebraical properties on the approximated relation.

In this section, a flexible model for learning the similarity relation from data is proposed (in Section 5). This model, called Rule-Based Similarity (RBS), aims at overcoming the issues related with the distance-based approaches. As a foundation, it uses Tversky's feature contrast model (Subsection 3.4). However, unlike the feature contrast model, it utilizes the rough set theory to automatically extract higher-level features of objects which are relevant for the assessment of the similarity and to estimate their importance. In the RBS model the aggregation of the similarities in local aspects is based on available data and takes into consideration dependencies between individual features. The flexibility of this model allows to apply it in a wide range of domains, including those in which objects are characterised by a huge number of attributes.

In the subsequent subsections some basic examples of similarity learning models are discussed. Subsection 4.1 explains the problem of similarity learning and points out desirable properties of a good similarity learning method. Subsection 4.2 is an overview of several approaches to similarity learning which mostly focus on tuning distance-based similarity functions. They utilize different techniques, such as attribute rankings, genetic algorithms or optimization heuristics, to select important attributes or to assign weights that express their relevance. On the other hand, Section 5 introduces the notion of Rule-Based Similarity whose focus is on constructing higher-level features of objects which are more suitable for expressing the similarity. Apart from explaining the motivation for the RBS model and its general construction scheme, some specialized modifications are presented. They adapt the model to tasks such as working with high dimensional data or learning a similarity function from textual data in an unsupervised manner.

4.1 Problem Statement

Similarity learning can be defined as a process of tuning a predefined similarity model or constructing a new one using available data. This task is often considered as a middle step in other data analysis assignments. If the main purpose for approximating a similarity relation is to better predict decision classes of new objects, facilitate planning of artificial agent actions or to divide a set of documents into semantically homogeneous groups, the resulting similarity model should help in obtaining better results than a typical baseline. Ideally, a process of learning the similarity should be characterised by a set features which indicate its practical usefulness.

The set of desirable similarity learning method properties include:

1. *Consistence with available data.*

   An ability to fit a similarity model to available data is the most fundamental feature of a similarity learning technique. It directly corresponds to an intuitive expectation that a trained model should be more likely to produce acceptable similarity evaluations than an a priori given one. An outcome of a perfect method should always be a proper similarity function (see Definition 15), regardless of a data set. Moreover, this property should hold even for new objects that were not available for learning. Unfortunately, such a perfect method does not exist. A good similarity learning model, however, should aim to fulfil this intuition and be consistent with available data.

2. *Consistence with a context of the similarity that is appropriate for a given task.*

   A trained similarity model should also be consistent with a given context. Hence, if the context is imposed by, for example, a classification task, the resulting similarity model should be more useful for assigning decision classes of new objects using one of the similarity-based decision rules (see Definitions 13 and 14 in Subsection 3.3) than the baseline. The verification of the precision of such a classifier can be treated as a good similarity learning evaluation method [16]. This particular approach is used in the experiments described in the next section.

3. *Ability to take into consideration an influence of objects from the data on evaluation of the similarity.*

   As it was mentioned in Subsection 3.2, similarity between two given objects often depends on the presence of other objects which are considered as a kind of a reference for comparison. Similarity learning methods that are able to construct a similarity model capable of capturing such a dependence are justifiable from the psychological point of view and are more likely to produce intuitive results [4, 21].

4. *Compliance with psychological intuitions (e.g. regarding object representations).*

   Another desirable property of similarity learning models is also related to intuitiveness of the resulting similarity evaluations. Assessments of similarity obtained using constructed similarity function should be comprehensible for domain experts. One way of ensuring this is to express the similarity in terms of meaningful higher-level features of objects. Such features can be extracted from data using standard feature extraction methods [88] as well as with specialized methods such as decision rules [22, 25, 27] or semantic text indexing tools [40, 129]. Not only can higher-level features help in capturing aspects of similarity that are difficult to grasp from lower-level sensory data but may also be used as a basis for a set representation of objects [4, 5]. Such representation can be more natural for objects that are difficult to represent in a metric space [5, 20, 21]. Moreover, by working with higher-level features the similarity evaluation can be associated with resolving conflicts between arguments for and against the similarity of given objects. Such an approach is usually more intuitive for human experts.

5. *Robustness for complex object domains (e.g. high dimensional data).*

   A good similarity learning method should be general enough to be possible to apply in many object domains. Usually, a similarity model is efficient for some data types while for others it yields unreliable results. The usage of a similarity learning technique for tuning parameters of a model can greatly extend the range of suitable data types. However, applications of a similarity learning method may also be confined. For instance, models with multiple parameters are more vulnerable to overfitting when there is a limited number of available instances in data (e.g. the few-objects-many-attributes problem [9, 130]).

6. *Computational feasibility.*

The last of the considered properties regards computational complexity of the similarity learning model. The complexity of a model can be considered in several aspects. A similarity learning method needs to be computationally feasible with regard to the size of a training data set, understood in terms of both, the number of available objects and the number of attributes. Either of those two sides can be more important in specific situations. Many models, however, are efficient in one of the aspects and inefficient in the other. The scalability of a similarity learning model often determines its practical usefulness.

Any similarity learning model can be evaluated with regard to the above characteristics. In particular, Rule-Based Similarity described in Section 5 was designed to possess all those properties.

4.2   Examples of Similarity Learning Models

The problem of similarity learning was investigated by many researchers from the field of data analysis [12, 15, 16, 18, 24, 61, 126, 131, 132]. In the applications discussed in Subsection 3.5, similarity functions which can be employed for a particular task can be adjusted to better fit the considered problem. The main aim of such an adjustment is to improve effectiveness of the algorithms which make use of the notion of similarity.

The commonly used similarity models (e.g. the distance-based models – see Subsection 3.4) neglect the context for similarity. However, the vast majority of similarity learning methods incorporate this context into the resulting model by, e.g., considering feedback from experts or by guiding the learning process using evaluations of the quality of the model computed on training data. Thus, in a typical case, the similarity learning can be regarded as a way of adaptation of a predefined similarity model to the context which is determined by a given task.

The process of learning the similarity relation may sometimes be seen as a supervised learning task. This is especially true when it can be described as a procedure in which an omniscient oracle is queried about similarities between selected pairs of objects to construct a decision system for an arbitrary classification algorithm [61, 126]. However, in many cases, direct assessments of the degrees of similarity which can be used as a reference are not available. In that situation, domain knowledge or some more general properties of the similarity in the considered context have to be used to guide the construction of the model. One example of such a property is stated in Definition 14. Since the later approach can be regarded as more practical, the following examples show similarity learning methods mainly designed to work in such a setting.

Feature Extraction and Attribute Ranking Methods. One of the most general methods for learning a similarity relation is to adjust the corresponding similarity function to a given data set by assigning weights, selecting relevant attributes or constructing new, more informative ones. Such weights can be used in combination with standard generalizations of similarity functions (e.g. a measure based on the generalized Minkowsky distance) to express the importance of particular local similarities.

Research on attribute selection techniques has always been in a scope of interest of the machine learning and data mining communities [81,82,89,90]. The dimensionality reduction allows to decrease the amount of computational resources needed for execution of complex analysis and very often leads to better quality of the final results [87,88,132]. The selection of a small number of meaningful features also enables better visualizations and can be crucial for human experts who want to gain insight into the data.

Feature selection methods can be categorised in several ways. One of those is the distinction between supervised and unsupervised algorithms. The unsupervised methods focus on measuring variability and internal dependencies of attributes in data. As an example of such a method one can give Principle Component Analysis [133] in which the representation of data is changed from original attributes to their representation in a space of eigenvectors, computed by eigenvalue decomposition of an attribute correlation matrix. The supervised methods information about decision classes to assess the relevance of particular attributes. They can be further divided into three categories, i.e. filter, wrapper and embedded methods [88,95,134].

The filter methods create rankings of individual features or feature subsets based on some predefined scoring function. Ranking algorithms can be divided into univariate and multivariate. The univariate rankers evaluate importance of individual attributes without taking into consideration dependencies between them. A rationale behind this approach is that a quality of an attribute should be related to its ability to discern objects from different decision classes. As an example of frequently used univariate algorithms one can give a simple correlation-based ranker [135], statistical tests [136] or rankers based on mutual information measure [137]. The multivariate attribute rankers try to assess the relevance in a context of other features. They explore dependencies among features by testing their usefulness in groups (e.g. the relief algorithm [138]) or by explicitly measuring relateness of pairs of attributes and applying the minimum-redundancy-maximum-relevance framework [137,139]. Another worth-noticing example of a multivariate feature ranker is the Breiman's relevance measure. It expresses the average increase of a classification error resulting from randomization of attributes that were used during construction of trees by the Random Forest algorithm [140,141].

In the second approach, subsets of features are ranked based on the performance of a predictive model constructed using those features. Attributes from the subset which achieved the highest score are selected. Usually, the same model is used for choosing the best feature set and making predictions for test data, because different classifiers may produce their best results using different features. Due to the fact that a number of all possible subsets of attributes is exponentially large, different heuristics are being used to search the attribute space. The most common heuristics include top-down search [142], bottom-up search [143] and Monte Carlo heuristics such as simulated annealing or genetic algorithms [144]. Although usually the wrapper approach yields better results than the filter approach, its computational complexity makes it difficult to apply for extremely high dimensional data.

Table 4. Summary of attribute selection methods

Filter methods:	Wrapper methods:	Embedded methods:
– attributes or attribute subsets receive scores based on some predefined statistic, – scores of individual attributes can be used as weights, – top ranked features can be selected as relevant.	– learning algorithms are evaluated on subsets of attributes, – many different subset generation techniques can be used, – the best subset is selected.	– feature selection can be an integral part of a learning algorithm, – irrelevant attributes may be neglected, – some new features may be constructed (internal feature extraction).

The embedded methods are integral parts of some learning algorithm. For instance, classifiers such as Support Vector Machine (SVM) [145, 146] can work in a space of higher dimensionality than the original data applying the kernel trick [147]. Moreover, efficient implementations of classifiers such as Artificial Neural Networks (ANN) [148, 149] automatically drop dimensions from the data representation if their impact on the solution falls below a predefined threshold.

The application of an attribute selection or ranking algorithm for learning a similarity may be dependent on its context. Practically all typical supervised feature selection algorithms can be employed for tuning a similarity function if the similarity is considered in the classification context. It is a consequence of the main feature of similarity in that context (see Definition 14). If the similarity needs to be consistent with decision classes, the more discriminative attributes are likely to be relevant in a similarity judgement. However, in the case of similarity in the context of "general appearance" unsupervised feature extraction methods need to be used.

Genetic Approaches. Genetic algorithms (GA) [150] are another popular tool for learning the parameters of similarity functions which are constructed using the *local-global* principle. The idea of GA was inspired by evolution process of living beings. In this approach parameters of the local similarities (e.g. their weights) and the aggregation function are treated as genes and are arranged into genotypes of the genome (also called chromosomes). In this nomenclature, the similarity learning process corresponds to searching for the most adapted genotype. The adaptation of a genotype to a given problem is measured using a fitness function. Since in applications to similarity learning a proper fitness function needs to be based on a similarity model evaluation method, such as those discussed in Subsection 3.3, the GA-based similarity learning may be regarded as a special case of the wrapper attribute ranking approach (see Subsection 4.2). However, the flexibility of GA makes it particularly popular among researchers from the case-base reasoning field [16, 131, 132].

In GA searching for the most adapted genotype is iterative. In each iteration, which is also called a life-cycle or a generation, chromosomes undergo four genetic operations, namely the replication (inheritance), mutation, crossover and elimination (selection).

1. Replication – a selected portion of genotypes survives the cycle and is carried out to the next one.
2. Mutation – a part of genotypes is carried out to the next generation with a randomly modified subset of genes.
3. Cross-over – some portion of genotypes exchange a part of their genetic code and generate new genotypes.
4. Elimination – a part of genotypes that were taken to the new population is removed based on their values of a fitness function.

Figure 8 presents a schema of an exemplary genetic optimization process (a genetic life-cycle). Initially, a random population of genotypes is generated, with each genotype coding a set of parameters of a similarity function that is being tuned. The genetic operations are repeatedly applied to consecutive populations until stop criteria of the algorithm are met.

Exact algorithms for performing the genetic operations may vary in different implementations of GA. However typically, the selection of genotypes to undergo the replication, mutation and crossover is non-deterministic. It usually depends on scores assigned by the fitness function. A common technique for selecting genotypes is called the roulette wheel selection – every member of a population receives a certain probability of being selected and the genotypes are chosen randomly from the resulting probability distribution. The selection of genotypes for different genetic operations is done independently, which means that a single genotype may undergo a few different genetic operations. It can also be chosen several times for the same operation type.

During the replication, selected genotypes are copied unchanged to the next generation. The mutation usually involves random selection of a relatively small subset of genes, which are then slightly modified and the resulting genotype is taken to the next cycle. The crossover operation is usually the most complex. Its simplest exemplary implementation may consist of swapping randomly selected genes between two genotypes. If all parameters of a similarity function are numeric, it may also be realized by computing two weighted averages of the

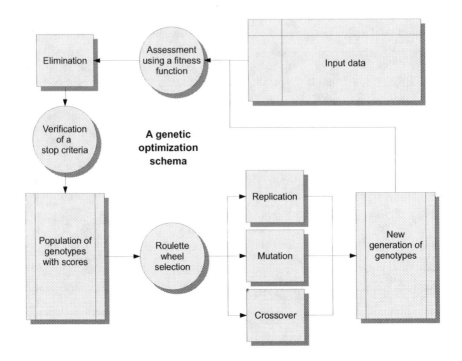

Fig. 8. A schema of a single genetic life-cycle

parent genotypes. One which gives more weight to genes from one parent and the second giving a higher weight to the other one. A more complicated variants of the crossover may include the construction of completely new features that are used for measuring the similarity or to define new local similarity measures [16]. The elimination of genotypes is performed to maintain a desired size of a population. Most commonly, genotypes with the lowest values of a fitness function are removed before starting a new life-cycle. In some implementations however, this last phase of the cycle is also done in a non-deterministic manner, using techniques such as the roulette wheel selection.

The most computationally intensive part of GA is the quality evaluation of genotypes belonging to the population. In the context of the similarity learning, the fitness function needs to assess a quality of a similarity function with parameters corresponding to each genotype in the population. Those assessments are usually performed using one of the methods discussed in Subsection 3.3 and they require a comparison of similarity values returned by the tested models on a training data set with some reference.

Relational Patterns Learning. A different model for learning the similarity relation from data was proposed among the relational patterns learning methods [15]. This approach also employs the *local-global principle* for defining approximations of the similarity. However, it differs from the previously discussed methods in that it tries to directly approximate the similarity in a local distance vector space. The learning in the context of classification is done through optimizing a set of parameters for an a priori given family of similarity approximations. Since the usage of distance-based local similarities enforces reflexivity and symmetry on the resulting approximation, this approach is highly related to searching for optimal approximations in tolerance approximation spaces [55, 56, 151].

Table 5. An exemplary data set describing a content of two vitamins in apples and pears (the data were taken from [15])

Vitamin A	Vitamin C	Fruit	Vitamin A	Vitamin C	Fruit
1.0	0.6	Apple	2.0	0.7	Pear
1.75	0.4	Apple	2.0	1.1	Pear
1.3	0.1	Apple	1.9	0.95	Pear
0.8	0.2	Apple	2.0	0.95	Pear
1.1	0.7	Apple	2.3	1.2	Pear
1.3	0.6	Apple	2.5	1.15	Pear
0.9	0.5	Apple	2.7	1.0	Pear
1.6	0.6	Apple	2.9	1.1	Pear
1.4	0.15	Apple	2.8	0.9	Pear
1.0	0.1	Apple	3.0	1.05	Pear

The first step in relational patterns learning algorithms is a transformation of data from an attribute value vector space into a local distance (or similarity) vector space. This process for an exemplary *fruit* data set (Table 5) taken from [15] is depicted on Figure 9.

For each pair of objects from a decision system $\mathbb{S}_d = (U, A \cup \{d\})$, all their local distances (or sometimes local similarities) are computed. Those values are used to represent the pairs in a new metric space, whose dimensionality is the same as the total number of original attributes. A new binary decision attribute is also constructed. It indicates whether the both of objects from the corresponding pair belong to the same decision class of original data. This new data representation can serve as an approximation space (see Subsection 2.2) for learning the similarity in the context of classification (see Definition 14).

The selection of the most suitable approximation from a given family is performed by searching for parameters that maximize the number of pairs with the same decision (white squares in the right plot of Figure 9) included in the approximation while maintaining the constraints resulting from Definition 14. One example of a family of approximations is the *parametrized conjunction* in a form:

$$(x, y) \in \tau_1^*(\epsilon_1, \ldots, \epsilon_{|A|}) \Leftrightarrow \bigwedge_{a_i \in A} \left[f_{a_i}(x, y) < \epsilon_i \right], \qquad (29)$$

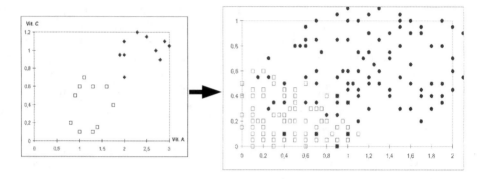

Fig. 9. An example of a transformation of the fruit data set (Table 5) from the attribute value vector space to the local distance vector space. In the plot on the left, the white squares correspond to apples and the black diamonds represent pears. In the plot on the right, the squares correspond to pairs of instances representing the same fruit, whereas the black circles are pairs of different fruits (an apple and a pear).

where A is a set of all conditional attributes, f_{a_i} are local distance functions and ϵ_i are parameters of the family. A different example of a useful family of approximations the *parametrized linear combination* form:

$$(x, y) \in \tau_2^*(w_0, \ldots, w_{|A|}) \Leftrightarrow \sum_{i=1}^{|A|} \left[w_i \cdot f_{a_i}(x, y) \right] + w_0 < 0, \qquad (30)$$

where w_i, $i = 0, \ldots, |A|$ are parameters. Those two families of approximations differ in geometrical interpretations of neighbourhoods they assign to the investigated objects. In the first one, the captured similar objects need to be in a rectangular-shaped area, whereas in the case of the second family, the neighbourhoods are diamond-like. Several heuristics for learning semi-optimal sets of parameters for different families of similarity relation approximations are shown in [15].

One disadvantage of the relational pattern learning approach is its computational complexity. The transformation of the original decision system into a local distance vector space requires $O(n^2)$ storage and computation time (n is the number of objects in the decision system). In order to avoid such a big computational cost, the transformation may be virtualized, i.e. it may not be physically performed but instead the computation of local distances and new decision value may be done "on demand" by the learning heuristic. Although such a solution decreases the space complexity, it usually leads to a significant increase in the time complexity of the method.

Another solution is to approximate similarity only to a selected small subset of objects from the decision system. This technique, called *local approximation* of the similarity, might be efficient, especially when it is possible to distinguish a small group of objects which are representative for the whole data.

Explicit Semantic Analysis. Many similarity learning models were proposed specifically to approximate a semantic similarity in corpora of textual data [115, 116, 129, 152]. One of the most successful approaches aims at improving a representation of documents, so that it better reflected a true meaning of the texts [129, 152]. With the new representation, the similarity of two documents is estimated using standard similarity measures, such as those described in Subsection 3.4.

One particularly interesting example of such a method is Explicit Semantic Analysis (ESA), proposed in [129]. It is based on an assumption that any document can be represented by predefined concepts which are related to the information that it carries (its semantics). Those concepts can be then treated as semantic features of documents. The process of choosing concepts that describe documents in the most meaningful way can be regarded as a feature extraction task [88].

In the ESA approach, natural language definitions of concepts from an external knowledge base, such as an encyclopaedia or an ontology, are matched against documents to find the best associations. A scope of the knowledge base may be general (like in the case of Wikipedia) or it may be focused on a domain related to the investigated text corpus, e.g. Medical Subject Headings (MeSH)[11] [153]. The knowledge base may contain some additional information on relations between concepts, which can be utilized during computation of the "concept-document" association indicators. Otherwise, it is regarded as a regular collection of texts with each concept definition treated as a separate document.

The associations between concepts from a knowledge base and documents from a corpus are treated as indicators of their relatedness. They are computed two-fold. First, after the initial preprocessing (stemming, stop words removal, identification of relevant terms), the corpus and the concept definitions are converted into the *bag-of-words* representation. Each of the unique terms in the texts is given a set of weights which express its association strength to different concepts.

Assume that after the initial processing of a corpus consisting of M documents, $D = \{T_1, \ldots, T_M\}$, there have been identified N unique terms (e.g. words, stems, N-grams) $w_1, \ldots, w_N$. Any text T_i in the corpus D can be represented by a vector $\langle v_1, \ldots, v_N \rangle \in \mathbb{R}_+^N$, where each coordinate $v_j(T_i)$ expresses a value of some relatedness measure for j-th term in the vocabulary (w_j) relative to this document. The most common measure for calculating $v_j(T_i)$ is the *tf-idf* (term frequency-inverse document frequency) index (see [115]) defined as:

$$v_j(T_i) = tf_{i,j} \cdot idf_j = \frac{n_{i,j}}{\sum_{k=1}^{N} n_{i,k}} \cdot \log \left(\frac{M}{|\{i : n_{i,j} \neq 0\}|} \right), \tag{31}$$

where $n_{i,j}$ is the number of occurrences of the term w_j in the document T_i.

---

[11] MeSH is a controlled vocabulary and thesaurus created and maintained by the United States National Library of Medicine. It is used to facilitate searching in life sciences related article databases.

In the second step, the bag-of-words representation of concept definitions is transformed to an inverted index which maps words into lists of K concepts, $c_1, \ldots, c_K$, described in a knowledge base. The inverted index is then used to perform a semantic interpretation of documents from the corpus. For each text, the semantic interpreter iterates over words that it contains, retrieves corresponding entries from the inverted index and merges them into a vector of concept weights (association strengths) that represent a given text.

Let $W_i = \langle v_j \rangle_{j=1}^N$ be a bag-of-words representation of an input text T, where v_j is a numerical weight of a word w_j expressing its association to the text T_i (e.g. its tf-idf). Let $inv_{j,k}$ be an inverted index entry for w_j, where $inv_{j,k}$ quantifies the strength of association of the term w_j with a knowledge base concept c_k, $k \in \{1, \ldots, K\}$. The new vector representation of T_i, called a *bag-of-concepts*, will be denoted by $C_i = (c_1(T_i), \ldots, c_K(T_i))$, where

$$c_k(T_i) = \sum_{j: w_j \in T_i} v_j \cdot inv_{j,k} \tag{32}$$

is a numerical association strength of k-th concept to the document T_i. In Subsection 6.3, texts will also be represented as a set of concepts with sufficiently high association level denoted by $F_i = \{f_k : c_k(T_i) \geq minAssoc_k\}$. Those concepts will be treated as binary semantic features of texts, such as those which are utilized by Tversky's contrast model [5] (see Subsection 3.4).

The representation of texts by sets of features can be easily transformed into an information system $\mathbb{S} = (D, F)$, where $F = \bigcup_{i=1}^{|D|} F_i$. Each possible feature of documents is treated as a binary attribute in $\mathbb{S}$. The semantic similarity between objects from $\mathbb{S}$ can be assessed using standard measures described in Subsection 3.4. However, due to sparsity and high dimensionality of this representation, usually spherical similarity functions, such as the *cosine similarity*, or set-oriented measures as *Jaccard index* and *Dice coefficient*, are employed. In several papers it is experimentally shown that this representation can yield better evaluations of the similarity than the standard bag-of-words [29, 40, 129].

If the utilized knowledge base contains additional information on semantic dependencies between the concepts, this knowledge can be used to further adjust the vector (32). Moreover, if experts could provide feedback in the form of manually labelled exemplary documents, some supervised learning techniques can also be employed in that task [38]. However, particular methods for automatic tagging of textual data are not in the scope of this research.

5   Rule-Based Similarity Learning Model

This section presents a similarity learning model which is the main contribution of this dissertation. The model, called Rule-Based Similarity (RBS), originally proposed in [25] and reared in [24, 26–29], was inspired by works of Amos Tversky. It can even be seen as a rough set extension of the psychologically plausible feature contrast model proposed in [5] (see also the discussion in Subsection 3.4).

Tversky's model is extended in a few directions. In RBS, some basic concepts from the rough set theory are used to automatically extract from available data, features that influence the judgement of similarity. Additionally, the proposed method for aggregation of local similarities and dissimilarities takes into consideration dependencies between the induced features that occur in data. This allows for a more reliable assessment of the importance of arguments for and against the similarity of investigated objects. Finally, the simplicity and flexibility of RBS makes it useful in a wide array of applications, including learning the similarity in a classification context from both regular and extremely high dimensional data. It can also be modified to allow learning the semantic similarity of textual documents.

The following subsections overview the construction of RBS in different application scenarios. Subsection 5.1 explains the main motivation behind the model and points out its relations to Tversky's feature contrast model. Next, Subsection 5.2 shows how the basic RBS model is constructed and then, Subsection 5.4 discusses an adaptation of RBS to the case when data describing considered objects are high dimensional. The last subsection (Subsection 5.5) shows how RBS can be adjusted to work in an unsupervised fashion, especially for learning a similarity measure appropriate for assessment of the similarity in a meaning of texts.

5.1   General Motivation for Rule-Based Similarity

The similarity learning models discussed in Subsection 4.2 allow to fit a parametrized similarity function or a family of approximation formulas to available data. This process can be understood as an adjustment of a similarity relation to a desired context. However, in case of the discussed methods this problem is reduced to tuning parameters of a preselected similarity model. An approach like that has to result in passing to the final model some properties which are not inferred from data and are potentially unwanted.

The approach to learning the similarity represented by the commonly used similarity models is usually based on an assumption that an expert is able to preselect at least a proper family of similarity models. This family is expected to contain a member which can sufficiently approximate the reality. Unfortunately, due to the complexity of the concept of similarity, this assumption may be false. Additionally, in some cases the family of possible approximations may be so large, that the extensive parameter tuning is likely to terminate at some relatively weak local optimum or even to overfit to training data while showing poor performance when used for new, previously unknown objects.

This problem is particularly conspicuous when the analysed data set is high dimensional. Typically, the number of parameters of a similarity learning model is at least linearly dependent on the number of attributes in data. Hence, dimensionality has a significantly adverse impact on a complexity of a model. Not only can very complex models suffer from overfitting but they are also unintuitive and difficult to interpret by experts.

Another important issue related with the similarity learning methods which utilize the *local-global principle* is a difficulty with modelling dependencies between local similarities/distances corresponding to different attributes. For instance, what weights should be assigned to a group of highly correlated attributes which are important for the similarity judgement individually? On one hand, each local similarity is important so it should have a high weight. On the other hand, if all of those local similarities are given a high weight, the final model can be biased towards a single aspect of a similarity while neglecting other, possibly as relevant factors.

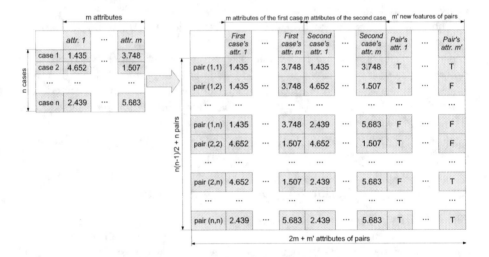

Fig. 10. A schema showing generalized tabular representations of an information system for the purpose of learning a similarity relation. An information system on the left is transformed into a relational pattern learning space represented on the right.

Moreover, in comparison to approximation of concepts, approximation of relations often requires extraction of some additional higher-level characteristics related to pairs of objects (see the discussion in Subsections 2.2 and 3.1). Figure 10 shows a general transformation schema of an information system that allows flexible learning of any similarity relation. Apart from attributes that describe each object, the transformed system may contain additional features that characterise two objects as a pair. Such features may correspond to a variety of statistics or different aggregation types of values of the original attributes. The models discussed in Subsection 4.2 simplifies this transformation by narrowing the set of new characteristics to predefined local similarities. Even though such a limitation is beneficial from the computational complexity point of view, it may severely deteriorate the model's ability to infer a semantically meaningful similarity relation. It was confirmed in a number of empirical studies that the introduction of higher-level features often significantly increases performance of similarity models [12, 23, 117, 120, 129].

The need for extraction of meaningful qualitative features of objects for the purpose of measuring their similarity was recognized by Tversky and motivated his contrast model [5] (see Subsection 3.4). Tversky argued that people rarely think in terms of exact numbers but instead they tend to operate on binary characteristics of objects, such as *an object is large* or *an object is round*. In his model, objects were represented by such higher-level features. For each pair, the features were divided into those which are arguments for the similarity (the common features) and those which constitute arguments against the similarity or, in other words, arguments for dissimilarity of the objects from the pair. The RBS model is in a large part inspired by this approach.

In Tversky's experiments relevant characteristics of the compared stimuli were usually defined by participants of the conducted study. However, for analysis of larger real-life data sets, meaningful features need to be extracted automatically. Their selection and influence on the final model needs to be dependent on a context of the similarity relation and, in particular, on other objects from the given data set. One of the main aims of the RBS model is to facilitate this task using a rough-set-motivated approach for approximation of relations (see Subsection 2.2).

Semantically meaningful higher-level features of objects can be extracted from data using a rule mining algorithm [22, 25]. Unlike in [22], however, in RBS such features are not only used for changing the representation of objects but are also utilized to construct approximations of similarities to each object in a training data set. A RBS similarity function value is derived from those approximations to allow convenient modelling of the dependencies imposed by the presence of different objects in the data (see the discussion in Subsection 3.2).

Another goal of RBS is to overcome the problem with selection of appropriate weights for the contrast model. Instead of assigning globally defined importance values to common and distinctive features of any pair, RBS aims at assessing strength of all arguments for and all arguments against the similarity, relative to an investigated pair. This approach allows RBS to better reflect the context in which the similarity of given objects is considered.

Finally, a good similarity learning model need to be scalable. The scalability of a model can be considered relative to a number of objects in the data as well as to a number of attributes. Construction of RBS does not require investigating all pairs of object during the learning, hence it is possible to approximate a similarity relation from larger data. By utilizing basic notions from the theory of rough sets, RBS can also be adapted to work with extremely high dimensional data.

In general, construction of a desired similarity learning model should include the following steps:

1. The selection of an appropriate context for the similarity.
2. The extraction of features which are relevant in the given context (definition of an approximation space).
3. The definition of a data-dependent similarity function that aggregates the features, while considering the preselected context and types of compared objects.

The next section shows how the RBS model implements these steps in order to incorporate the properties discussed in Subsection 4.1.

5.2   Construction of the Rule-Based Similarity Model

The Rules-based Similarity (RBS) model was developed as an alternative to the distance-based approaches [25]. It may be seen as a rough set extension to the psychologically plausible feature contrast model proposed by Tversky [5]. As in the case of the contrast model, in RBS the similarity is assessed by examining whether two objects share some binary higher-level features. Unlike in Tversky's approach, however, in RBS features that are relevant for a considered similarity context are automatically extracted from data. Their importance is also assessed based on available data, which allows to model the influence of information about other objects on the similarity judgement (see the discussion in Subsection 3.1).

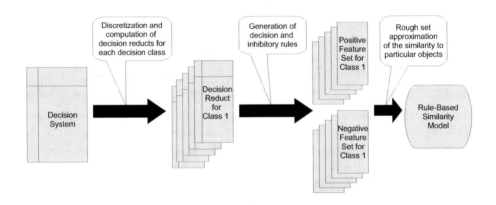

Fig. 11. A construction schema of the RBS model

The construction of RBS is performed in three steps. The schema from Figure 11 shows those steps in a case when the similarity is learnt in a classification context (the context is defined by a decision attribute in a data set). Since originally the notion of the RBS model was proposed for a classification purpose, that specific context will be assumed for the remaining part of this section.

The first step involves transformation of raw attribute values, which often are numerical, into a more abstract symbolic representation that resembles basic qualitative characteristics of objects. As discussed in Subsections 3.1 and 3.4, such characteristics are more likely to be used by humans and are more suitable for an assessment of a local similarity from a psychological point of view [1, 3–5, 11, 21]. For example, values of an attribute expressing a length of a car can be transformed into intervals labelled as *short, medium* and *long*. Those new values are easier to comprehend and utilize by humans in their judgements. Of course, semantics of each of those values can be different for different people. It will also be dependent on particular cars represented in the data.

For the purpose of a practical data analysis, however, it is often sufficient to apply a heuristic discretization technique to divide numerical attribute values into intervals representing meaningful qualitative symbols.

Algorithm 2. The calculation of a decision reduct from numerical data

---

Input: a decision system $\mathbb{S}_d = (U, A \cup \{d\})$;

Output: a decision reduct $DR \subset A$ coupled with sets of cuts for each attribute from DR;

```
 1  begin
 2      DR = empty list;                  // an empty list of attributes
 3      SC = empty list;                  // an empty list of selected cuts
 4      CC = ∅;                           // a set of cut candidates
 5      foreach a ∈ A do
 6          Compute cut candidates CC_a for the attribute a using
            the guidelines from [154];
 7          CC = CC ∪ CC_a;
 8          SC[a] = ∅;
 9      end
10      i = 1; while there are conflicts in S'_d = (U, DR', d) do
11          Q_max = −∞;
12          foreach cut ∈ CC do
13              Q(cut) = quality of cut;
14              if Q(cut) > Q_max then
15                  Q_max = Q(cut);
16                  best_cut = cut;
17                  best_a = attribute a which corresponds to best_cut;
18              end
19          end
20          DR[i] = best_a;
21          SC[a] = SC[a] ∪ best_cut;
22      end
23      foreach a ∈ DR do
24          if there are no conflicts in S'_d = (U, (DR \ {a})', d) then
25              DR = DR \ {a};
26              SC[a] = ∅;
27          end
28      end
29      return DR and SC;
30  end
```

The discretization can be combined with dimensionality reduction, e.g. by using a discernibility-based discretization method described in [154] to compute a set of symbolic attributes that discern all objects in the data (or nearly all in the approximate case). In this approach a subset of attributes with a corresponding set of cuts is selected from a larger attribute set in a greedy fashion.

It is done using a simple deterministic heuristic which starts with an empty set and iteratively adds the most promising attributes with corresponding cuts until the decision determination criterion is satisfied [34, 154]. Since the resulting set of discretized attributes discern all or sufficiently many[12] instances from different decision classes of the original decision system, it can be easily adjusted to become a desired type of a reduct (definitions of several types of reducts can be found in Subsection 2.3). For this purpose cuts that are abundant need to be eliminated. Therefore such a method can be viewed as a simultaneous supervised discretization and computation of decision reducts [34, 155]. This approach to the dimensionality reduction does not only boost the construction of RBS, but is also helpful in identification of truly relevant local features. For those reasons it was used in all experiments with RBS on numerical data presented in this dissertation (see Subsections 6.1 and 6.2). Algorithm 2 shows the procedure for classical reducts which in this case are understood as irreducible sets of *discretized* attributes that discern all objects from different decision classes[13] [34]. The algorithm assumes that due to the presence of numerical attributes there is no inconsistency in the original data table (i.e. there are no indiscernible objects).

In Algorithm 2, DR' denotes a set of attributes from DR discretized using the corresponding cuts from the list SC. To facilitate computations for high dimensional data some randomness can be introduced to the generation of candidate cut. In this way the algorithm can be employed for finding a diverse set of good quality decision reducts [33, 34]. This approach was used in the extension to RBS which is discussed in Subsection 5.4. The resulting set of attributes can also be a super-reduct by skipping the attribute elimination phase in order to capture more, potentially important similarity aspects.

Because a class or a type of an object may have a significant impact on its similarity assessments to other objects in data [5, 11, 21], different sets of important features need to be extracted for different decision classes. For this reason, in a case when a decision attribute in data has more than two values, the discretization and attribute reduction in RBS need to be performed separately for each decision class, using the one-vs-all approach.

In the second step, higher-level features that are relevant for the judgement of similarity are derived from data using a rule mining algorithm. Each of those features is defined by the characteristic function of the left-hand side (the antecedent) of a rule. In RBS, two types of rules are generated – decision rules (see Definition 3) that form a set of candidates for relevant positive features, and inhibitory rules (see Definition 4) which are regarded as relevant distinctive features. Depending on a type of a rule, the corresponding feature can be useful either as an argument for or against the similarity to a matching object.

The induction of rules in RBS may be treated as a process of learning aggregations of local similarities from data. Features defined by antecedents of

---

[12] A desired number of discerned instances can be treated as a parameter that governs the approximation quality.

[13] A discretized attribute corresponds to a pair consisting of the original attribute and a set of cuts that define nominal values (intervals).

the rules express higher-level properties of objects. For instance, a characteristic indicating that a car is big may be expressed using a formula:

$$car\_length = high \wedge car\_width = high \wedge car\_height = high \ . \qquad (33)$$

The feature defined in this way approximates the concept of a big car. Such a concept is more likely to be used by a person who assesses the similarity between two cars in a context of their appearance, than the exact numerical values of lengths, widths and heights. It can be noticed, for example, when people are explaining why they think that two objects are similar. It is more natural to say that two cars are similar because they are both big, rather than saying that one of them has $5,034mm$ length, $1,880mm$ width and $1,438mm$ height; the other is $5,164mm$ long, $1,829mm$ wide and $1,415mm$ high, and the differences in the corresponding parameters are small.

The choice of the higher-level features in RBS is not unique. Different heuristics for computation of reducts and different parameter settings of rule induction algorithms lead to the construction of different feature sets. As a consequence, the corresponding similarity approximation space changes along with the representation of the objects. The new representation may define a family of indiscernibility classes which is better fitted to the approximation of similarities to particular objects. In this context, it seems trivial to say that some approximation spaces are more suitable for approximating the similarities than others. Therefore the problem of learning the similarity relation in RBS is closely related to searching for a relevant approximation space [56, 151] (see also the discussion in Subsection 2.2).

More formally, let $F_{(i)}^{+}$ and $F_{(i)}^{-}$ be the sets of binary features derived from the decision and the inhibitory rules (see Definitions 3 and 4), respectively, generated for i-th decision class:

$$F_{(i)}^{+} = \left\{\phi : \left(\phi \rightarrow (d = i)\right) \in RuleSet_i\right\} ,$$

$$F_{(i)}^{-} = \left\{\phi : \left(\phi \rightarrow \neg(d = i)\right) \in RuleSet_i\right\} .$$

$RuleSet_i$ is a set of rules derived from a reduct DR_i associated with the i-th decision class. The rule set may be generated using any rule mining algorithm but it is assumed, that if not stated otherwise, $RuleSet_i$ consists of rules that are true in $\mathbb{S}$ (their *confidence factor* is equal to 1 – see Subsection 2.1) and cover all available training data, i.e. for every $u \in U$ there exists $\pi \in RuleSet_i$ such that $u \vDash lhs(\pi)$. Moreover, for efficiency in practical applications of the model it may be necessary to require that the generated sets of rules $RuleSet_i$ be minimal. It means that there is no rule $\pi \in RuleSet_i$ that could be removed without reducing the set of covered objects or, in other words, for every $\pi \in RuleSet_i$ there exists $u \in lhs(\pi)(U)$ which is not covered by any other rule from $RuleSet_i$.

A feature ϕ is also a decision logic formula, i.e. a conjunction of descriptors defined over discretized attributes, that corresponds to an antecedent of some rule (see the notation introduced in Subsection 2.1). We will say that an object u, described in a decision system $\mathbb{S} = (U, A)$, has a feature ϕ iff $u \vDash \phi$. A set of all objects from U that have the feature ϕ (the meaning of ϕ in $\mathbb{S}$) will be denoted by $\phi(U)$.

In RBS a similarity relation is approximated by means of approximating *multiple concepts* of being similar to a specific object. In the rough set setting, a similarity to a specific object is a well-defined concept. In the proposed model, it consists of those object from U which share with u at least one feature from the set $F_{(i)}^+$, where i is *assumed* to be the decision class of u ($d(u) = i$):

$$SIM_{(i)}(u) = \bigcup_{\phi \in F_{(i)}^+ \wedge u \models \phi} \phi(U) \tag{34}$$

Analogically, the approximation of the dissimilarity to u is a set of objects from U which have at least one feature from $F_{(i)}^-$ that is *not in common* with u:

$$DIS_{(i)}^0(u) = \bigcup_{\phi \in F_{(i)}^- \wedge u \not\models \phi} \phi(U) \tag{35}$$

For convenience, the set of objects that have at least one feature from $F_{(i)}^-$ that is *in common* with u will be denoted by:

$$DIS_{(i)}^1(u) = \bigcup_{\phi \in F_{(i)}^- \wedge u \models \phi} \phi(U) \tag{36}$$

To abbreviate the notation only $SIM(u)$ and $DIS(u)$ will be written when the decision for an object u is known:

$$SIM(u) = SIM_{(d(u))}(u); \qquad DIS(u) = DIS_{(d(u))}^0(u) \tag{37}$$

It is worth noticing that within the theory of rough sets the set $SIM(u)$ can be seen as an outcome of an uncertainty function $SIM : U \to \mathbb{P}(U)$ (see Definition 5). A proof of this fact is quite trivial. From the definition of the set $SIM(u)$ it follows that $u \in SIM(u)$. Moreover, if $u_1 \in SIM(u_2)$, then there exists $\phi \in F_{(d(u_2))}^+$ such that $u_1 \in \phi(U) \wedge u_2 \models \phi$. If so, then $u_1 \models \phi$, thus $d(u_1) = d(u_2)$ and $u_2 \in SIM(u_1)$.

Analogically, the set $DIS(u)$ is an outcome of a function $DIS : U \to \mathbb{P}(U)$ which can be seen as an opposite of SIM. The function SIM induces a tolerance relation in U, whereas DIS induces a relation that can be called an *intolerance relation*. From the definition, $\forall_{u \in U} u \notin DIS(u)$, i.e. the relation induced by DIS is anti-reflexive. Moreover, this relation is asymmetric since for every $u_1, u_2 \in U$, if $u_1 \in DIS(u_2)$ then $u_2 \notin DIS_{(d(u_2))}^0(u_1)$.

The functions SIM and DIS are used for the approximation of the similarity and the dissimilarity to objects from U. In the RBS model, the assessment of a degree in which an object u_1 is similar and dissimilar to u_2 is done using two functions:

$$Similarity(u_1, u_2) = \mu_{sim}(u_1, SIM_{d(u_1)}(u_2)) = \frac{|SIM(u_1) \cap SIM_{d(u_1)}(u_2)|}{|SIM(u_1)| + C_{sim}}, \tag{38}$$

$$Dissimilarity(u_1, u_2) \; = \; \hat{\mu}_{dis}\big(u_1, DIS^1_{d(u_1)}(u_2)\big) \; = \; \frac{\left| DIS(u_1) \cap DIS^1_{d(u_1)}(u_2) \right|}{|DIS(u_1)| + C_{dis}} .$$

$$(39)$$

In the above formulas C_{sim} and C_{dis} are positive constants which can be treated as parameters of the model. The function $\mu_{sim} : U \times \mathbb{P}(U) \rightarrow [0, 1)$ can be seen as a membership function from the rough set theory (see Definition 6). It measures a degree in which an object u_1 fits to the concept of the similarity to u_2. The function $\hat{\mu}_{dis} : U \times \mathbb{P}(U) \rightarrow [0, 1)$ may be regarded as an *anti-membership* function since it measures a degree in which u_1 is not similar to u_2 (i.e. is dissimilar to u_2). It is also worth noticing that if the assumptions regarding the consistency and the coverage of the utilized rules are true, then for every $u \in U$, $|SIM(u)| > 0$ and $|DIS(u)| > 0$, and the functions $Similarity$ and $Dissimilarity$ are well-defined for every pair $(u_1, u_2) \in U \times \Omega$, even in a case when $C_{sim} = C_{dis} = 0$.

The similarity function of the RBS model combines values of $Similarity$ and $Dissimilarity$ for a given pair of objects. It can be expressed as:

$$Sim_{RBS}(u_1, u_2) = F\Big(Similarity(u_1, u_2), Dissimilarity(u_1, u_2)\Big) \qquad (40)$$

where $F : \mathbb{R} \times \mathbb{R} \rightarrow \mathbb{R}$ can be any function that is monotonically increasing with regard to its first argument (i.e. a value of $Similarity$) and monotonically decreasing with regard to its second argument (a value of $Dissimilarity$). One example of such a function can be:

$$Sim_{RBS}(u_1, u_2) = \frac{Similarity(u_1, u_2) + C}{Dissimilarity(u_1, u_2) + C} \qquad (41)$$

where $C > 0$ is a small constant, which is introduced to avoid division by zero and to ensure that $Sim_{RBS}(u_1, u_2) = 1$ for u_1, u_2 which are neither similar nor dissimilar (i.e. $Similarity(u_1, u_2) = Dissimilarity(u_1, u_2) = 0$). In this particular form the RBS similarity function was used in experiments described in Subsections 6.1 and 6.2.

Alternatively, a similarity degree in RBS could also be expressed as a simple difference between the similarity and dissimilarity of two objects, as in the case of Tversky's model:

$$Sim'_{RBS}(u_1, u_2) = Similarity(u_1, u_2) - Dissimilarity(u_1, u_2) \qquad (42)$$

In this form, the RBS function takes values between -1 and 1, with its neutral value equal 0. An advantage of this function is that it does not need the additional constant C. It can be easily shown that all the mathematical properties of Sim_{RBS}, which are discussed in Subsection 5.3, are independent of the exact form of the function F as long as the requirement regarding its monotonicity is met.

Depending on the type and parameters of a rule mining algorithm utilized for the creation of the feature sets $F^+_{(i)}$ and $F^-_{(i)}$, the sets $SIM(u)$ and $DIS(u)$ can have different rough set interpretations (Figure 12). If all the rules are true

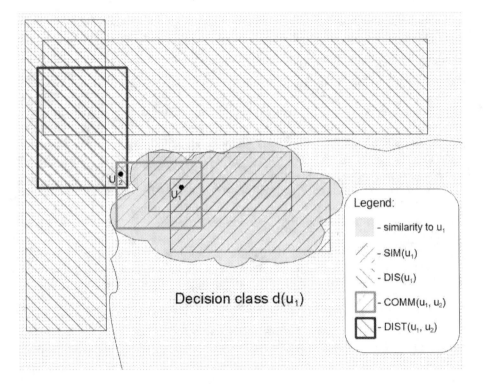

Fig. 12. A graphical interpretation of an approximation of similarity to a single object in RBS

in $\mathbb{S}$, then $SIM(u)$ and $DIS(u)$ would be equivalent to lower approximations of the concepts of similarity and dissimilarity to u in U, respectively. Otherwise, if the rules with a lower confidence coefficient were allowed, $SIM(u)$ and $DIS(u)$ would correspond to upper approximations of the similarity and the dissimilarity to u. Their properties and granulation may be treated as parameters of the model. In applications they can be tuned to boost the quality of the induced relation. This tuning process can be regarded as searching for the optimal approximation space (see Subsection 2.2).

Figure 12 shows a simplified graphical interpretation of the RBS model. The grey area in the picture represents a concept of similarity to object u_1 from the decision class $d(u_1)$. The rectangles inside this region correspond to an approximation of the concept of being similar to u_1. They are defined by indiscernibility classes of training objects that share at least one feature from $F^+_{(d(u_1))}$ with u_1. Analogically, the rectangles outside the decision class approximate the concept of the dissimilarity to u_1 and they contain instances from the set $DIS^0_{d(u_1)}(u_1)$. The local similarity value of u_2 to u_1 in this example would be calculated as a ratio between a fraction of the similarity approximation shared by u_1 and u_2,

Table 6. An exemplary decision table displaying one's preferences regarding general appearance of selected cars. $F^+_{Nice} = \{\phi^+_1, \phi^+_2, \phi^+_3, \phi^+_4\}$ and $F^-_{Nice} = \{\phi^-_1, \phi^-_2, \phi^-_3, \phi^-_4\}$.

Object:	ϕ^+_1	ϕ^+_2	ϕ^+_3	ϕ^+_4	ϕ^-_1	ϕ^-_2	ϕ^-_3	ϕ^-_4	Decision
Ford Mustang	1	0	1	0	0	0	0	0	Nice
Toyota Avensis	0	0	0	0	1	1	0	1	notNice
Audi A4	0	0	0	0	1	0	1	0	notNice
Porsche Carrera	0	1	0	1	0	0	0	0	Nice
Mercedes S-Class	0	0	0	0	0	1	0	1	notNice
Chevrolet Camaro	0	1	1	0	0	0	0	0	Nice
Volkswagen Passat	0	0	0	0	0	1	1	0	notNice
Mitsubishi Eclipse	1	0	1	1	0	0	0	0	Nice

and a fraction of the dissimilarity approximation which is characteristic only to u_2. In Figure 12, areas corresponding to those fractions are highlighted in blue and red, respectively.

The function Sim_{RBS} can be employed for the classification of objects from unknown decision classes as it only uses information about the class of the first object from the pair. New objects can be classified in a cased-based fashion, analogically to the *k-nearest neighbors* algorithm. Exemplary similarity-based classification functions are presented in Subsection 3.3.

5.3   Properties of the Rule-Based Similarity Function

To illustrate the evaluation of the similarity in RBS, let us consider the decision system from Table 6. Assume that we want to evaluate the similarity of *Ford Mustang* to *New_Car* in a context of their appearance, which is judged by a given person. We know preferences of this person regarding cars (the classes of objects) from our decision table but we have no information regarding the classification of *New_Car*. During the construction of the RBS model, the data set describing the selected cars was discretized and some consistent decision rules[14] were induced for each of the two possible classes. Since the decision for *Ford Mustang* is *Nice*, we choose the positive features from the rules pointing at this class (i.e. rules in a form of $\phi \rightarrow Nice$). The negative features are chosen among the rules indicating the *notNice* decision.

Suppose that from the set of antecedents of the rules induced for the decision *Nice*, two were matching *New_Car*: ϕ^+_1 and ϕ^+_4. Additionally, there was one feature derived from a rule classifying objects as *notNice*, that matched the tested car: ϕ^-_1. From the decision table we know that *Ford Mustang* has in common with *New_Car* only the feature ϕ^+_1, so this feature is an argument for their similarity. In addition, the feature ϕ^-_1 does not match *Ford Mustang* therefore this feature provide an argument for dissimilarity of the compared cars. Although the rule ϕ^+_4 does not match *Ford Mustang*, it is not considered as an

---

[14] Since there are only two decisions, inhibitory rules for one class correspond to decision rules for the other.

argument for the dissimilarity of the two cars because the features from the set F_{Nice}^+ may only become arguments for the similarity. Since two out of three cars which match to the features of *Ford Mustang* have the feature ϕ_1^+ and three out of four cars with decision *notNice* have the feature ϕ_1^-, if we set $C_{sim} = C_{dis} = 1$, the RBS value equals:

$$Sim_{RBS}(FordMustang, New\_Car) = \left(\frac{2}{3 + C_{sim}} + C \right) \bigg/ \left(\frac{3}{4 + C_{dis}} + C \right)$$
$$= \left(\frac{2 + 4C}{4} \right) \bigg/ \left(\frac{3 + 5C}{5} \right) = \frac{5 + 10C}{6 + 10C} \ .$$

For a very small value of C we get the value $\approx \frac{5}{6}$. Since this value is lower than 1, *Ford Mustang* should be considered dissimilar to *New_ Car*.

The RBS model shares many properties with Tversky's contrast model of the similarity. In both models the evaluation of the similarity is seen as a feature matching process. Objects from the data are represented by sets of qualitative features rather than by vectors in an attribute space [5]. Furthermore, both models consider features as possible arguments for or against the similarity and aggregate those arguments during the similarity assessment.

The construction of RBS makes the resulting model flexible and strongly data-dependent. As in the contrast model, in RBS the similarity function is likely to be not symmetric, especially when the compared objects are from different decision classes. Moreover, in a case of inconsistency of a data set (see Definition 2), a relation induced using the RBS similarity function may be even not reflexive. This fact is in accordance with the main feature of the similarity for the classification (Definition 14). It also reflects a phenomena, that availability of information about decision classes (types or predefined labels) of examined stimuli impacts human judgements of the similarity [20, 21].

The similarity functions of RBS and the contrast model also have in common a number of mathematical properties, such as the maximality of marginal values and the monotonicity with regard to the inclusion of the feature sets:

Proposition 1. *Let $\mathbb{S}_d = (U, A \cup \{d\})$ be a consistent decision system, $U \subseteq \Omega$ and let $Sim_{RBS} : U \times \Omega \to \mathbb{R}$ be a similarity function of the RBS model, constructed for $\mathbb{S}_d$ using rules that are true in $\mathbb{S}_d$ and cover all objects from U. The following inequity holds for every $u \in U$ and $u' \in \Omega$:*

$$Sim_{RBS}(u, u) \geq Sim_{RBS}(u, u') \ . \tag{43}$$

Proof. To prove this inequity it is sufficient to show that $Similarity(u, u)$ is maximal and $Dissimilarity(u, u) = 0$ for every $u \in U$. Since the RBS model is constructed from rules that cover all objects from U, $|SIM(u)| > 0$ and for any $X \subseteq U$ we have:

$$Similarity(u, u) = \frac{|SIM(u) \cap SIM(u)|}{|SIM(u)| + C_{sim}} \geq \frac{|SIM(u) \cap X|}{|SIM(u)| + C_{sim}} \ . \tag{44}$$

Analogically, $|DIS(u)| > 0$ and since the rules are true in $\mathbb{S}_d$, $DIS^1_{d(u)}(u) = \emptyset$. If so, then

$$Dissimilarity(u, u) = \frac{\left| DIS(u) \cap DIS^1_{d(u)}(u) \right|}{|DIS(u)| + C_{dis}} = 0 \ . \tag{45}$$

$\square$

Proposition 2. *Let* $\mathbb{S}_d = \left(U, A \cup \{d\} \right)$ *be a consistent decision system,* $U \subseteq \Omega$ *and let* $Sim_{RBS} : U \times \Omega \to \mathbb{R}$ *be a similarity function of the RBS model, constructed for* $\mathbb{S}_d$ *using rules that cover all objects from* U. *In addition, let us consider objects* $u \in U$ *and* $u', u'' \in \Omega$, *such that* u' *and* u'' *are represented by feature sets* $\{\Phi^+_{(d(u))}, \Phi^-_{(d(u))}\}$ *and* $\{\Psi^+_{(d(u))}, \Psi^-_{(d(u))}\}$, *respectively. The following implication holds for every* $u \in U$:

$$\left(\Phi^+_{(d(u))} \supseteq \Psi^+_{(d(u))} \wedge \Phi^-_{(d(u))} \subseteq \Psi^-_{(d(u))} \right) \Rightarrow Sim_{RBS}(u, u') \geq Sim_{RBS}(u, u'') \ . \tag{46}$$

Proof. To prove the above implication it is sufficient to show that for objects considered in the proposition we have $Similarity(u, u') \geq Similarity(u, u'')$ and $Dissimilarity(u, u') \leq Dissimilarity(u, u'')$.

Let us consider the sets $\Phi^+_{(d(u))}$ and $\Psi^+_{(d(u))}$:

$$\Phi^+_{(d(u))} \supseteq \Psi^+_{(d(u))} \Rightarrow SIM_{(d(u))}(u') \supseteq SIM_{(d(u))}(u'')$$
$$\Rightarrow \left| SIM(u) \cap SIM_{(d(u))}(u') \right| \geq \left| SIM(u) \cap SIM_{(d(u))}(u'') \right|$$

This and the fact that $\forall_{u \in U} |SIM(u)| > 0$ implies that $Similarity(u, u') \geq Similarity(u, u'')$.

Analogically, if $\Phi^-_{(d(u))} \subseteq \Psi^-_{(d(u))}$ then $DIS^1_{d(u)}(u') \subseteq DIS^1_{d(u)}(u'')$ and as a consequence $Dissimilarity(u, u') \leq Dissimilarity(u, u'')$.

$\square$

Proposition 3. *Let* $\mathbb{S}_d = \left(U, A \cup \{d\} \right)$ *be a consistent decision system,* $U \subseteq \Omega$ *and let* $Sim_{RBS} : U \times \Omega \to \mathbb{R}$ *be a similarity function of the RBS model, constructed for* $\mathbb{S}_d$ *using rules that are true in* $\mathbb{S}_d$ *and cover all objects from* U. *In addition, let us consider objects* u, u', *such that* $d(u) = d(u') = i$ *and* u, u' *are represented by feature sets* $\{\Phi^+_{(i)}, \Phi^-_{(i)}\}$ *and* $\{\Psi^+_{(i)}, \Psi^-_{(i)}\}$, *respectively. The following implication holds for any such* $u, u' \in U$:

$$\left(\Phi^+_{(i)} \supseteq \Psi^+_{(i)} \wedge \Phi^-_{(i)} \subseteq \Psi^-_{(i)} \right) \Rightarrow Sim_{RBS}(u, u') \leq Sim_{RBS}(u', u) \ . \tag{47}$$

Proof. It is sufficient to show that for all objects $u, u' \in U$ considered in the proposition, $Similarity(u, u') \leq Similarity(u', u)$ and $Dissimilarity(u, u') \geq Dissimilarity(u', u)$.

The second inequity is trivial due to the fact that $d(u) = d(u')$ and the rules are true in $\mathbb{S}_d$. In such a case $DIS^1(u) = DIS^1(u') = \emptyset$ and $Dissimilarity(u, u') = Dissimilarity(u', u) = 0$. To show the validity of the first inequity let us consider $u, u' \in U$ described by feature sets $\Phi^+_{(i)}$ and $\Psi^+_{(i)}$, respectively. We have:

$$\Phi^+_{(i)} \supseteq \Psi^+_{(i)} \Rightarrow SIM(u) \supseteq SIM(u')$$
$$\Rightarrow SIM(u) \cap SIM(u') = SIM(u') \qquad \text{and}$$
$$|SIM(u)| \geq |SIM(u')| \quad .$$

If so, then:

$$Similarity(u, u') = \frac{|SIM(u) \cap SIM(u')|}{|SIM(u)| + C_{sim}} = \frac{|SIM(u')|}{|SIM(u)| + C_{sim}}$$
$$\leq \frac{|SIM(u')|}{|SIM(u')| + C_{sim}} = Similarity(u', u) \quad .$$

That concludes the proof. $\qquad \square$

The next proposition shows that the RBS similarity function is suitable for constructing approximations of similarity relations in the context of classification. Fundamental properties of such relations were discussed in Subsection 3.2. However, before we can formulate this proposition we first need to prove a simple lemma:

Lemma 1. *Let Π be a set of decision rules generated for a consistent decision system $\mathbb{S}_d = (U, A \cup \{d\})$ and let Π_1, Π_2 denote two subsets of Π. Additionally, let $supp(\Pi_1) = \bigcup_{\pi \in \Pi_1} lhs(\pi)(U)$ and $supp(\Pi_2) = \bigcup_{\pi \in \Pi_2} lhs(\pi)(U)$. If Π covers all objects from U and is minimal in U, then*

$$supp(\Pi_1) \subseteq supp(\Pi_2) \Leftrightarrow \Pi_1 \subseteq \Pi_2 \quad . \tag{48}$$

Proof. The implication $\Pi_1 \subseteq \Pi_2 \Rightarrow supp(\Pi_1) \subseteq supp(\Pi_2)$ is trivial. To prove the second implication, for a moment let us assume that the conditions from Lemma 1 are met and $supp(\Pi_1) \subseteq supp(\Pi_2)$ but there exists a rule $\pi \in \Pi_1$ such that $\pi \notin \Pi_2$. In such a case, $supp(\{\pi_1\}) \subseteq supp(\Pi_1) \subseteq supp(\Pi_2)$, so $\forall_{u \in lhs(\pi)(U)} \exists_{\pi' \in \Pi_2} u \models lhs(\pi')$. This, however, contradicts with the assumption that Π is minimal. $\qquad \square$

A direct consequence of Lemma 1 is that $supp(\Pi_1) = supp(\Pi_2) \Leftrightarrow \Pi_1 = \Pi_2$.

In the following proposition there will be an additional assumption regarding the sets of rules $RuleSet_i$ used in the construction of the RBS model. Namely, apart from the consistency, coverage and minimality of the rule sets, it will be assumed that each $RuleSet_i$ is sufficiently rich to ensure the uniqueness of a representation by the sets of new features of all objects which are discernible in the original decision system $\mathbb{S}_d = (U, A \cup \{d\})$. More formally, we will assume that for every $u, u' \in U$ represented by new feature sets $\{\Phi^+_{(i)}, \Phi^-_{(i)}\}$ and $\{\Psi^+_{(i)}, \Psi^-_{(i)}\}$, respectively, $u' \notin [u]_A \Leftrightarrow (\Phi^+_{(i)} \neq \Psi^+_{(i)} \vee \Phi^-_{(i)} \neq \Psi^-_{(i)})$. This property corresponds to the solvability assumption in Tversky's contrast model [5]. It is worth noticing that for any consistent decision system $\mathbb{S}_d$ (see Definition 2) it is always possible to construct sets $RuleSet_i$ that meet all of the above requirements. In the simplest case, it is sufficient to take the rules whose predecessors correspond to descriptions of indiscernibility classes in $\mathbb{S}_d$ and successors point out the corresponding decisions.

Proposition 4. *Let τ be a similarity relation in a context of classification in a universe Ω. Additionally, let $\mathbb{S}_d = (U, A \cup \{d\})$ be a consistent decision system, $U \subseteq \Omega$ and let $Sim_{RBS} : U \times \Omega \to \mathbb{R}$ be a similarity function of the RBS model, constructed for $\mathbb{S}_d$ using rule sets, which have the properties of consistency, coverage, minimality and uniqueness of a representation. The function Sim_{RBS} is a proper similarity function for the relation τ within the set U.*

Proof. Let us denote by $\tau_{(\epsilon)}^{Sim_{RBS}}$ a set of all pairs $(u, u') \in U \times U$ for which $Sim_{RBS}(u, u') \geq \epsilon$. To show that the function Sim_{RBS} has the property of being a proper similarity function (Definition 15) for the relation τ within the set U we will give values of ϵ_1 and ϵ_2 such that for any $u, u' \in U$ we have:

$$Sim_{RBS}(u, u') \geq \epsilon_1 \Rightarrow (u, u') \in \tau \tag{49}$$

$$(u, u') \in \tau \Rightarrow Sim_{RBS}(u, u') \geq \epsilon_2 \tag{50}$$

and the sets $\tau_{(\epsilon_1)}^{Sim_{RBS}}$ and $U \setminus \tau_{(\epsilon_2)}^{Sim_{RBS}}$ are not empty.

We will start the proof by showing that if

$$Sim_{RBS}(u, u') = F\big(Similarity(u, u'), Dissimilarity(u, u')\big)$$

for F that is increasing with regard to its first argument and decreasing with regard to the second, then the implication 49 is true for $\epsilon_1 = F(sim_{max}, 0)$, where $sim_{max} = \max\limits_{u \in U}\big(Similarity(u, u)\big)$. In particular, we will show that

$$Sim_{RBS}(u, u') \geq F(sim_{max}, 0) \Leftrightarrow (u' \in [u]_A \land u \in U_{max}),$$

where $U_{max} = \{u \in U : u = \operatorname*{argmax}\limits_{u \in U} |SIM(u)|\}$.

Since all utilized rules are consistent and they cover all objects from U, for any $u, u' \in U$ we have $Dissimilarity(u, u') = 0 \Leftrightarrow d(u) = d(u')$. Moreover, due to the fact that $\mathbb{S}_d$ is consistent and the utilized rules uniquely represent the objects from U, for any $u \in U$ and $u' \in [u]_A$ we have $SIM(u') = SIM(u)$. If so, then $u \in U_{max} \Rightarrow [u]_A \subseteq U_{max}$ and

$$Similarity(u, u') = Similarity(u', u) = \frac{|SIM(u)|}{|SIM(u)| + C_{sim}} . \tag{51}$$

Thus, the inequity $Sim_{RBS}(u, u') \geq F(sim_{max}, 0)$ holds for every pair (u, u') such that $u \in U_{max}$ and $u' \in [u]_A$.

On the other hand, let us imagine that there exist objects $u, u' \in U$ such that $u \notin U_{max} \lor u' \notin [u]_A$ and

$$Sim_{RBS}(u, u') \geq F(sim_{max}, 0) \lor Sim_{RBS}(u', u) \geq F(sim_{max}, 0)$$

or, equivalently,

$$Similarity(u, u') \geq sim_{max} \lor Similarity(u', u) \geq sim_{max}.$$

If $u \notin U_{max}$ but $u' \in [u]_A$ we get an inconsistency, because all $u \in U$ for which $u' \in [u]_A$ and $Similarity(u, u')$ is maximal, by definition must belong to U_{max}. Now, if it is true that $u \in U_{max} \land u' \notin [u]_A$ and $Similarity(u, u') \geq sim_{max}$, then we have:

$$Similarity(u, u') \geq Similarity(u, u) \Leftrightarrow |SIM(u) \cap SIM_{d(u)}(u')| \geq |SIM(u)|$$
$$\Leftrightarrow d(u) = d(u') \land SIM(u) = SIM(u') \ .$$

That also results in an inconsistency because, based on the assumption regarding the minimality of the rule sets and Lemma 1, the objects u and u' must have the same representation by new features, and thus $(u, u') \in IND_A$ (by the uniqueness of a representation). Hence, the only possibility left is that $u \in U_{max} \land u' \notin [u]_A$ and $Similarity(u', u) \geq sim_{max}$. In such a case we would have:

$$\frac{|SIM(u') \cap SIM_{d(u')}(u)|}{|SIM(u')| + C_{sim}} \geq sim_{max} \Leftrightarrow d(u) = d(u') \land SIM(u') \subseteq SIM(u) \land$$

$$|SIM(u')| \geq |SIM(u)|$$
$$\Leftrightarrow SIM(u) = SIM(u') \ ,$$

which again contradicts with the assumption about the uniqueness of a representation and proves that $Sim_{RBS}(u, u') \geq F(sim_{max}, 0) \Leftrightarrow (u' \in [u]_A \land u \in U_{max})$. Since a similarity relation in the context of a classification is assumed to be reflexive, it shows that the implication 49 is true for $\epsilon_1 = F(sim_{max}, 0)$. Moreover, due to the fact that U is finite, the maximum value of the function $Similarity$ has to be taken by at least one pair $(u, u') \in U \times U$, and thus $\tau_{(\epsilon_1)}^{Sim_{RBS}} \neq \emptyset$.

To show that there exists ϵ_2 for which the implication 50 is true we will use the fact that τ is assumed to have the main feature of the similarity for the classification (see Definition 14). As we already noticed, due to the consistency and coverage of the utilized rules we have $Dissimilarity(u, u') = 0 \Leftrightarrow d(u) = d(u')$, and $d(u) \neq d(u') \Rightarrow Similarity(u, u') = 0$. If so, then for $\epsilon_2 = F(0, 0)$ we get $\tau_{(\epsilon_2)}^{Sim_{RBS}} \supseteq_U IND_{\{d\}} \supseteq_U \tau$. Moreover, since $Dissimilarity(u, u') > 0$ for any pair $(u, u') \in U \times U$ such that $d(u) \neq d(u')$, we have $U \setminus \tau_{(\epsilon_2)}^{Sim_{RBS}} \neq \emptyset$. Thus it is sufficient to take $\epsilon_2 = F(0, 0)$. □

5.4 Rule-Based Similarity for High Dimensional Data

In the Rule-Based Similarity model the notion of decision reduct is used for finding a concise set of attributes which can serve as building blocks for constructing higher-level features. Nevertheless, it has been noted that a single reduct may fail to capture all critical aspects of the similarity in a case when there are many important "raw" attributes. To overcome this problem, an extension to RBS called

---

Algorithm 3. The computation of (ϵ, δ)-dynamic reducts in DRBS

Input: a decision system $\mathbb{S}_d = (U, A \cup \{d\})$;
  a parameter $NoOfAttr << |A|$;
  parameters $\epsilon, \delta \in [0, 1)$;
  integers $MaxDDR, MaxTry, NSets$;
Output: a set of (ϵ, δ)-dynamic reducts DDR_{set};

1 **begin**
2     $DDR_{set} = \emptyset$;
3     $i = 0$;
4   **while** $|DDR_{set}| < MaxDDR \wedge i < MaxTry$ **do**
5   Randomly draw $NoOfAttr$ attributes from A and construct $A' \subset A$, $|A'| = NoOfAttr$;
6   Compute a decision reduct DR of $\mathbb{S}'_d = (U, A', d)$;
7        $k = 0$;
8   **for** $j = 1$ **to** $NSets$ **do**
9   Randomly draw $\lfloor (1 - \epsilon) \cdot |U| \rfloor$ objects from U (without repetition) and create $\mathbb{S}''_d = (U', DR, d)$;
10   **if** $DR \in RED(\mathbb{S}''_d)$ **then**
11             $k = k + 1$;
12   **end**
13   **end**
14   **if** $k/NSets > 1 - \delta$ **then**
15          $DDR_{set} = DDR_{set} \cup \{DR\}$;
16   **end**
17       $i = i + 1$;
18   **end**
19   **return** DDR_{set};
20 **end**

Dynamic Rule-Based Similarity (DRBS) was proposed [27,28]. The main aim of the DRBS model is to extend the original model by taking into consideration a wider spectrum of possibly important aspects of the similarity.

During construction of the DRBS model, many independent sets of rules are generated from heterogeneous subsets of attributes. In this way, the resulting higher-level features are more likely to cover the factors that can influence similarity or dissimilarity of objects (the positive and negative feature sets) from a domain under scope. Within the model, the attributes that are used to induce the rules are selected by computation of multiple decision reducts from random subsets of data. This method can be seen as an analogy to the Random Forest algorithm [140], in which multiple decision trees are constructed. In DRBS however, the rules derived in this manner are not directly employed for classification but they are utilized to define multiple RBS similarity functions. Those local models are then combined in order to construct a single function which can yield a better approximation of a similarity relation in the context of a classification.

Although in all experiments described in this dissertation DRBS was implemented using the (ϵ, δ)-dynamic decision reducts [100, 101] (see Definition 12), any kind of an efficient dimensionality reduction technique, such as approximate reducts [98, 99] or decision bireducts [36] could be used (see the definitions in Subsection 2.3). The dynamic decision reducts, however, tend to be reliable even in a case when only a few hundreds of objects are available for the learning and thus are suitable for coping with the *few-objects-many-attributes* problem [26, 130].

Algorithm 3 shows an efficient procedure for computing (ϵ, δ)-dynamic decision reducts. Although the algorithm does not give any guarantee as to the number of returned dynamic reducts, in practical experiments with real-life data sets (see Subsection 6.2) it has always successfully generated a sufficient number of reducts for constructing a reliable DRBS model. Its advantage for the similarity learning is that it naturally adjusts the number of generated local RBS models to the available data. In particular, for reasonable values of ϵ and δ, the number of produced reducts for data sets describing objects with many important similarity aspects is likely to be higher than for those which describe simpler problems, characterised with fewer potentially important features.

The DRBS similarity function combines values of the local similarity functions. Due to a partially randomized reduct construction process, the individual RBS models represent more independent aspects of the similarity. That in turn results in a better performance of their ensemble [36, 102, 147]. This particular characteristic makes the DRBS model akin to the Random Forest algorithm where the final classification is done by combining decisions of multiple decision trees, constructed from random subsets of attributes and objects [140]. Unlike in the Random Forest, however, the classification results which are based on DRBS do not lose their interpretability. For each tested object we can explain our decision by indicating the examples from our data set which were used in the decision-making process (i.e. the k most similar cases). Equation 52 shows a basic form of a DRBS similarity function which averages outputs of the N local RBS models:

$$Sim_{DRBS}(u_1, u_2) = \frac{1}{N} \cdot \sum_{j=1}^{N} \left(Sim_{RBS}^{(j)}(u_1, u_2) \right), \qquad (52)$$

where $Sim_{RBS}^{(j)}(u_1, u_2)$ is the value of the RBS similarity function for the j-th decision reduct. This function can be easily modified to reflect relative importances of individual RBS models:

$$Sim_{wDRBS}(u_1, u_2) = \frac{\omega^{(j)} \cdot \sum_{j=1}^{N} \left(Sim_{RBS}^{(j)}(u_1, u_2) \right)}{\sum_{j=1}^{N} \omega^{(j)}}. \qquad (53)$$

In the above equation, weights $\omega^{(j)}$ correspond to quality of RBS models, which can be estimated using some of the methods described in Subsection 3.3. For this purpose, usually a part of objects from a learning set needs to be held back as a validation set.

DRBS introduces a few new parameters to the similarity model, of which the most important are $NoOfAttr$ and $MaxDDR$. They both govern the process of randomized computation of reducts. The first one tells how many attributes are randomly drawn from data for computation of a single reduct. The second one sets maximal number of the reducts to be generated. Together, those parameters influence the thickness of a coverage of truly important similarity aspects. Knowing their values, it is possible to estimate a chance of an attribute to be considered for inclusion into at least one reduct and the expected number of its occurrences within the final set of reducts.

If by p_{attr} we denote the ratio between $NoOfAttr$ and the total number of attributes in data ($p_{attr} = \frac{NoOfAttr}{|A|}$), the occurrence probability of an attribute $attr$ in at least one reduct and the expected number of its occurrences are equal:

$$p(attr) = 1 - MaxDDR \cdot (1 - p_{attr})^{MaxDDR} \quad and \quad E(attr) = MaxDDR \cdot p_{attr},$$

respectively. In practice, these two quantities can be used to set reasonable values of $NoOfAttr$ and $MaxDDR$ for a given data set.

Another two important parameters are ϵ and δ which have a significant impact on properties of generated dynamic reducts. The higher ϵ and lower δ, the more robust are the resulting dynamic decision reducts. However, too restrictive values of those parameters may cause a serious deterioration in a computational efficiency of the algorithm or even prevent its completion.

Alternatively, if instead of dynamic reducts, the new feature sets were defined using decision bireducts, many parameters of the DRBS model could be replaced by a single ratio that governs the generation of random permutations (for more details refer to [36]). In practical experiments with DRBS, however, only the approximations derived from dynamic decision reducts have been used so far.

5.5   Unsupervised Rule-Based Similarity for Textual Data

The idea behind the RBS model can also be applied to carry out unsupervised similarity learning [29]. In particular, the RBS model was extended to facilitate an approximation of a semantic similarity of scientific articles.

The construction of the model starts with assigning concepts from a chosen knowledge base to a training corpus of documents. This can be done in an automatic fashion with the use of methods such as ESA [129] (see Subsection 4.2). The associations to the key concepts assigned to the documents can be transformed to binary features and therefore, are suitable to use with the contrast model of similarity. However, a direct application of this model would not take into consideration data-based relations between concepts from the knowledge base and a potentially different meaning of those relations for different documents. The problem of finding appropriate values of parameters of the Tversky's model would also remain unsolved. The proposed extension of RBS aims to overcome those issues [29]. It is called the *unsupervised RBS* model, since it can be seen as a continuation of the research on the similarity learning model for high dimensional data [28].

Let F be a set of all possible semantic features of texts from a corpus D and let F_i be a set of the most important concepts related to the document T_i, $F = \bigcup_{i=1}^{|D|} F_i$. The documents from D can be represented in an information system $\mathbb{S} = (D, F)$, as explained in Subsection 4.2. An example of such a system is shown in Table 7. We can say that two documents described by this table have a common feature if they both have value 1 in the corresponding column (e.g. the documents T_1 and T_2 have three common features: f_2, f_3 and f_{10}). The binary attributes in this system may correspond to tags assigned by experts or by discretizing numeric weights of ontological entities generated using methods such as ESA.

In many practical applications, the numerical values of an association strength between a concept and a document may be discretized into more than two intervals in order to precisely model their bond. In this case, it is reasonable to define a few binary features that represent consecutive intervals and remain dependent, in a sense that if a feature is *"highly related"* to a document then it is also *"weakly related"*, but not the opposite. Such an approach is popular in Formal Concept Analysis [156] and allows to model a psychological phenomena that usually simpler objects are more similar to the more complex ones than the other way around.

In order to find out which combinations of independent concepts comprise the informative aspects of similarity, we could compute information reducts of $\mathbb{S}$ [80, 91] or, to obtain more compact and robust subsets of F, some form of approximate information reducts [34, 99]. However, during the research on the unsupervised RBS model the information bireducts were proposed [29] in order to limit its bias toward common concepts and objects of negligible importance.

Information bireducts can be defined similarly to the decision bireducts (see Subsection 2.3), however their interpretation is slightly different.

Definition 17 (Information bireduct)
Let $\mathbb{S} = (D, F)$ be an information system. A pair (B, X), where $B \subseteq F$ and $X \subseteq D$, is called an information bireduct, iff B discerns all pairs of objects in X and the following properties hold:

1. *There is no proper subset $C \subsetneq B$ such that C discerns all pairs of objects in X.*
2. *There is no proper superset $Y \supsetneq X$ such that B discerns all pairs of objects in Y.*

Just as in the case of decision bireducts, information bireducts do not allow any inconsistencies in X. In a context of information bireducs, however, consistence is understood as an ability to distinguish between any pair of objects in the selected set.

It is interesting to compare information bireducts with templates studied in the association rule mining [96, 97] or concepts known from the formal concept analysis [156, 157]. Templates aim at describing a maximum number of objects with the same (or similar enough) values on a maximum number of attributes. Similarly, concepts are defined as non-extendable subsets of objects that are indiscernible with respect to non-extendable subsets of attributes. On the other hand, information bireducts describe non-extendable subsets of objects that are discernible using irreducible subsets of attributes. The templates and concepts might be seen as corresponding to the most regular areas of data, while the information bireducts correspond to the most irregular, chaotic or one might even claim – the most informative data. Hence, information bireducts can be also called anti-templates or anti-concepts.

In a context of similarity learning, information bireducts can also be intuitively interpreted as artificial agents that try to assess the similarity between given objects. Each of such agents can be characterised by its experience and preferences. In a bireduct, the experience of an agent is explicitly expressed by the set X – the set of cases that the agent knows. The preferences of an agent are modelled in an information bireduct by the set of attributes which are the factors taken into account when the agent makes a judgement. Such an interpretation makes information bireducts become an interesting tool for constructing similarity models from data.

Table 7. An information system $\mathbb{S}$ representing a corpus of nine documents, with three exemplary bireducts

	f_1	f_2	f_3	f_4	f_5	f_6	f_7	f_8	f_9	f_{10}
T_1	1	1	1	0	0	0	0	0	1	1
T_2	0	1	1	1	1	0	1	0	0	1
T_3	1	1	0	0	0	0	1	0	0	0
T_4	0	0	0	0	1	0	0	1	0	0
T_5	1	0	1	0	1	1	0	0	0	0
T_6	1	0	1	0	0	0	0	0	0	0
T_7	0	1	1	0	0	1	1	1	0	0
T_8	0	0	0	0	1	1	1	1	1	0
T_9	1	1	0	0	1	0	0	0	1	0

Exemplary information bireducts:

$BR_1 = (\{f_2, f_3, f_8, f_9\},$
$\quad \{T_1, T_2, T_3, T_4, T_6, T_7, T_8, T_9\})$

$BR_2 = (\{f_3, f_5, f_7, f_9\},$
$\quad \{T_1, T_2, T_3, T_4, T_5, T_6, T_7, T_8, T_9\})$

$BR_3 = (\{f_1, f_2, f_5\},$
$\quad \{T_2, T_3, T_5, T_6, T_7, T_8, T_9\})$

For each bireduct $BR = (B, X)$, $B \subseteq F$, $X \subseteq D$, we can define a commonality relation in D with regard to BR. One example of such a relation can be $\varsigma|_{BR}$ which is defined as follows:

$$(T_i, T_j) \in \varsigma|_{BR} \iff T_j \in X \wedge \left| F_{i|_{BR}} \cap F_{j|_{BR}} \right| \geq p, \tag{54}$$

where $p > 0$, $T_i, T_j \in D$ and $F_{i|_{BR}}$ is a representation of T_i restricted to features from B. Intuitively, two documents are in the commonality relation $\varsigma|_{BR}$ if and only if one of them is covered by the bireduct BR and they have at least p common concepts. The commonality class of a document T with regard to BR will be denoted by $I_{BR}(T)$ since it can be regarded as a specific type of an uncertainty function in the theory of rough sets (see Definition 5). For instance, if we consider the information system from Table 7 and the commonality relation defined by the formula (54) with $p = 2$, then $I_{BR_1}(T_1) = \{T_1, T_2, T_7, T_9\}$ and $I_{BR_1}(T_5) = \emptyset$.

It is important to realize that a commonality of two objects is something conceptually different than indiscernibility. For example, the documents T_5 and T_6 are indiscernible with regard to the features from the bireduct BR_1 but they are not in the commonality relation since they have only one feature f_3 in common.

A similarity model needs to have a functionality which allows it to be applied for analysis of new documents. Typically, we would like to assess their similarity to the known documents (those available during the learning phase) in order to index, classify or assign them to some clusters. For this reason, in the definition of the commonality relation only T_j needs to belong to X. This also makes it more convenient to utilize information bireducts that explicitly define the set of reference cases for which the comparison with the new ones is well defined.

The commonality relation (54) can be used to locally estimate the real significance of arguments for and against the similarity of documents which are being compared. Those arguments, i.e. sets of higher-level features of documents, can be aggregated analogously to the case of the regular RBS similarity function (40). In particular, the similarity of T_i to T_j with regard to a bireduct BR can be computed using the following formula:

$$Sim_{BR}(T_i, T_j) = \frac{|I_{BR}(T_i) \cap I_{BR}(T_j)|}{|I_{BR}(T_i)| + C} - \frac{|(X \setminus I_{BR}(T_i)) \cap I_{BR}(T_j)|}{|X \setminus I_{BR}(T_i)| + C}. \qquad (55)$$

As in the case of the functions *Similarity* and *Dissimilarity* of the RBS model (see Subsection 5.2), the constant $C > 0$ is added to avoid division by zero.

Since each information bireduct is a non-extendable subset of documents, coupled with an irreducible subset of features that discern them, it carries maximum information on a diverse set of reference documents. Due to this property, the utilization of bireducts nullifies the undesired effect which common objects (or usual features) would impose on sizes of the commonality classes and thus, on the similarity function value. Moreover, such a use of the information bireducts in combination with the commonality relation (54) substitutes the need for manual tuning of additional parameters. Instead, the relative intersection size of the commonality classes locally expresses the relevance of arguments for similarity without a need for considering additional parameters. By analogy, the importance of arguments against the similarity is reflected by the relative size of a set that comprises those documents which are not in the commonality class of the first document and are sufficiently compliant with the second text.

Following the example from Table 7, the formula (55) can be used to compute the similarity between any two documents from $\mathbb{S}$ with regard to a chosen bireduct BR_i. For instance, for a very small c, $Sim_{BR_1}(T_1, T_2) \approx 3/4 - 0 = 0.75$, $Sim_{BR_1}(T_1, T_5) = 0 - 0 = 0$ and $Sim_{BR_1}(T_1, T_8) \approx 0 - 1/4 = -0.25$. It is worth noting that the proposed approach keeps the flexibility of the original RBS model and does not impose any properties on the resulting similarity function. Depending on the data and on the selection of $\tau|_{BR}$, the function Sim_{BR} may be not symmetric ($Sim_{BR_1}(T_2, T_1) \approx 0.8 \neq Sim_{BR_1}(T_1, T_2)$), and even not reflexive ($Sim_{BR_1}(T_3, T_3) = 0$). In this case the lack of the reflexivity is a consequence of the fact that $T_3 \notin X$, thus a meaningful assessment of the similarity to this document is not possible. This flexibility of the model makes it consistent with observations made by psychologists [5].

The utilization of information bireducts allows to conveniently model different aspects of similarity. By analogy to the initial experiments with decision bireducts [36], a set of information bireducts will cover much broader aspects of data than an equally sized set of the regular information reducts. This allows to capture approximate dependencies between features which could not be discovered using classical methods and may contribute to the overall performance of the model. The algorithm proposed in [36] for computation decision bireducts can be easily adjusted to the case of information bireducts (Algorithm 4). The randomization of the algorithm guarantees that its multiple executions will produce a diverse set of bireducts.

Algorithm 4. The calculation of an information bireduct of $\mathbb{S} = (D, F)$

Input: an information system $\mathbb{S} = (D, F)$;
         a random permutation $\sigma : \{1, ..., |D| + |F|\} \to \{1, ..., |D| + |F|\}$;
Output: an information bireduct (B, X), $B \subseteq F$, $X \subseteq D$;

1 **begin**
2     $B = F$;
3     $X = \emptyset$;
4   **for** $i = 1$ **to** $|D| + |F|$ **do**
5   **if** $\sigma(i) \leq |F|$ **then**
6   **if** $B \setminus \{F_{\sigma(i)}\}$ *discerns all pairs in* X **then**
7              $B \leftarrow B \setminus \{F_{\sigma(i)}\}$
8   **end**
9   **end**
10   **else**
11   **if** B *discerns all pairs in* $X \cup \{T_{\sigma(i)-K}\}$ **then**
12              $X \leftarrow X \cup \{T_{\sigma(i)-K}\}$
13   **end**
14   **end**
15   **end**
16   **return** (B, X);
17 **end**

To robustly evaluate similarity of two documents the agents need to interact by combining their assessments. The simplest method of such an interaction is to average votes of all agents. In such a case, the final similarity of T_i to T_j can be computed using the following formula:

$$Sim(T_i, T_j) = \frac{\sum_k Sim_{BR_k}(T_i, T_j)}{\#extracted\ bireducts}. \tag{56}$$

For example, if for the information system $\mathbb{S}$ from Table 7 we consider information bireducts BR_1, BR_2 and BR_3, the final similarity of T_1 to T_2 would be equal to $Sim(T_1, T_2) = (0.75 + 0.05 + 0.3)/3 \cong 0.37$.

The design of such a similarity function is computationally feasible and does not require tuning of unintuitive parameters. It also guarantees that the resulting similarity function keeps the flexibility and psychologically plausible properties. Moreover, this kind of an ensemble significantly reduces the variance of similarity judgements in a case when the available data set changes over time (e.g. new documents are added to the repository) and increases model robustness.

However, some more sophisticated methods can also be employed for carrying out the interaction between the agents (bireducts), in order to improve performance in a given task or to reduce similarity computation costs. For instance, properties of extracted bireducts can be used to select only those which will most likely contribute to the performance of the model. The considered properties may include, e.g. a number of selected features, a size of the reference document subset or an average intersection with other bireducts [36]. Using such statistics in combination with general knowledge about the data it is possible to decrease the number of bireducts required for making consistent similarity assessments.

5.6   Summary of the Rule-Based Similarity Models

The construction of the RBS model makes it flexible and allows to apply it in many object domains. By its design, the model tries to incorporate all the plausible properties of a similarity learning method listed in Subsection 4.1. For instance, if the rules which are used for constructing the RBS and DRBS models are consistent, the resulting similarity function is guaranteed to respect the fundamental feature of a similarity relation in a classification context (see Definition 14). Hence, the models are consistent with the training data. For unsupervised RBS this property is difficult to verify in a general case. However, if the semantic concepts which represent the documents are properly assigned (e.g. by experts or a well-trained supervised algorithm), the consistence with data is a natural consequence. Moreover, the similarity function of the RBS model is a proper similarity function if only the data set is consistent and the utilized sets of rules meet a few general requirements (i.e. they are consistent with the data, cover all objects, are minimal and allow to uniquely identify all objects that originally were discernible – see Subsection 5.3).

By its design RBS takes into consideration the context for evaluation of the similarity. A value of the resulting similarity function depends on a decision class of a referent object. The similarity values are also influenced by a presence of other objects in the data. Due to the utilization of the rough sets (i.e. the use of notions such as a reduct, an uncertainty and membership function, as well as the overall approach which resembles searching for appropriate similarity approximation space), the model is capable of automatically adapting itself to the data at hand. This characteristic contributes to good performance of the RBS models in tasks such as a supervised classification. This fact is confirmed by experiments on real-life data which are described in the next Section 6.

The proposed model can be more intuitive for domain experts than typical distance-based approaches. Unlike distance-based metrics, RBS does not enforce any undesirable properties on the induced similarity relation. The set representation, originally borrowed from Tversky's feature contrast model, is more natural for complex objects than the vector representation in a metric space. It is particularly important in situations when the vectors representing objects would have to be high dimensional and possibly sparse (e.g. typical bag-of-words representation of textual documents). The set representation also allows to conveniently model the phenomenon that the lack of some important characteristics in both of compared objects is not an argument for their similarity. Moreover, RBS treats the evaluation of similarity as a problem of resolving conflicts between arguments for and against the similarity, which has an intuitive interpretation.

An important aspect of RBS models is their computational complexity. The construction time of the models depends on particular algorithms used for extracting higher level features. Thanks to the proposed extensions the model can be efficiently built even for very high dimensional data sets. A bigger issue is related to a time cost of a similarity assessment between a single pair of objects. Since the model considers influence of other objects on the context, the computation cost of the RBS similarity function can be in the worst case linear with regard to the number of objects in the data. Since the corresponding cost for typical distance-based similarity functions is constant, such models are easer to apply for analysis on data sets with many objects. On the other hand, the bounded computation cost and robustness with regard to the number of attributes (the sizes of higher-level feature sets can be limited by applying simple filters on rule induction algorithms) makes RBS a useful tool for solving the few-objects-many-attributes problem.

6   Experimental Evaluation of the Rule-Based Similarity Model

This section presents the results of experiments in which Rule-Based Similarity was used for constructing similarity models from various types of data. The aim of those experiments was to demonstrate feasibility of the rule-based approach to the similarity learning problem. Quality of the proposed model was evaluated using methods briefly described in Subsection 3.3. Depending on the context in which a given similarity model was meant to be applied (i.e. an object classification or a semantic similarity of texts), its quality was judged based on a performance of the 1-nearest neighbour classifier or on a conformity of the similarity function to feedback provided by domain experts.

The performance of the RBS model was additionally compared to several other similarity models as well as to the state-of-the-art classifiers in the investigated domain. For the sake of an in-depth analysis of the results, not only are the raw evaluation values presented but also their statistical significance is given. Although most of those results were already published and presented at respectful conferences [24–29], some new views at those tests are shown as well.

All the experiments described in this section were implemented and executed in R System [158]. RBS and its extensions were coded in a native R language with an exception of the discretization algorithm (see Algorithm 2 in Subsection 5.2), which was supported by a C++ code. This code was executed through the *.C* interface provided by R. The whole experimental environment, including the code, data sets and documentation, that allows to conveniently repeat a major part of the conducted experiments is available on request[15].

The section is divided into three subsections. The first one discusses the performance of the original RBS model. In that subsection (Subsection 6.1), RBS was constructed for several benchmark data sets from the UCI repository[16] [159] and compared with several common distance-based models. Next, Subsection 6.2 shows the evaluation of the proposed model on microarray data sets which are an example of high dimensional data. Finally, in Subsection 6.3 a case-study of semantic similarity learning from biomedical texts is presented.

6.1 Performance of Rule-Based Similarity in a Classification Context

The original RBS model was tested on a range of benchmark data sets and compared to several commonly used similarity models. Its performance was also verified on a few high dimensional data sets to check its usefulness for learning a similarity relation characterised by multiple possible aspects. This subsection describes the methodology and presents the results of those test.

Description of the Benchmark Data Sets. The first series of experiments with RBS was conducted on a set of six benchmark data tables, from which five were downloaded from the UC Irvine Machine Learning Repository [159] and one was taken from an R System library *MASS* [158] (the *Cars93* data). They concern domains such as classification of cars, handwritten digits recognition, breast tumour diagnosis and recurrence risk assessment, automatic assessment of nursery applications and Internet advertisements recognition.

A few basic characteristics of the utilized data sets are shown in Table 8. They significantly differ in both, the number of objects and attributes. Three of the selected data sets contain nominal attributes, whereas numeric attributes are present in five tables. The *Nursery* data set was the only one containing purely nominal features. The number of decision classes for each set ranged from two (the *WDBC*, *WPBC* and *InternetAds* data sets) to ten (the *Pendigits* data).

Additional experiments were performed to assess usefulness of the RBS model for the similarity learning from high dimensional data. For this series of tests, four microarray data sets[17] were selected along with the *InternetAds* data table which was already used in the initial experiments. Two of the chosen microarray data sets (*PTC* and *Barrett*) are smaller benchmark tables, whereas the

---

[15] *janusza@mimuw.edu.pl*

[16] UC Irvine Machine Learning Repository: http://archive.ics.uci.edu/ml/

[17] For more information on microarray data see Subsection 6.2.

Table 8. A brief summary of the benchmark datasets used in the experiments with the original RBS model

Data set:	no. instances	no. attributes	numeric attributes	nominal attributes	no. decision classes
Cars93	93	27	Yes	Yes	6
Pendigits	10992	17	Yes	No	10
WDBC	569	31	Yes	No	2
WPBC	198	33	Yes	No	2
Nursery	12958	9	No	Yes	4
InternetAds	3279	1559	Yes	Yes	2

Table 9. A brief summary of microarray data sets used in the experiments

Data set:	no. samples (instances)	no. genes (attributes)	no. decision classes
PTC	51	16502	2
Barrett	90	22277	3
HepatitisC	124	22277	4
SkinPsoriatic	180	54675	3

two other (*HepatitisC* and *SkinPsoriatic*) were obtained from the ArrayExpress repository[18] [160] and were created as a result of larger research projects (experiment accession numbers E-GEOD-14323 and E-GEOD-13355, respectively).

A microarray analysis is an important source of high dimensional data. A case-based approach to knowledge discovery from collections of microarrays is popular due to scarcity of available data samples [14,19,161]. The construction of a reliable similarity model for such a data type is usually a challenging task. Since each data sample is described by numerous attributes (from a few thousands to a few hundred thousands), it is difficult to select those relevant in the considered context. The evaluation of RBS on microarray data was performed to check whether the reduct-based construction of relevant features is effective for high dimensional data.

Compared Similarity Models. In the experiments, the RBS model was constructed for the classification context (see Subsection 3.2 and Subsection 5.2), which was defined by the decision attributes in the data sets. Several other similarity models were also constructed for each of the decision tables. Some of them, such as the Euclidean distance-based model[19] (the Gower distance-based similarity, see Subsection 3.4), were unsupervised, whereas the others utilized the information about classification of objects to adapt to the data.

Among the supervised similarity models used in this comparison, the most important one was the distance-based model combined with a genetic algorithm for learning parameters of local distances. This model will be called Genetic-Based

---

[18] www.ebi.ac.uk/arrayexpress

[19] For the data sets containing nominal attributes the Gower distance was used.

Similarity (GBS). This approach was described in more details in Subsection 4.2. The genetic algorithm was coded in the native R language [158]. As the local distances it used an absolute difference for numeric attributes and the equivalence test for the nominal ones. The local distance values were aggregated using the Euclidean norm (the Gower metric – see Subsection 3.4).

The value of the parameter that governs the population size (i.e. the number of chromosomes) was set to 1000 for the smaller data sets and to 250 for the larger. The probabilities of the replication, mutation and crossover operations for a particular chromosome were computed using the roulette wheel selection technique, based on a distribution of scores (fitness values) in the population. The exact copies of the chromosomes chosen for replication were taken to the next generation. The chromosomes which were chosen for mutation were randomly modified on a small number of genes (the genes were also chosen at random) and added to the new generation. Next, the chromosomes chosen to crossover were randomly matched in pairs to produce two offspring. The new chromosomes were computed as a weighted averages of the parent chromosomes. Finally, scores of the new generation members were computed and the chromosomes with lower scores were eliminated so that the size of the population did not exceed the starting value. In this way the selection of a chromosome was not directly dependent on its fitness but instead, it was conditioned on its ranking in the population.

Additionally, two different similarity learning models were implemented for the experiments on the high dimensional data sets. Both of those models represented the feature selection approach to similarity learning (see Subsection 4.2). The first one, denoted by Gover+FS, was a combination of the Gower distance-based similarity function with a filter attribute selection method. Relevant features were selected using a t-statistic filter. The attributes were ranked according to average p-values of a t-test (the lower the average, the higher the rank) that check equity of attributes' values within pairs of decision classes. The final number of top-ranked attributes for the model was decided using the leave-one-out cross-validation [162, 163] on the available training data. This number was chosen within the range of 2 to 1000. The second model, called Minkowski+FS, extended the first one by allowing to tune the local distance aggregation function (the p parameter in the Minkowski's aggregation function, see Subsection 3.4). To increase the performance of all the distance-based models, numeric attributes in the data sets were scaled before the experiments.

The RBS model was designed for each of the data sets as it was described in Section 5. The relevant higher-level features were constructed from the attributes constituting decision superreducts[20]. The attributes were selected and discretized using a supervised greedy heuristic [34, 154] which was modified so that instead of selecting only one cut at a time, the algorithm was able to simultaneously choose cuts on several attributes that discern most of the samples from different decision classes. The rules which define the higher-level features were discovered using the *decision apriori* algorithm implemented in the *arules*

---

[20] A decision superreduct is a set of attributes that discern all objects from different decision classes but does not need to be minimal. See Subsection 2.3.

Table 10. A summary of the models used in the experiments from Section 6.1

Name:	A short description of a model:
Gower	A standard Gower distance-based similarity function.
Gower + FS	A Gower distance-based similarity function with a t-statistic filter for selecting relevant attributes.
Minkowski + FS	A Minkowski distance-based similarity function combined with a t-statistic attribute filter and a metric parameter learning wrapper.
GBS	A genetic algorithm-based similarity function learning.
RBS	The original Rule-Based Similarity model.

R System library. Only consistent rules[21] were considered with a minimal support factor set to minimum from 5 and 1% of a total number of objects in a training set. Table 10 summarizes the similarity models used in the experiments described in this subsection.

Evaluation Method and Discussion of the Results. The quality of the compared similarity models was evaluated indirectly by measuring classification performance of 1-NN classification rule (Definition 13) applied to the corresponding similarity functions. This similarity model evaluation method was discussed in Subsection 3.3. The classification accuracy (ACC), defined as:

$$ACC = \frac{|\{u \in TestSet : \hat{d}(u) = d(u)\}|}{|TestSet|}, \tag{57}$$

where $TestSet$ is a set of test objects and $\hat{d}(u)$ is a prediction of a decision class for an object u, was estimated using the 10-fold cross-validation technique [164]. The cross-validation was repeated 12 times with different partitioning of data sets into folds. Although in each cross-validation run the division of data was random, the same partitioning was used for every tested similarity model in order to facilitate the comparison of the evaluation results. The mean and standard deviations of model accuracies were computed and the significance of differences in results was assessed using the paired t-test with a 0.99 confidence level.

Table 12 shows the mean and standard deviation of accuracy obtained by similarity models described in this section for the regular data sets. Figure 13 also conveniently visualizes those results.

The classification accuracies of the similarity models on the benchmark data are comparable. The RBS model achieved significantly better results on the data sets containing nominal attributes, with an exception of the *InternetAds* data. Although the accuracy of the RBS model for the most of datasets was slightly higher than the accuracy of the GBS model, the difference was significant (p-value of a t-test was lower than 0.01) only for *Cars93*, *Nursery* in favour of the RBS and *Pendigits* in favour of the GBS. However, it is worth noticing that the time needed to perform the tests was much shorter for the rule-based approach.

---

[21] A rule is called consistent or true if its confidence equals 1. See Subsection 2.1.

Table 11. A comparison of the classification accuracy (ACC) of the tested models

Dataset:	Gower acc. (%)	GBS acc. (%)	RBS acc. (%)
Cars93	63.44 ± 2.41	87.96 ± 1.11	89.25 ± 1.10
Pendigits	97.46 ± 0.21	98.57 ± 0.26	97.30 ± 0.55
WDBC	95.20 ± 0.31	95.66 ± 0.64	95.53 ± 0.48
WPBC	73.13 ± 1.25	76.25 ± 0.82	76.79 ± 0.85
Nursery	76.28 ± 0.39	78.35 ± 0.31	97.02 ± 0.05
InternetAds	96.52 ± 0.09	96.06 ± 0.44	96.07 ± 0.14

Comparing to the simple Gover distance-based approach, RBS turned out to be more reliable for all data tables, except *Pendigits* and *InternetAds*. The average classification accuracy of RBS was statistically higher (p-value of a t-test ≤ 0.01) for the *Cars93*, *WPBC* and *Nursery* data. Interestingly, the performance of RBS was significantly lower than the performance of the Gower model for the *InternetAds* data set. This fact can be treated as an argument for a hypothesis that the RBS model may fail to capture all relevant aspects of similarity when the dimensionality of a data set is high.

To further investigate this problem the second series of experiments was conducted, in which the performance of RBS was compared to several distance-based models on high dimensional data sets. Table 11 shows the results of those tests. They are also displayed in Figure 14.

Table 12. A comparison of the classification accuracy (ACC) of several similarity models for high dimensional data sets

Dataset:	Gower	Gower+FS	Minkowski+FS	GBS	RBS
InternetAds	96.52±0.09	96.79±0.14	96.75±0.12	96.06±0.44	96.07±0.14
PTC	84.31±1.41	96.08±1.67	98.04±0.77	95.74±1.95	98.04±1.31
Barrett	51.67±2.23	55.11±1.86	59.78±1.43	55.55±1.97	62.56±2.12
HepatitisC	86.36±1.66	84.54±1.58	85.08±1.06	84.83±1.38	86.58±0.83
SkinPsoriatic	71.17±1.50	70.83±2.13	69.50±2.39	72.17±1.56	79.00±1.12

The results seem to confirm a hypothesis that similarity learning may have a significant impact on a quality of a similarity model for high dimensional data. For the *PTC* and *Barrett* data the basic Gower distance-based model, which does not adapt to particular data sets, achieved much lower accuracies than all other similarity models. Moreover, the accuracy of the Gover model was lower than the accuracy of RBS on every data table except *InternetAds* and the difference was statistically significant for the *PTC*, *Barrett* and *SkinPsoriatic* data. On the other hand, its results on the *InternetAds* and *HepatitisC* data sets show that even such a simple model may be sufficient to obtain comparable, if not better, results to much more sophisticated approaches, like the genetic algorithm-based similarity learning (GBS) or RBS.

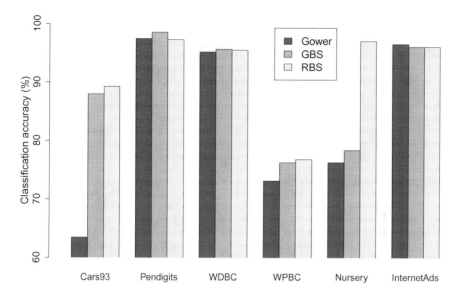

Fig. 13. A visualization of classification accuracy obtained by the compared similarity models on the benchmark data sets

Accuracy scores achieved by RBS were significantly higher (p-value < 0.01) than those of other similarity learning models for the *Barrett*, *HepatitisC* and *SkinPsoriatic* data. In particular, on average RBS turned out to be more reliable than GBS for all data sets except *InternetAds*. The genetic approach does not work very well for high dimensional data. Its probably due to the over-fitting problem which is likely to happen when extensive supervised tuning is performed for models with many parameters [7]. It has been observed, however, that for data sets with many potentially important attributes (i.e. *InternetAds*, *HepatitisC*) the results of RBS are comparable to those of the much simpler models (Gower, Gower+FS). This, perhaps, can be explained by the fact that RBS was using a much lower number attributes than the other models (the total number of attributes used by RBS for a single data set never exceeded 70, whereas for other models it often was more than ten times greater). On one hand, this characteristic can be advantageous since it facilitates interpretability of the model. On the other hand, however, it may deteriorate the performance of the model for complex classification problems, when the number of important features is usually high.

6.2 Evaluation of the Dynamic Rule-Based Similarity Model on Microarray Data

The construction of a similarity model for a high dimensional data may require incorporation of numerous characteristics or factors that have an impact on

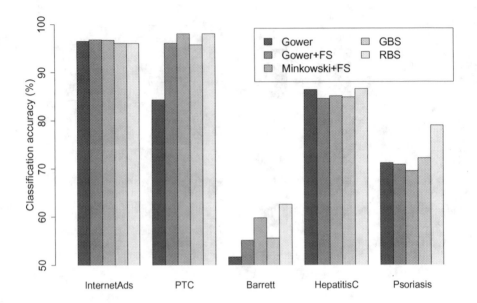

Fig. 14. A visualization of classification accuracy obtained for the high dimensional data sets

similarity judgements in a given context. The DRBS model was proposed in order to enable working on multiple features during the construction of the model, while keeping the reliability and flexibility of RBS which are provided by utilization of notions from the theory of rough sets. This subsection shows some applications of DRBS to analysis of several microarray data sets. It also presents the comparison of classification results of the 1-NN algorithm which uses a DRBS-induced similarity function, and a few state-of-the-art classifiers that are commonly employed for microarray data.

Microarrays as an Example of Real-Life High Dimensional Data. The microarray technology allows researchers to simultaneously monitor thousands of genes in a single experiment. In a microarray data set, specific microarray experiments are treated as objects (e.g. tissue samples). The attributes of those objects correspond to different genes and their values correspond to *expression levels* – the intensity of a process in which information coded in a gene is transformed into a specific gene product. Figure 15 visualizes a single microarray chip after an experiment and its representation in a decision table.

In recent years, a lot of attention of researchers has been put into investigation of this kind of data. That growing interest is largely motivated by numerous practical applications of knowledge acquired from microarray analysis in medical diagnostics, treatment planning, drugs development and many more [165]. When dealing with microarrays, researchers have to overcome the problem of insufficient availability of data. Due to very high costs of microarray processing, usually the number of examples in data sets is limited to several dozens. This

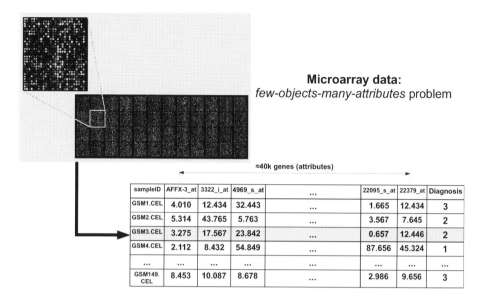

Microarray data:
few-objects-many-attributes problem

≈40k genes (attributes)

sampleID	AFFX-3_at	3322_i_at	4969_s_at	...	22095_s_at	22379_at	Diagnosis
GSM1.CEL	4.010	12.434	32.443	...	1.665	12.434	3
GSM2.CEL	5.314	43.765	5.763	...	3.567	7.645	2
GSM3.CEL	3.275	17.567	23.842	...	0.657	12.446	2
GSM4.CEL	2.112	8.432	54.849	...	87.656	45.324	1
...	...	...	...	...	...	...	...
GSM149.CEL	8.453	10.087	8.678	...	2.986	9.656	3

Fig. 15. A visualization of a microarray chip after an experiment (the top left corner) and its representation in a decision system. The intensity of a colour of spots at the chip reflects expression levels of the genes.

fact, combined with a large number of examined genes, makes many of the classic statistical or machine learning models unreliable and encourages researchers to develop specialized methods for solving the *few-objects-many-attributes* problem [130].

Thorough experiments have been conducted to test the performance of the DRBS model on 11 microarray data sets. All the data samples were downloaded from the ArrayExpress[22] repository. All the data available in the repository are in the MIAME[23] standard [166]. To find out more about this open repository refer to [160]. Each of the used data sets was available in a partially processed form as two separate files. The first one was a data table which contained information about expression levels of genes in particular samples and the second was a *SDRF*[24] file storing meta-data associated with samples (e.g. decision classes). Entries in those files had to be matched during the preprocessing phase. Figure 16 shows a standard microarray data preprocessing schema.

The data sets used in experiments were related to different medical domains and diverse research problems (the ArrayExpress experiment accession numbers are given in parentheses):

1. *AcuteLymphoblasticLeukemia* (ALL) – a recognition of acute lymphoblastic leukemia genetic subtypes (E-GEOD-13425).

---

[22] www.ebi.ac.uk/arrayexpress
[23] **M**inimal **I**nformation **A**bout **M**icroarray **E**xperiment
[24] **S**ample and **D**ata **R**elationship **F**ile

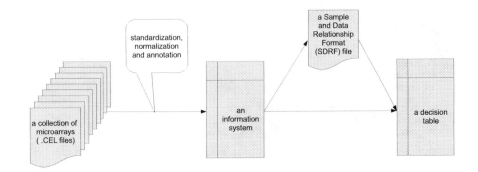

Fig. 16. A standard preprocessing schema for microarray data sets

2. *AnthracyclineTaxaneChemotherapy* (ATC) – a prediction of response to anthracycline/ taxane chemotherapy (E-GEOD-6861).
3. *BrainTumour* (BTu) – diagnosis of human gliomas (E-GEOD-4290).
4. *BurkittLymphoma* (BLy) – a diagnostics of human Burkitts lymphomas (E-GEOD-4475).
5. *GingivalPeriodontits* (GPe) – transcription profiling of human healthy and diseased gingival tissues (E-GEOD-10334).
6. *HeartFailurFactors* (HFF) – transcription profiling of human heart samples with different failure reasons (E-GEOD-5406).
7. *HepatitisC* (HeC) – an investigation of a role of the chronic hepatitis C virus in the pathogenesis of HCV-associated hepatocellular carcinoma (E-GEOD-14323).
8. *HumanGlioma* (HGl) – a recognition of genomic alterations that underlie brain cancers (E-GEOD-9635).
9. *OvarianTumour* (OTu) – a recognition of the ovarian tumour genetic subtypes (E-GEOD-9891).
10. *SepticShock* (SSh) – profiling of critically ill children with the systemic inflammatory response syndrome (SIRS), sepsis, and septic shock spectrum (E-GEOD-13904).
11. *SkinPsoriatic* (SPs) – an investigation of genetic changes related to the skin psoriasis (E-GEOD-13355).

Apart from matching the decisions to samples some additional preprocessing was needed to remove those decision classes which were represented by less than 3 instances. The first 10 data sets were previously used in RSCTC'2010 Discovery Challenge [37]. The eleventh set was previously used for the comparison of the original RBS with distance-based similarity learning models in [26] (see Subsection 6.1). A part of those data sets was also used in the preliminary experiments, in which a developing version DRBS was compared to the original RBS model ([27]). Table 13 presents some basic characteristics of the data sets. They differ in the number of samples (from 124 to 284), the number of examined genes (it varies between 22276 and 61358) and decision classes (2 to 5). Only

Table 13. A brief summary of microarray data sets used in the experiments

Data set:	no. samples	no. genes	no. classes (& class distribution)
ALL	190	22276	5 $(0.28, 0.23, 0.19, 0.23, 0.07)$
ATC	160	61358	2 $(0.59, 0.41)$
BTu	180	54612	4 $(0.28, 0.13, 0.14, 0.45)$
BLy	221	22282	3 $(0.20, 0.58, 0.22)$
GPe	247	54674	2 $(0.74, 0.26)$
HFF	210	22282	3 $(0.51, 0.41, 0.08)$
HeC	124	22276	4 $(0.14, 0.38, 0.15, 0.33)$
HGl	186	59003	5 $(0.57, 0.18, 0.08, 0.07, 0.10)$
OTu	284	54620	3 $(0.87, 0.06, 0.07)$
SSh	227	54674	5 $(0.47, 0.23, 0.12, 0.08, 0.10)$
SPs	180	54675	3 $(0.32, 0.36, 0.32)$

data sets which contained more than 100 samples were used in the experiments with DRBS.

Some of the data sets have significantly uneven class distribution, with one dominant class represented by majority of samples and a few minority classes represented by a small number of objects. Typically, in microarray data, the minority classes are more interesting than the dominant one and this fact should be reflected by the quality measure used to assess the performance of classification algorithms. For this reason, the quality of the models employed in the experiments was evaluated using the *balanced accuracy* (BAC) measure. This is a modification of the standard classification accuracy (Eq. 57) which is insensitive to imbalanced frequencies of decision classes. It is calculated by computing standard classification accuracies ($Accuracy_i$) for each decision class and then averaging the result over all classes ($d = 1, \ldots, l$). In this way, every class has the same contribution to the final result, no matter how frequent it is:

$$ACC_i = \frac{|\{u \in TestSet : \hat{d}(u) = d(u) = i\}|}{|\{u \in TestSet : d(u) = i\}|},$$

$$BAC = \Big(\sum_{i=1}^{l} ACC_i \Big)/l \quad, \tag{58}$$

where l is a total number of decision classes, $TestSet$ is a set of test samples and $\hat{d}(u)$ is a prediction for a sample u. In a case of a 2-class problem with no adjustable decision threshold, balanced accuracy is equivalent to *Area Under the ROC Curve* (AUC). Thus, it may be viewed as a generalization of AUC for multiclass classification problems. Balanced accuracy is insensitive to imbalanced class distribution. This particular measure was used during RSCTC'2010 Discovery Challenge [37] to evaluate solutions of participants and it is also used in the experiments described further in this subsection.

Comparison with the State-of-the-Art in the Microarray Data Classification. The performance of DRBS for microarray data sets was evaluated in two series of experiments. In the first one, DRBS was compared to the original RBS model and three distance-based approaches. The distance-based models used different feature selection techniques combined with a Minkowski distance-based similarity measure (see Subsection 4.2) whose parameter p was automatically tuned on available training data. The utilized feature selection methods were based on *correlation test* [135], *t-test* [88, 136] and the *relief* algorithm [138], respectively. Table 14 shows the results of this comparison for six data tables from the basic track of RSCTC'2010 Discovery Challenge [37]. The results are also visualized in Figure 17.

Table 14. Results of different similarity models for microarray data sets. For each table, the best score is marked in red and the second best is in blue. Mean and standard deviation values are given.

Data set:	1-NN+$corTest$	1-NN+$tTest$	1-NN+$relief$	RBS	DRBS
ALL	0.894±.024	0.936±.023	0.927±.017	0.862 ± .017	0.929 ± .008
BTu	0.548±.010	0.548±.028	0.633±.021	0.613 ± .027	0.687 ± .010
GPe	0.744±.019	0.779±.016	0.785±.025	0.795 ± .018	0.885 ± .016
HFF	0.509±.023	0.532±.029	0.550±.019	0.541 ± .011	0.706 ± .022
HGl	0.509±.023	0.512±.033	0.516±.018	0.464 ± .019	0.648 ± .013
SSh	0.434±.032	0.457±.022	0.458±.024	0.424 ± .023	0.478 ± .017

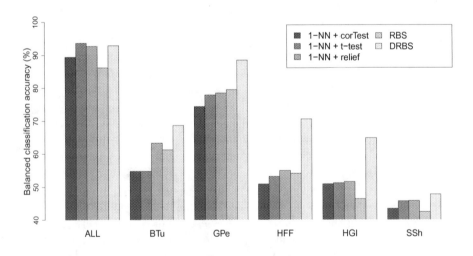

Fig. 17. Balanced classification accuracies of the compared similarity models

From the results of those experiments it is clearly visible that DRBS is a viable improvement over RBS for high dimensional microarray data. Not only did DRBS achieve better balanced accuracy than RBS for each of the data sets but also the differences in their results were always statistically significant. Comparing to the distance-based models, DRBS performed better for five out of six data sets. Only for the *AcuteLymphoblasticLeukemia* data (ALL) the average BAC score of 1-NN model combined with the t-test filter turned out to be higher, yet even in that case, the difference could not be marked as statistically significant. Unexpectedly, the most reliable gene selection method for the distance-based models was the *relief* algorithm, which was ranked second for four tables.

In the second series of experiments, the classification performance achieved with a combination of DRBS and the simple classification rule from Definition 13 was compared to the results of the Random Forest [140, 167] and SVM [136, 145] algorithms which are considered as the state-of-the-art in the biomedical data domain. All the models were implemented in R System ([158]). The DRBS model consisted of $(0.9, 0.95)$–dynamic reducts (see Definition 12) constructed from 250 randomly selected subsets of $5 * \lfloor \sqrt{|A|} \rfloor$ genes, were $|A|$ is a total number of genes (attributes) in a data set. These values guaranteed that a probability of an inclusion of any particular gene to at least one random subset was greater than 0.95 (see Subsection 5.4). Those particular parameter values were chosen as a trade off between computational requirements and robustness of the model. No tuning of the parameters was performed during the experiments due to computation complexity reasons, but it was observed that, for several different data sets, an increase in the number of random subsets of genes usually leads to a slightly better classification quality.

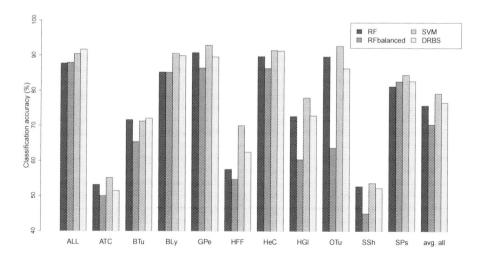

Fig. 18. A visualization of the classification accuracies obtained by the compared algorithms

Table 15. Results of the tests evaluated using the classification accuracy (ACC) measure. For each data set, the best score is marked in red and the second best is in blue. Mean and standard deviation values are given.

Data set:	RF	RF *balanced*	SVM	DRBS
ALL	87.98 ± 0.97	88.77 ± 1.19	90.39 ± 0.96	91.71 ± 0.60
ATC	53.28 ± 2.90	49.64 ± 3.72	55.73 ± 2.78	51.35 ± 3.45
BTu	71.30 ± 1.26	66.44 ± 1.68	71.44 ± 1.45	72.08 ± 1.16
BLy	86.01 ± 1.65	86.05 ± 1.17	90.54 ± 1.79	89.89 ± 1.74
GPe	90.69 ± 1.02	86.50 ± 0.55	92.95 ± 0.90	89.57 ± 1.35
HFF	59.29 ± 1.75	56.03 ± 2.35	70.28 ± 1.81	62.54 ± 2.45
HeC	89.92 ± 1.52	87.16 ± 1.40	91.60 ± 1.80	91.26 ± 1.57
HGl	72.45 ± 1.91	61.74 ± 2.11	77.96 ± 1.23	72.76 ± 1.13
OTu	89.61 ± 0.41	64.91 ± 1.72	92.66 ± 0.52	86.27 ± 1.07
SSh	52.57 ± 1.53	44.49 ± 3.24	53.71 ± 2.48	52.31 ± 1.41
SPs	81.16 ± 1.47	82.64 ± 0.82	84.77 ± 1.45	82.69 ± 1.29
avg. ACC	75.84 ± 14.98	70.40 ± 16.44	79.27 ± 14.67	76.58 ± 15.42

Apriori algorithm from the *arules* package was used for the generation of the rule sets for DRBS. The implementation of Random Forest from the package *randomForest* was used with parameter settings recommended in [167]. Additionally, a balanced version of RF model was checked in which empirical probabilities of decision classes (computed on a training set) were used during the voting as a cut-off values. Support Vector Machine was trained with a linear kernel. The implementation from the package *e1071* was used. Other parameters of SVM were set to values used by the winners of the advanced track of RSCTC'2010 Discovery Challenge [37]. No gene selection method was used for any of the compared models.

The quality of the compared models was assessed using two different quality measures – mean accuracy (ACC, Eq. 57) and balanced accuracy (BAC, Eq. 58). Those measures highlight different properties of a classification model. By its definition, the balanced accuracy gives more weight to instances from minority classes, whereas the standard mean accuracy treats all objects alike and, as a consequence, usually favours the majority class. Depending on applications, each of those properties can be useful, thus, a robust classification model should be able to achieve a high score regardless of the quality measure used for the assessment. The tests were performed using 5-fold cross validation technique. The experiments were repeated 12 times for each of the data sets and models. This testing methodology has been proved to yield reliable error estimates in terms of bias and standard deviation (see [162–164]). The results in terms of the accuracy and the balanced accuracy are given in Tables 15 and 16, respectively.

As expected, there were significant differences between performances of the models depending on the quality measure used for the assessment. In terms of the accuracy, SVM turned out to be the most reliable. It achieved the best score on 9 data sets, whereas DRBS scored the best on 2 data tables. Different results were noted in terms of the balanced accuracy – DRBS and Random Forest (the

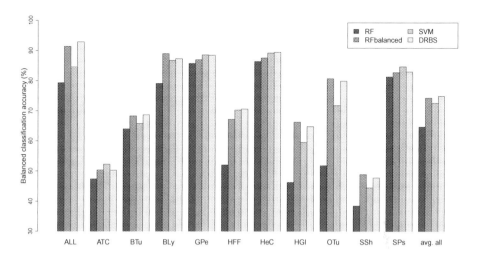

Fig. 19. A visualization of the balanced classification accuracies obtained by the compared algorithms.

Table 16. Results of the tests evaluated using the balanced accuracy (BAC) measure. For each data set, the best score is marked in red and the second best is in blue. Mean and standard deviation values are given.

Data set:	RF	RF*balanced*	SVM	DRBS
ALL	79.34 ± 2.08	91.40 ± 1.25	84.68 ± 2.68	92.93 ± 0.77
ATC	47.28 ± 2.92	50.46 ± 4.08	52.92 ± 2.97	50.33 ± 3.39
BTu	63.93 ± 1.80	68.49 ± 1.83	65.88 ± 1.82	68.71 ± 1.04
BLy	79.15 ± 2.30	89.08 ± 1.32	86.65 ± 2.52	87.30 ± 2.17
GPe	85.88 ± 1.50	87.04 ± 0.70	88.76 ± 1.30	88.52 ± 1.59
HFF	51.98 ± 3.13	67.17 ± 1.72	70.62 ± 2.10	70.64 ± 2.18
HeC	86.42 ± 1.78	87.55 ± 1.36	89.28 ± 1.57	89.52 ± 1.55
HGl	46.35 ± 2.83	66.46 ± 1.43	59.49 ± 2.10	64.76 ± 1.32
OTu	51.79 ± 1.67	80.75 ± 2.23	71.8 ± 1.80	79.91 ± 2.25
SSh	38.08 ± 1.98	48.98 ± 2.94	44.51 ± 2.13	47.77 ± 1.68
SPs	81.32 ± 1.43	82.80 ± 0.80	84.87 ± 1.42	82.95 ± 1.29
avg. BAC	64.68 ± 18.16	74.56 ± 15.22	72.68 ± 15.59	74.85 ± 15.71

balanced version) had the highest mean score on 4 sets, whereas SVM ranked first on 3 data sets. Pairwise comparisons of the tested models are summarized in Tables 17 and 18. For each pair, a number of data sets for which the model named in a column achieved a higher average score is given.

The statistical significance of differences in the results between each of models was verified using the paired Wilcoxon test. This particular statistical test was used instead of the standard t-test because balanced accuracies of different classifiers are not likely to have a normal distribution with equal variances. A

Table 17. A pairwise comparison of accuracies (ACC) of the tested models. Tables show the number of data sets for which the model named in a column achieved a higher score. The number of statistically significant wins is given in parentheses.

Method name:	RF	RF*balanced*	SVM	DRBS
lower\higher	(higher)	(higher)	(higher)	(higher)
RF (lower)	–	3 (1)	11 (9)	7 (4)
RF*balanced* (lower)	8 (8)	–	11 (11)	11 (9)
SVM (lower)	0 (0)	0 (0)	–	2 (1)
DRBS (lower)	4 (1)	0 (0)	9 (6)	–

null hypothesis was tested that the true performance measurements obtained for the particular data set have equal means. Due to a large number of the required comparisons a Bonferroni correction was applied and each test was conducted on 0.9999 confidence level. Differences in means were marked as significant (i.e. the null hypothesis was rejected and a statistical proof was found that performance of one of the model is higher) if the p-value[25] of the test was lower than 0.01. The results of this comparison are also shown in Tables 17 and 18 (in parentheses).

It is worth noticing that DRBS turned out to be the most stable classification model – differences in its score in terms of the accuracy and the balanced accuracy were the smallest of the tested models. For example, although SVM achieved the highest average accuracy on all data sets (79.27), the average difference between its accuracy and the balanced accuracy was 6.59. The value of the same statistic for DRBS was 1.73, with average accuracy of 76.58 (it ranked second in terms of the accuracy measure). DRBS achieved the highest average balanced accuracy of 74.85. This score was only slightly higher than the result of the second algorithm – balanced Random Forest (74.56). The results of the Random Forest algorithms, however, significantly differed with regard to the quality measures. The absolute differences between average values of the two utilized indicators for the Random Forest and balanced Random Forest models were 11.16 and 4.16, respectively. Those results clearly show that DRBS can successfully compete with the state-of-the-art in the microarray data classification.

6.3   Unsupervised Similarity Learning from Textual Data

This subsection demonstrates an application of the unsupervised RBS model for computation of semantic similarity of texts. Reliable semantic similarity assessment is crucial for numerous practical problems, such as information retrieval [113,114], clustering of documents or search results [46,116], or multi-label classification of textual data [32]. The usefulness of unsupervised RBS in one of those tasks, namely document grouping, is verified on a corpus of scientific articles related to biomedicine. The notion of information bireducts (see Subsection 2.3) is combined with Explicit Semantic Analysis (ESA) (see Subsection 4.2) in order to extract important features of the texts, and the performance of unsupervised RBS is compared to the cosine similarity model.

---

[25] The p-value of a statistical test is the probability of obtaining a test statistic value as extreme as the observed one, assuming that the null hypothesis of the test is true.

Table 18. A pairwise comparison of balanced accuracies (BAC) of the models. Tables show the number of data sets for which the model in a column achieved a higher score. The number of statistically significant wins is given in parentheses.

Method name:	RF	RF*balanced*	SVM	DRBS
lower\higher	(higher)	(higher)	(higher)	(higher)
RF (lower)	–	11 (8)	11 (10)	11 (10)
RF*balanced* (lower)	0 (0)	–	5 (3)	6 (3)
SVM (lower)	0 (0)	6 (6)	–	8 (5)
DRBS (lower)	0 (0)	5 (2)	3 (1)	–

Testing Methodology. The experiments were performed on a document corpus consisting of 1000 research papers related to biomedicine which were downloaded from PubMed Central repository [168]. The ESA algorithm, which was used for extracting semantic features of the texts, was adapted to work with the MeSH ontology [153] and implemented in R System [158]. Prior to the experiments, documents from the corpus were processed with natural language processing tools from the *tm* and *RStem* libraries, and the associations between the documents and the MeSH headings were precomputed[26]. The documents were represented by *bags-of-concepts* (see Subsection 4.2) to construct the unsupervised RBS model described in Subsection 5.5, as well as the other models used for comparison.

Additionally, all of the documents were manually labelled by experts from the U.S. National Library of Medicine (NLM) with the MeSH subheadings [153]. Those labels represent a topical content of the documents and as such, they can serve as means for evaluation of truly semantic relatedness of the texts (see Subsection 3.3). In the presented experiments, they were used for computation of a semantic proximity between the analysed documents, which is treated as a reference for the compared similarity functions.

The semantic proximity was measured using F_1-*distance*, defined as:

$$F_1\text{-}distance(T_i, T_j) = 1 - 2 \cdot \frac{precision(S_i, S_j) \cdot recall(S_i, S_j)}{precision(S_i, S_j) + recall(S_i, S_j)}, \qquad (59)$$

where S_i and S_j are sets of labels (MeSH subheadings) assigned by experts, that represent documents $T_i, T_j \in D$, respectively, and

$$precision(S_i, S_j) = \frac{|S_i \cap S_j|}{|S_i|}, \qquad recall(S_i, S_j) = \frac{|S_i \cap S_j|}{|S_j|}. \qquad (60)$$

This particular measure was chosen since it is often used for evaluation of results in the information retrieval setting [115, 127]. Although the evaluation of a similarity measure by a distance metric may fail to capture some of the psychologically important properties and underestimate its real quality, in this way

---

[26] The corpus used in the experiments is a subset of a data set prepared for JRS'2012 Data Mining Competition [32,38]. For more details on the contest and data preprocessing refer to http://tunedit.org/challenge/JRS12Contest)

it was possible to quantitatively assess many similarity models and use those assessments to objectively compare them (see Subsection 3.3).

The semantic proximity between two sets of documents is defined as an average of F_1-*distance* between each pair of texts from different sets:

$$semDist(D_1, D_2) = \frac{\displaystyle\sum_{T_i \in D_1, T_j \in D_2} F_1\text{-}distance(T_i, T_j)}{|D_1| \cdot |D_2|} \ . \tag{61}$$

In experiments, three types of comparisons between the similarity models were made. In the first one, for each of the models, its correlation with the semantic proximity values (Eg. 59) was computed. This kind of evaluation is often used in psychological studies where researchers try to measure the dependence between values returned by their models and assessments made by human subjects [5,21, 169]. It is also commonly utilized in studies on semantic similarity of texts [129].

One disadvantage of this evaluation method is that the linear correlation does not necessarily indicate the usefulness of the measure in practical applications such as the information retrieval or clustering. For this reason, the second series of tests was performed, which aimed at measuring semantic homogeneity of clusterings resulting from the use of different similarity models. To each document in the corpus there was assigned its average semantic proximity (Eg. 61) to other documents from the same cluster and to the remaining texts. If for a division of data into k groups we denote documents from a corpus D belonging to the same cluster as T_i by $cluster_{T_i}$, then we can define a semantic homogeneity of T_i with regard to the semantic proximity function $semDist$ as:

$$homogeneity(T_i) = \frac{B(T_i) - A(T_i)}{\max\big(A(T_i), B(T_i)\big)}, \qquad \text{where}$$

$$A(T_i) = semDist(T_i, cluster_{T_i} \setminus T_i) \qquad \text{and}$$

$$B(T_i) = semDist(T_i, D \setminus cluster_{T_i}).$$

If $cluster_{T_i} \setminus T_i = \emptyset$, then it is assumed that $homogeneity(T_i) = 1$. The average semantic homogeneity of all documents can be used as a measure of clustering quality. Since useful similarity models should lead to meaningful clustering results, the average semantic homogeneity can be employed to intuitively evaluate the usefulness of the compared similarity models for the clustering task.

Finally, in the last series of tests, it was measured how clustering separability is influenced by different similarity models. Two hierarchical clustering algorithms, agnes (AGglomerative NESting) and diana (DIvisive ANAlysis), were used in the experiments. They are described in [127]. Those algorithms differ in the way they form a hierarchy of data groups. Agnes starts by assigning each observation to a different (singleton) group. In the consecutive steps, the two *nearest* clusters are combined to form one larger cluster. This operation is repeated until there remains only a single cluster. The distance between two clusters is evaluated using a linkage function (see the brief discussion in Subsection 3.5). To maximize the semantic homogeneity of the clusters, in the experiments the *complete linkage* method was used.

The diana algorithm starts with a single cluster containing all observations. Next, the clusters are successively divided until each cluster contains a single observation. At each step, only a single group, with the highest internal dissimilarity is split. Two different algorithms were used in the experiments to verify the stability of the compared similarity models and avoid the bias towards a single clustering method.

Apart from a clustering hierarchy, those algorithms return agglomerative (AC) and divisive coefficients (DC), respectively. These coefficients express conspicuousness of a clustering structure in a clustering tree [127]. Although they are internal measures and their value does not necessarily correspond to the semantic relatedness of objects within the clusters, they can give an intuition on interpretability of a clustering solution.

Compared Similarity Models. Four similarity models were implemented in R System [158] for the purpose of the experiments. The unsupervised RBS was constructed as described in Subsection 5.5. The documents from the corpus were given associations to MeSH headings using ESA. An information system $\mathbb{S} = (D, F)$ was constructed consisting of 1000 documents described by a total of $25,640$ semantic features. During preprocessing, the features which were not present in at least one document from the corpus D were filtered out from further analysis. Numerical association values of each term were transformed into four distinct symbolic values. The discretization was based on general knowledge of the data (e.g. for each feature possible association values ranged from 0 to 1000) and the cut thresholds were constant for every feature (i.e. they were set to $= 0$, ≥ 300, ≥ 700 and ≥ 1000).

From the discretized information system, 500 information bireducts (see Subsection 2.3) were computed using random permutations (see Algorithm 4). As expected, they significantly differ in selection of features and reference documents. On average, a bireduct consisted of 210 attributes (min $= 173$, max $= 246$), with each attribute belonging on average to 9 bireducts (min $= 1$, max $= 42$). The average number of documents in a single bireduct was 995 (min $= 988$, max $= 1,000$), and each document appeared on average in 498 bireducts (min $= 489$, max $= 500$). All of the computed information bireducts were used for assessment of similarity by the unsupervised RBS model.

Apart from the unsupervised RBS, for the sake of comparison three other similarity models were implemented. The first one was the standard cosine similarity. In this model, for documents T_i and T_j, represented by vectors C_i, C_j of numerical association strengths to headings from the MeSH ontology (i.e. the vector representation of *bag-of-concepts* described in Subsection 4.2), the cosine similarity is:

$$Sim_{cos}(T_i, T_j) = 1 - Dist_c(C_i, C_j) \ , \tag{62}$$

where $Dist_c$ is the cosine distance function (see Subsection 3.4). This particular measure is very often used for the comparison of texts due to its robust behaviour in high dimensional and sparse data domains.

The second reference model used in the comparison was also based on the cosine similarity measure. However, unlike in the first one, its similarity judgements were not based on the entire data but were ensembled from 500 local evaluations. Each of those local assessments was made by the cosine similarity restrained to the features selected by a corresponding information bireduct (the same as those used in the construction of the unsupervised RBS model). The similarity function of this model was:

$$Sim_{ens}(T_i, T_j) = \sum_{l=1}^{500} Sim_{cos}(T_i|_{BR_l}, T_j|_{BR_l}), \tag{63}$$

were $T|_{BR}$ is a document T represented only by features from BR. This model will be called *cosine ensemble*. It was included in the experiments to investigate the impact of the similarity aggregation technique utilized in unsupervised RBS, on the overall quality of metric-based similarity.

The last reference model, which is called *single RBS*, was constructed using the notion of a commonality relation (Formula 54) and the same aggregation method as in the unsupervised RBS (Formula (55)). The only difference was that it did not use bireducts to create multiple local sub-models but instead, it made similarity assessments using the whole data set. Such a model can be interpreted as a super-agent whose experience covers all available documents and who takes into consideration all possible factors at once. It was used to verify whether the bireduct-based ensemble approach is beneficial for the unsupervised RBS model.

Results of Experiments. In the experiments, the similarity models described in the previous subsection were used to assess similarities between every pair of documents from the corpus. This allowed to construct four similarity matrices, in which a value at an intersection of i-th row and j-th column expressed similarity of the document T_i to T_j. The reference semantic proximity matrix was also constructed using Formula (59), just as it is described in previous section.

Correlations measurements between the values from the similarity matrices obtained for each similarity model and the semantic proximity values are displayed in Table 19. Since similarity assessments made using different measures are likely to come from different distributions, the Spearman's rank correlation [169] was utilized in this test to increase its reliability.

The result of the unsupervised RBS in this test is much higher than results of other models. It is interesting that the correlation of the third of the reference models (the single RBS) with the semantic proximity was the lowest. This highlights the benefit from considering multiple similarity aspects in the RBS approach. On the other hand, the difference between the two cosine-base similarity models is negligible which suggests that the ensemble approach may be ineffective for spherical similarity measures.

Table 19. The correlations between the tested similarity models and the semantic proximity

cosine	cosine ensemble	single RBS	unsupervised RBS
0.155	0.153	0.144	0.186

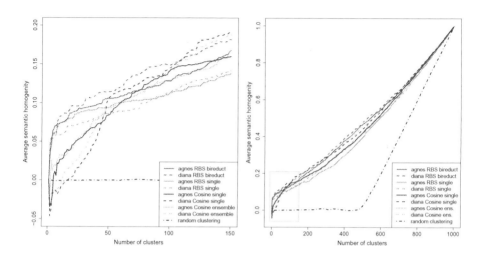

Fig. 20. The comparison of the average semantic homogeneity of clusterings into a consecutive number of groups using different similarity models and clustering algorithms. The plot on the left is a close up of the most interesting area from the plot on the right. The clustering based on a randomly generated dissimilarity matrix is given as the black dot-dashed line.

The second test involved the computation of two clustering trees for each of the models using the agnes and diana algorithms [127]. Since their implementations from the *cluster* library [158]) can work only with symmetric dissimilarity matrices, the similarity matrix of each examined model had to be transformed using Formula (64):

$$dissMatrix = 1 - (simMatrix + simMatrix^T)/2 \ , \qquad (64)$$

where $simMatrix$ is a square matrix with similarity values, 1 is a square matrix containing 1 at every position and $*^T$ is the transposition operation.

Figure 20 presents average semantic homogeneities (62) of clusterings into a consecutive number of groups made using the compared similarity models. The plot on the left is a close up to the area in the plot on the right which is marked by a rectangle with dotted edges. This area corresponds to the most interesting part of the plot because a clustering of documents into a large number of groups produces small individual clusters and is very often difficult to interpret.

The results of this test show evident superiority of the unsupervised RBS similarity over other models for grouping into 2 to 50 groups. Interestingly, in this interval, the semantic homogeneity of the single RBS is also much higher than in the case of the cosine-based measures. The maximum difference between the unsupervised RBS and the cosine similarity for the agnes algorithm is visible when the clustering is made into 4 groups and is equal to 0.083. For the diana algorithm the difference is even higher – when clustering is made into 10 groups it reaches 0.097. For clustering into 51 to approximately 150 groups the results, especially for the agnes algorithm, change slightly in favour of the cosine similarity. The highest loss of the unsupervised RBS was to the cosine ensemble model and reached 0.015 for division of data into 101 groups. Going further, the unsupervised RBS takes the lead again but the difference is not that apparent. In Figure 20 there are also results of a clustering made using a random dissimilarity matrix (as the black dot-dashed line). They can serve as a reference since they clearly show that all of the investigated similarity models led to much better results than the random clusterings.

The compared models differ also in the results of the internal clustering measures. Table 20 shows the agglomerative (AC) and divisive (DC) coefficients of the clustering trees obtained for each similarity model.

Table 20. Values of the internal clustering separability measures

measure:	cosine	cosine ensemble	single RBS	unsupervised RBS
AC	0.33	0.37	0.55	0.58
DC	0.28	0.31	0.51	0.54

In this test, the clustering for the both RBS-based models significantly outperformed the cosine similarity approaches. Higher values of the coefficients indicate that the clusterings resulting from the use of the proposed model are clearer (the groups of documents are better separated), thus, they are more likely to be easier to interpret for experts and end-users. It is also worth noticing that the both ensemble-based measures achieved higher internal scores than their single-model counterparts.

Finally, some additional tests were performed to check how some additional information about generated bireducts can be used for selecting relevant local similarity models during the construction of unsupervised RBS. This can be seen as a way of learning an optimal interaction scheme between artificial agents that try to assess the similarity of given documents. In those experiments, the local RBS models were sorted by the decreasing size of the corresponding bireducts[27]. Next, the correlations between the semantic proximity matrix and the similarity assessments (made using an unsupervised RBS model constructed from the first k bireducts) were computed with k ranging from 1 to 500. The highest score was obtained for a model consisting only of 10 bireducts – it reached 0.203 comparing to 0.186 when all the bireducts were used (see Table 19). It seems that by using an

---

[27] A size of a bireduct is understood as a sum of cardinalities of its attribute set and its document set.

additional validation document set it would be possible to estimate the optimal number of bireducts to be included into the model, and to increase its overall performance. Moreover, considering a lower number of local models would have a positive impact on the scalability of the proposed similarity learning process.

7   Concluding Remarks

This section concludes the dissertation and summarises the presented research on similarity learning from high dimensional data. It also indicates some possible research directions for future development of the described models and points out some interesting application areas.

7.1   Summary

The dissertation discusses the problem of learning a similarity relation that reflects human perception and is based on information about exemplary objects represented in an information system. A special focus is on a situation when the considered objects are described by many attributes and thus their typical representation in a metric space would be extremely high dimensional. For such a case, the typically used distance-based similarity models often fail to capture true resemblance between compared objects [5,9].

Following the research of Amos Tversky on general properties of a similarity relation, a similarity model is proposed in which the metric representation of objects is shifted to a representation by sets of features. In this model, which is called Rule-Based Similarity (RBS), assessments of a similarity degree of a pair of objects depend not only on a context in which the similarity is considered, but also on other objects in the available data. This property remains consistent with observations made by numerous psychologists [4,5,10,20,21].

The proposed similarity model utilizes notions from the theory of rough sets, which is briefly discussed in Section 2. In fact, similarity learning in RBS can be seen as a process of adjusting a similarity approximation space [51,55,56] to better fit the desired context. Apart from the fundamental concepts of rough sets, Section 2 outlines the rough set approach to selecting relevant attributes (i.e. attribute reduction) and forming rules that represent knowledge about a given data set. Those techniques are later applied in the RBS model for discovering sets of higher-level features that influence similarity judgements.

The third section of this dissertation is devoted to the concept of similarity and its general properties. A special emphasis is put on the necessity of fixing a context in which the similarity of two objects is to be considered as it may greatly influence the evaluation outcome (Subsection 3.2). An attempt is also made to form a definition of a similarity function that would meet intuitive expectations for a natural resemblance measure. As a result, the definition of a *proper similarity function* is proposed in Subsection 3.3, which is followed by a discussion of methods for the evaluation of a similarity function quality. Additionally, Section 3 includes an overview of similarity models that are typically employed for solving real-life problems and highlights the essential differences between common distance-based similarity metrics and Tversky's feature contrast model [5]. It also shows application examples of similarity models in a variety of machine learning tasks.

Section 4 focuses on techniques that allow learning a similarity relation or a similarity function from data. It starts with an overview of desirable properties of a similarity learning model and a presentation of several approaches to the problem of adjusting a given distance-based similarity function to better fit a data set at hand, by exploiting *the local-global principle* (Subsection 4.2). Then, Section 5 presents the RBS model, which is the main contribution of this dissertation.

The motivation for RBS comes from observations of psychologists who noticed that properties of similarity do not necessarily correspond to those of distance-based models. In fact, in a specific circumstances every basic property of a distance-based similarity relation can be questioned [5, 20]. On the other hand, some practitioners noticed that non-metric representations of objects require defining their higher-level characteristics [4, 50, 120] which often are not present in the original data. For this reason, the construction procedure of RBS, described in Subsection 5.2, includes an automatic feature extraction step that uses decision and inhibitory rules to form sets of arguments for and against the similarity of given objects. During assessment of the similarity, those arguments are aggregated analogically to the contrast model. Unlike in that model, however, weights of the feature sets need not to be set manually, but are derived from available data. Subsection 5.3 discusses several plausible properties of the proposed model and shows that under certain conditions its similarity function meets the definition of the proper similarity function for a similarity in the context of a classification problem.

The original RBS model was extended in order to facilitate its application to two different types of problems that typically involve dealing with high dimensional data. The first extension, which is described in Subsection 5.4, is designed for learning a similarity function in a context of a classification problem from data containing tens thousands of attributes and possibly only a few hundreds of objects. Dynamic Rule-Based Similarity (DRBS) utilizes the notion of dynamic decision reducts for constructing multiple sets of features that may robustly represent different views or aspects of the similarity. Those aspects are then aggregated using DRBS similarity function by an analogy with the Random Forest algorithm [140].

The main purpose of the second extension, called unsupervised RBS (Subsection 5.5), is unsupervised rule-based learning of a semantic resemblance between texts. In order to make it possible, the higher-level features of textual documents that represent relevant aspects of their semantics are extracted using a combination of Explicit Semantic Analysis (ESA) [38, 129] and a novel notion of information bireduct [29, 36, 170]. Due to the utilization of the information bireducts, the evaluation of similarity in the unsupervised RBS model can be interpreted as an interaction between artificial agents who are characterised by different experience and preferences, and thus have different views on semantics of the compared documents.

Finally, Section 6 presents the results of experimental evaluation of different RBS models for a wide array of data types. The performance of the original RBS was compared to several distance-based similarity learning models on well-known benchmark data tables acquired from UCI machine learning repository [159]. The empirical quality evaluation of 1-nearest-neighbour classification revealed that RBS can successfully compete with popular similarity models on standard data sets. For high dimensional microarray data from ArrayExpress [160], not only did DRBS significantly outperform other similarity models but it also achieved better classification results in terms of the balanced accuracy measure than the Random Forest and SVM algorithms, which are considered the state-of-the-art. Unsupervised RBS was also tested and its usefulness for practical applications, such as document clustering, was verified. Groupings constructed using this model turned out to be more semantically homogeneous than those obtained from clustering using standard methods.

7.2   Future Works

There are several possible directions for the future research on the rule-based models of similarity. One idea is to focus on the incorporation of domain knowledge into the model. For example, by using a dedicated similarity ontology it would be possible to model similarity of complex objects or even behavioural patterns changing over time [23, 45, 120, 171]. This kind of a domain knowledge may be effectively used to learn the local similarity relations as well as to create even better higher-level features, e.g. by merging those rules which are semantically similar. Moreover, the method for aggregating arguments for and against the similarity of given objects that is used in RBS is just one of many possibilities. In the future some other aggregation functions could be tried. Such functions could even be learnt from data based on some auxiliary knowledge or interactions with experts.

RBS may also serve as a means for extending notions of rough sets and rough approximations. Currently, there exist several generalizations of rough sets that are based on the notion of similarity [57, 58]. It might be interesting to combine similarity-based rough sets with rough set-based similarity due to the conforming philosophy of those two models. Such a combination can help in obtaining approximations which are more intuitive for human experts and thus can be more useful for real-life data analysis.

Another possible direction in research on RBS is to focus on scalability of the model. In order to facilitate its practical applications in a wide array of domains, scalability of RBS needs to be further enhanced. The scalability can be considered in several aspects, e.g. in terms of a number of training and test objects or in terms of a total number of attributes. Currently, the computational cost of RBS models strongly depends on particular algorithms used for the discretization, attribute reduction and extraction of rules. Having constructed an

RBS model, the evaluation of similarity between a single pair of objects can be done in a linear time with regard to the number of extracted rules and objects. Moreover, since a value of RBS similarity function can be computed by a single SQL query, even a sub-linear time complexity would be possible to achieve by utilization of modern analytical database technologies [172, 173]. Therefore, an implementation of RBS that would be able to make use of contemporary Relational Database Management Systems (RDBMS) would definitely be helpful in real-life applications of the model.

An important factor in the scalability context is also the method for computation of reducts that represent different aspects of the approximated similarity relation. This problem is closely related to an efficient construction of reduct ensembles [36, 103]. The results of the recent research in this topic suggest that an incorporation of auxiliary knowledge about clusterings of original attributes in data can greatly speed up the computation of diverse sets of reducts [33].

Finally, it would be very useful to come up with a unified framework for developing and testing similarity learning methods. Although there exist systems for data analysis that make use of rough set methods for a feature subset selection and extraction of rules, e.g. RSES and RSESlib [174] or Rosetta [175], there is no environment allowing to conveniently combine those tools for the construction of higher-level similarity models. Such an extension, for example in a form of a library for increasingly popular R System [158], would definitely bring benefit to the rough set community, as well as to other data mining researchers. Algorithms used in the construction of RBS models combined with discretization and rule induction methods implemented for the described experiments may serve as a starting point for this task.

Any further progress in the field of learning similarity relation from data would be beneficial to researchers from many domains. This problem is especially important in domains such as biomedicine, where efficient and more accurate models could lead to discovering more effective and safer drugs or better planing of treatments [37, 120, 145, 165]. The classical distance-based approach is often unable to deal with the few-objects-many-attributes problem and the rule-based approach appears to be a promising alternative.

Acknowledgements. I wish to thank my supervisor, Prof. Hung Son Nguyen, for his guidance and invaluable support in writing this dissertation. I am thankful to Prof. Andrzej Skowron for countless inspiring conversations and remarks. They have always been a motivation for my research and writing. I am very grateful to Dominik Ślęzak whose enthusiasm and excellent ideas have taught me to stay open-minded. I would also like to sincerely thank others from Institute of Mathematics, who I consider not only my colleagues but my friends. I have learnt a lot from their knowledge and experience.

My deepest gratitude, however, goes to my family, especially to my parents and two sisters who have supported me through my entire life. Non of the things that I achieved would be possible without them. Thank You!

References

1. Pinker, S.: How the mind works. W. W. Norton (1998)
2. Schank, R.C.: Dynamic Memory: A Theory of Learning in Computers and People. Cambridge University Press, New York (1982)
3. Thagard, P.: 10. In: Mind: Introduction to Cognitive Science. Segunda edn. MIT Press, Cambridge (2005)
4. Hahn, U., Chater, N.: Understanding similarity: A joint project for psychology, case based reasoning, and law. Artificial Intelligence Review 12, 393–427 (1998)
5. Tversky, A.: Features of similarity. Psychological Review 84, 327–352 (1977)
6. Aamodt, A., Plaza, E.: Case-based reasoning: Foundational issues, methodological variations, and system approaches. Artificial Intelligence Communications 7(1), 39–59 (1994)
7. Mitchell, T.M.: Machine Learning. McGraw Hill series in computer science. McGraw-Hill (1997)
8. Tan, P.N., Steinbach, M., Kumar, V.: Introduction to Data Mining. Addison-Wesley, Boston (2006)
9. Beyer, K., Goldstein, J., Ramakrishnan, R., Shaft, U.: When is "nearest neighbor" meaningful? In: Beeri, C., Bruneman, P. (eds.) ICDT 1999. LNCS, vol. 1540, pp. 217–235. Springer, Heidelberg (1998)
10. Krantz, D.H., Tversky, A.: Similarity of rectangles: An analysis of subjective dimensions. Journal of Mathematical Psychology 12(1), 4–34 (1975)
11. Tversky, A., Krantz, D.H.: The dimensional representation and the metric structure of similarity data. Journal of Mathematical Psychology 7(3), 572–596 (1970)
12. Chopra, S., Hadsell, R., LeCun, Y.: Learning a similarity metric discriminatively, with application to face verification. In: Proceedings of the 2005 IEEE Computer Society Conference on Computer Vision and Pattern Recognition (CVPR 2005), pp. 539–546. IEEE Computer Society, Washington, DC (2005)
13. Hechenbichler, K., Schliep, K.: Weighted k-Nearest-Neighbor Techniques and Ordinal Classification (October 2004), a Discussion paper
14. Martín-Merino, M., De Las Rivas, J.: Improving k-NN for human cancer classification using the gene expression profiles. In: Adams, N.M., Robardet, C., Siebes, A., Boulicaut, J.-F. (eds.) IDA 2009. LNCS, vol. 5772, pp. 107–118. Springer, Heidelberg (2009)
15. Nguyen, S.H.T.: Regularity analysis and its applications in data mining. PhD thesis, Warsaw University, Faculty of Mathematics, Informatics and Mechanics, Part II: Relational Patterns (1999)
16. Stahl, A., Gabel, T.: Using evolution programs to learn local similarity measures. In: Ashley, K.D., Bridge, D.G. (eds.) ICCBR 2003. LNCS, vol. 2689, pp. 537–551. Springer, Heidelberg (2003)
17. Wojna, A.: Analogy-based reasoning in classifier construction. PhD thesis, Warsaw University, Faculty of Mathematics, Informatics and Mechanics (2004)
18. Xing, E.P., Ng, A.Y., Jordan, M.I., Russell, S.J.: Distance metric learning with application to clustering with side-information. In: Becker, S., Thrun, S., Obermayer, K. (eds.) Advances in Neural Information Processing Systems 15, NIPS 2002, December 9-14, pp. 505–512. MIT Press, Vancouver (2002)
19. Xiong, H., Chen, X.W.: Kernel-based distance metric learning for microarray data classification. BMC Bioinformatics 7(299) (2006) (online)
20. Gati, I., Tversky, A.: Studies of similarity. In: Rosch, E., Lloyd, B. (eds.) Cognition and Categorization, pp. 81–99. L. Erlbaum Associates, Hillsdale (1978)

21. Goldstone, R., Medin, D., Gentner, D.: Relational similarity and the noninde-pendence of features in similarity judgments. Cognitive Psychology 23, 222–262 (1991)
22. Sebag, M., Schoenauer, M.: A rule-based similarity measure. In: Wess, S., Richter, M., Althoff, K.-D. (eds.) EWCBR 1993. LNCS, vol. 837, pp. 119–130. Springer, Heidelberg (1994)
23. Janusz, A.: Similarity relation in classification problems. In: Chan, C.-C., Grzymala-Busse, J.W., Ziarko, W.P. (eds.) RSCTC 2008. LNCS (LNAI), vol. 5306, pp. 211–222. Springer, Heidelberg (2008)
24. Janusz, A.: Learning a Rule-Based Similarity: A comparison with the Genetic Ap-proach. In: Proceedings of the Workshop on Concurrency, Specification and Pro-gramming (CS&P 2009), Kraków-Przegorzaly, Poland, September 28-30, vol. 1, pp. 241–252 (2009)
25. Janusz, A.: Rule-based similarity for classification. In: Proceedings of the WI/IAT 2009 Workshops, Milan, Italy, September 15-18, pp. 449–452. IEEE Computer Society, Los Alamitos (2009)
26. Janusz, A.: Discovering rules-based similarity in microarray data. In: Hüllermeier, E., Kruse, R., Hoffmann, F. (eds.) IPMU 2010. LNCS, vol. 6178, pp. 49–58. Springer, Heidelberg (2010)
27. Janusz, A.: Utilization of dynamic reducts to improve performance of the rule-based similarity model for highly-dimensional data. In: Proceedings of the 2010 IEEE/WIC/ACM International Conference on Web Intelligence and International Conference on Intelligent Agent Technology - Workshops, pp. 432–435. IEEE (2010)
28. Janusz, A.: Dynamic rule-based similarity model for DNA microarray data. In: Peters, J.F., Skowron, A. (eds.) Transactions on Rough Sets XV. LNCS, vol. 7255, pp. 1–25. Springer, Heidelberg (2012)
29. Janusz, A., Ślęzak, D., Nguyen, H.S.: Unsupervised similarity learning from tex-tual data. Fundamenta Informaticae 119(3)
30. Janusz, A.: Combining multiple classification or regression models using genetic algorithms. In: Szczuka, M., Kryszkiewicz, M., Ramanna, S., Jensen, R., Hu, Q. (eds.) RSCTC 2010. LNCS, vol. 6086, pp. 130–137. Springer, Heidelberg (2010)
31. Janusz, A.: Combining multiple predictive models using genetic algorithms. In-telligent Data Analysis 16(5), 763–776 (2012)
32. Janusz, A., Nguyen, H.S., Ślęzak, D., Stawicki, S., Krasuski, A.: JRS'2012 Data Mining Competition: Topical Classification of Biomedical Research Papers. In: Yao, J., Yang, Y., Słowiński, R., Greco, S., Li, H., Mitra, S., Polkowski, L. (eds.) RSCTC 2012. LNCS, vol. 7413, pp. 422–431. Springer, Heidelberg (2012)
33. Janusz, A., Ślęzak, D.: Utilization of attribute clustering methods for scalable computation of reducts from high-dimensional data. In: Ganzha, M., Maciaszek, L.A., Paprzycki, M. (eds.) Proceedings of Federated Conference on Computer Science and Information Systems - FedCSIS 2012, Wrocław, Poland, September 9-12, pp. 295–302 (2012)
34. Janusz, A., Stawicki, S.: Applications of approximate reducts to the feature selec-tion problem. In: Yao, J., Ramanna, S., Wang, G., Suraj, Z. (eds.) RSKT 2011. LNCS, vol. 6954, pp. 45–50. Springer, Heidelberg (2011)
35. Kurach, K., Pawłowski, K., Romaszko, Ł., Tatjewski, M., Janusz, A., Nguyen, H.S.: An ensemble approach to multi-label classification of textual data. In: Zhou, S., Zhang, S., Karypis, G. (eds.) ADMA 2012. LNCS, vol. 7713, pp. 306–317. Springer, Heidelberg (2012)

36. Ślęzak, D., Janusz, A.: Ensembles of bireducts: Towards robust classification and simple representation. In: Kim, T.-H., Adeli, H., Slezak, D., Sandnes, F.E., Song, X., Chung, K.-I., Arnett, K.P. (eds.) FGIT 2011. LNCS, vol. 7105, pp. 64–77. Springer, Heidelberg (2011)

37. Wojnarski, M., Janusz, A., Nguyen, H.S., Bazan, J., Luo, C., Chen, Z., Hu, F., Wang, G., Guan, L., Luo, H., Gao, J., Shen, Y., Nikulin, V., Huang, T.-H., McLachlan, G.J., Bošnjak, M., Gamberger, D.: RSCTC'2010 discovery challenge: Mining DNA microarray data for medical diagnosis and treatment. In: Szczuka, M., Kryszkiewicz, M., Ramanna, S., Jensen, R., Hu, Q. (eds.) RSCTC 2010. LNCS, vol. 6086, pp. 4–19. Springer, Heidelberg (2010)

38. Janusz, A., Świeboda, W., Krasuski, A., Nguyen, H.S.: Interactive document indexing method based on explicit semantic analysis. In: Yao, J., Yang, Y., Słowiński, R., Greco, S., Li, H., Mitra, S., Polkowski, L. (eds.) RSCTC 2012. LNCS, vol. 7413, pp. 156–165. Springer, Heidelberg (2012)

39. Ślęzak, D., Janusz, A., Świeboda, W., Nguyen, H.S., Bazan, J.G., Skowron, A.: Semantic analytics of pubMed content. In: Holzinger, A., Simonic, K.-M. (eds.) USAB 2011. LNCS, vol. 7058, pp. 63–74. Springer, Heidelberg (2011)

40. Szczuka, M., Janusz, A., Herba, K.: Clustering of rough set related documents with use of knowledge from dBpedia. In: Yao, J., Ramanna, S., Wang, G., Suraj, Z. (eds.) RSKT 2011. LNCS, vol. 6954, pp. 394–403. Springer, Heidelberg (2011)

41. Pawlak, Z.: Information systems, theoretical foundations. Information Systems 3(6), 205–218 (1981)

42. Pawlak, Z., Skowron, A.: Rough sets and boolean reasoning. Information Sciences 177(1), 41–73 (2007)

43. Pawlak, Z., Skowron, A.: Rough sets: Some extensions. Information Sciences 177(1), 28–40 (2007)

44. Pawlak, Z., Skowron, A.: Rudiments of rough sets. Information Sciences 177(1), 3–27 (2007)

45. Bazan, J.: Hierarchical classifiers for complex spatio-temporal concepts. In: Peters, J.F., Skowron, A., Rybiński, H. (eds.) Transactions on Rough Sets IX. LNCS, vol. 5390, pp. 474–750. Springer, Heidelberg (2008)

46. Ngo, C.L., Nguyen, H.S.: A tolerance rough set approach to clustering web search results. In: Boulicaut, J.-F., Esposito, F., Giannotti, F., Pedreschi, D. (eds.) PKDD 2004. LNCS (LNAI), vol. 3202, pp. 515–517. Springer, Heidelberg (2004)

47. Pawlak, Z.: Rough sets, rough relations and rough functions. Fundamenta Informaticae 27(2-3), 103–108 (1996)

48. Peters, G., Lingras, P., Ślęzak, D., Yao, Y.: Rough Sets: Selected Methods and Applications in Management and Engineering. In: Advanced Information and Knowledge Processing. Springer, London (2012)

49. Sikora, M., Sikora, B.: Rough natural hazards monitoring. In: Peters, G., Lingras, P., Ślęzak, D., Yao, Y. (eds.) Selected Methods and Applications of Rough Sets in Management and Engineering. Advanced Information and Knowledge Processing, pp. 163–179. Springer, London (2012)

50. Nguyen, S.H., Bazan, J., Skowron, A., Nguyen, H.S.: Layered learning for concept synthesis. In: Peters, J.F., Skowron, A., Grzymała-Busse, J.W., Kostek, B.z., Swiniarski, R.W., Szczuka, M.S. (eds.) Transactions on Rough Sets I. LNCS, vol. 3100, pp. 187–208. Springer, Heidelberg (2004)

51. Skowron, A., Stepaniuk, J.: Approximation of relations. In: RSKD 1993: Proceedings of the International Workshop on Rough Sets and Knowledge Discovery, pp. 161–166. Springer, London (1994)

52. Szczuka, M.S., Skowron, A., Stepaniuk, J.: Function approximation and quality measures in rough-granular systems. Fundamenta Informaticae 109(3), 339–354 (2011)
53. Gomolinska, A.: Approximation spaces based on relations of similarity and dissimilarity of objects. Fundamenta Informaticae 79(3-4), 319–333 (2007)
54. Greco, S., Matarazzo, B., Słowiński, R.: Fuzzy similarity relation as a basis for rough approximations. In: Polkowski, L., Skowron, A. (eds.) RSCTC 1998. LNCS (LNAI), vol. 1424, pp. 283–289. Springer, Heidelberg (1998)
55. Polkowski, L.T., Skowron, A., Zytkow, J.M.: Rough foundations for rough sets. In: Lin, T.Y. (ed.) Rough Sets and Soft Computing. Conference Proceedings, pp. 142–149. San Jose State University, San Jose (1994)
56. Skowron, A., Stepaniuk, J.: Tolerance approximation spaces. Fundamenta Informaticae 27(2/3), 245–253 (1996)
57. Słowiński, R., Vanderpooten, D.: Similarity relation as a basis for rough approximations. In: Wang, P. (ed.) Advances in Machine Intelligence and Soft-Computing, vol. IV, pp. 17–33. Duke University Press, Durham (1997)
58. Słowiński, R., Vanderpooten, D.: A generalized definition of rough approximations based on similarity. IEEE Transactions on Data and Knowledge Engineering 12, 331–336 (2000)
59. Yao, Y.: Semantics of fuzzy sets in rough set theory. In: Peters, J.F., Skowron, A., Dubois, D., Grzymała-Busse, J.W., Inuiguchi, M., Polkowski, L. (eds.) Transactions on Rough Sets II. LNCS, vol. 3135, pp. 297–318. Springer, Heidelberg (2004)
60. Hu, X., Cercone, N.: Rough sets similarity-based learning from databases. In: KDD, pp. 162–167 (1995)
61. Maurer, A.: Learning similarity with operator-valued large-margin classifiers. Journal of Machine Learning Research 9, 1049–1082 (2008)
62. Komorowski, J., Pawlak, Z., Polkowski, L., Skowron, A.: Rough sets: A tutorial (1998)
63. Dubois, D., Prade, H.: Rough fuzzy sets and fuzzy rough sets. International Journal of General Systems 17(2-3), 191–209 (1990)
64. Pal, S.K.: Soft data mining, computational theory of perceptions, and rough-fuzzy approach. Information Sciences 163(1-3), 5–12 (2004)
65. Pal, S.K., Meher, S.K., Dutta, S.: Class-dependent rough-fuzzy granular space, dispersion index and classification. Pattern Recognition 45(7), 2690–2707 (2012)
66. Zadeh, L.A.: Fuzzy sets. Information and Control 8(3), 338–353 (1965)
67. Świeboda, W., Nguyen, H.S.: Rough Set Methods for Large and Sparse Data in EAV Format. In: 2012 IEEE RIVF International Conference on Computing & Communication Technologies, Research, Innovation, and Vision for the Future (RIVF), Ho Chi Minh City, Vietnam, February 27-March 1, pp. 1–6. IEEE (2012)
68. Greco, S., Matarazzo, B., Słowiński, R.: Handling missing values in rough set analysis of multi-attribute and multi-criteria decision problems. In: Zhong, N., Skowron, A., Ohsuga, S. (eds.) RSFDGrC 1999. LNCS (LNAI), vol. 1711, pp. 146–157. Springer, Heidelberg (1999)
69. Latkowski, R.: Flexible indiscernibility relations for missing attribute values. Fundamenta Informaticae 67(1-3), 131–147 (2005)
70. Stefanowski, J., Tsoukiàs, A.: Incomplete information tables and rough classification. Computational Intelligence 17(3), 545–566 (2001)
71. Grzymala-Busse, J.W.: Rough set strategies to data with missing attribute values. In: Lin, T.Y., Ohsuga, S., Liau, C.J., Hu, X. (eds.) Foundations and Novel Approaches in Data Mining. SCI, vol. 9, pp. 197–212. Springer, Heidelberg (2006)

72. Grzymala-Busse, J.W., Rzasa, W.: Local and global approximations for incomplete data. In: Greco, S., Hata, Y., Hirano, S., Inuiguchi, M., Miyamoto, S., Nguyen, H.S., Słowiński, R. (eds.) RSCTC 2006. LNCS (LNAI), vol. 4259, pp. 244–253. Springer, Heidelberg (2006)

73. Skowron, A., Stepaniuk, J., Świniarski, R.W.: Modeling rough granular computing based on approximation spaces. Information Sciences 184(1), 20–43 (2012)

74. Pawlak, Z.: Decision logik. Bulletin of the EATCS 44, 201–225 (1991)

75. Delimata, P., Moshkov, M.J., Skowron, A., Suraj, Z.: Inhibitory Rules in Data Analysis: A Rough Set Approach. SCI, vol. 163. Springer (2009)

76. An, A., Cercone, N.: Rule quality measures for rule induction systems: Description and evaluation. Computational Intelligence 17(3), 409–424 (2001)

77. Dean, P., Famili, A.: Comparative performance of rule quality measures in an induction system. Applied Intelligence 7, 113–124 (1997)

78. Lavrač, N., Flach, P.A., Zupan, B.: Rule Evaluation Measures: A Unifying View. In: Džeroski, S., Flach, P.A. (eds.) ILP 1999. LNCS (LNAI), vol. 1634, pp. 174–185. Springer, Heidelberg (1999)

79. Džeroski, S., Cestnik, B., Petrovski, I.: Using the m-estimate in rule induction. Journal of Computing and Information Technology 1(1), 37–46 (1993)

80. Pawlak, Z.: Rough sets - Theoretical Aspects of Reasoning about Data. Kluwer Academic Publishers (1991)

81. Modrzejewski, M.: Feature selection using rough sets theory. In: Brazdil, P.B. (ed.) ECML 1993. LNCS, vol. 667, pp. 213–226. Springer, Heidelberg (1993)

82. Nguyen, H.S., Skowron, A.: Boolean reasoning for feature extraction problems. In: Raś, Z.W., Skowron, A. (eds.) ISMIS 1997. LNCS, vol. 1325, pp. 117–126. Springer, Heidelberg (1997)

83. Zhong, N., Dong, J., Ohsuga, S.: Using rough sets with heuristics for feature selection. Journal of Intelligent Information Systems 16(3), 199–214 (2001)

84. Katzberg, J.D., Ziarko, W.: Variable precision rough sets with asymmetric bounds. In: Proceedings of the International Workshop on Rough Sets and Knowledge Discovery, RSKD 1993, pp. 167–177. Springer, London (1994)

85. Ziarko, W.: Variable precision rough set model. Journal of Computer and System Sciences 46, 39–59 (1993)

86. Pawlak, Z.: Rough sets: present state and the future. Foundations of Computing and Decision Sciences 18(3-4), 157–166 (1993)

87. Guyon, I., Elisseeff, A.: An introduction to variable and feature selection. Journal of Machine Learning Research 3, 1157–1182 (2003)

88. Guyon, I., et al.: Feature Extraction: Foundations and Applications. Studies in Fuzziness and Soft Computing. Springer (August 2006)

89. Nguyen, H.S., Nguyen, S.H., Skowron, A.: Searching for features defined by hyperplanes. In: Michalewicz, M., Raś, Z.W. (eds.) ISMIS 1996. LNCS, vol. 1079, pp. 366–375. Springer, Heidelberg (1996)

90. Valdés, J., Barton, A.: Relevant attribute discovery in high dimensional data: Application to breast cancer gene expressions, 482–489 (2006)

91. Skowron, A., Rauszer, C.: The Discernibility Matrices and Functions in Information Systems, pp. 331–362. Kluwer, Dordrecht (1992)

92. Nguyen, H.S.: On the decision table with maximal number of reducts. Electronic Notes in Theoretical Computer Science 82(4), 198–205 (2003)

93. Ślęzak, D.: Various approaches to reasoning with frequency based decision reducts: a survey, pp. 235–285. Physica-Verlag GmbH, Heidelberg (2000)

94. Ślęzak, D.: Rough sets and functional dependencies in data: Foundations of association reducts. In: Gavrilova, M.L., Kenneth Tan, C.J., Wang, Y., Chan, K.C.C. (eds.) Transactions on Computational Science V. LNCS, vol. 5540, pp. 182–205. Springer, Heidelberg (2009)

95. Kohavi, R., John, G.H.: Wrappers for feature subset selection. Artif. Intell. 97, 273–324 (1997)

96. Nguyen, H.S.: Approximate boolean reasoning: Foundations and applications in data mining. In: Peters, J.F., Skowron, A. (eds.) Transactions on Rough Sets V. LNCS, vol. 4100, pp. 334–506. Springer, Heidelberg (2006)

97. Nguyen, H.S., Ślęzak, D.: Approximate reducts and association rules. In: Zhong, N., Skowron, A., Ohsuga, S. (eds.) RSFDGrC 1999. LNCS (LNAI), vol. 1711, pp. 137–145. Springer, Heidelberg (1999)

98. Ślęzak, D.: Approximate reducts in decision tables. In: Proceedings of IPMU 1996 (1996)

99. Ślęzak, D.: Approximate entropy reducts. Fundamenta Informaticae 53(3-4), 365–390 (2002)

100. Bazan, J.G., Skowron, A., Synak, P.: Dynamic reducts as a tool for extracting laws from decisions tables. In: Raś, Z.W., Zemankova, M. (eds.) ISMIS 1994. LNCS, vol. 869, pp. 346–355. Springer, Heidelberg (1994)

101. Bazan, J.G.: A comparison of dynamic and non-dynamic rough set methods for extracting laws from decision tables. In: Polkowski, L., Skowron, A. (eds.) Rough Sets in Knowledge Discovery 2: Applications, Case Studies and Software Systems, pp. 321–365. Physica Verlag (1998)

102. Wróblewski, J.: Ensembles of classifiers based on approximate reducts. Fundamenta Informaticae 47(3-4), 351–360 (2001)

103. Ślęzak, D., Widz, S.: Is it important which rough-set-based classifier extraction and voting criteria are applied together? In: Szczuka, M., Kryszkiewicz, M., Ramanna, S., Jensen, R., Hu, Q. (eds.) RSCTC 2010. LNCS, vol. 6086, pp. 187–196. Springer, Heidelberg (2010)

104. Bauer, E., Kohavi, R.: An empirical comparison of voting classification algorithms: Bagging, boosting, and variants. Machine Learning 36(1-2), 105–139 (1999)

105. Dietterich, T.G.: An experimental comparison of three methods for constructing ensembles of decision trees: Bagging, boosting, and randomization. Machine Learning 40(2), 139–157 (2000)

106. Stefanowski, J.: An experimental study of methods combining multiple classifiers - diversified both by feature selection and bootstrap sampling. In: Atanassov, K.T., Kacprzyk, J., Krawczak, M., Szmidt, E. (eds.) Issues in the Representation and Processing of Uncertain and Imprecise Information, pp. 337–354. Akademicka Oficyna Wydawnicza EXIT, Warsaw (2005)

107. Smyth, B., McClave, P.: Similarity vs. diversity. In: Aha, D.W., Watson, I. (eds.) ICCBR 2001. LNCS (LNAI), vol. 2080, pp. 347–361. Springer, Heidelberg (2001)

108. Husserl, E.: The Crisis of European Sciences and Transcendental Phenomenology. Northwestern University Press, Evanston (1970) German original written in 1937

109. Schütz, A.: The Phenomenology of the Social World. Northwestern University Press, Evanston (1967)

110. Coomans, D., Massart, D.: Alternative k-nearest neighbour rules in supervised pattern recognition: Part 1. k-nearest neighbour classification by using alternative voting rules. Analytica Chimica Acta 136, 15–27 (1982)

111. Patrick, E.A., Fischer III, F.P.: A generalized k-nearest neighbor rule. Information and Control 16(2), 128–152 (1970)

112. Basu, S.: Semi-supervised Clustering: Probabilistic Models, Algorithms and Experiments. PhD thesis, The University of Texas at Austin (2005)
113. Hliaoutakis, A., Varelas, G., Voutsakis, E., Petrakis, E.G.M., Milios, E.: Information retrieval by semantic similarity. Int. Journal on Semantic Web and Information Systems (IJSWIS). Special Issue of Multimedia Semantics 3(3), 55–73 (2006)
114. Rinaldi, A.M.: An ontology-driven approach for semantic information retrieval on the web. ACM Transactions on Internet Technology 9, 10:1–10:24 (2009)
115. Feldman, R., Sanger, J. (eds.): The Text Mining Handbook. Cambridge University Press (2007)
116. Ho, T.B., Nguyen, N.B.: Nonhierarchical document clustering based on a tolerance rough set model. International Journal of Intelligent Systems 17, 199–212 (2002)
117. Janusz, A.: A similarity relation in machine learning. Master's thesis, University Warsaw, Faculty of Mathematics, Informatics and Mechanics (2007) (in Polish)
118. Beals, R., Krantz, D.H., Tversky, A.: Foundations of multidimensional scaling. Psychological Review 75(2), 127–142 (1968)
119. Bazan, J.: Behavioral pattern identification through rough set modeling. Fundamenta Informaticae 72(1–3), 37–50 (2006)
120. Bazan, J., Kruczek, P., Bazan-Socha, S., Skowron, A., Pietrzyk, J.J.: Automatic planning of treatment of infants with respiratory failure through rough set modeling. In: Greco, S., Hata, Y., Hirano, S., Inuiguchi, M., Miyamoto, S., Nguyen, H.S., Słowiński, R. (eds.) RSCTC 2006. LNCS (LNAI), vol. 4259, pp. 418–427. Springer, Heidelberg (2006)
121. Kumar, N., Lolla, N., Keogh, E., Lonardi, S., Ratanamahatana, C.A.: Time-series bitmaps: a practical visualization tool for working with large time series databases. In: SIAM 2005 Data Mining Conference, pp. 531–535. SIAM (2005)
122. Strong, G., Gong, M.: Similarity-based image organization and browsing using multi-resolution self-organizing map. Image Vision Comput. 29(11), 774–786 (2011)
123. Borg, I., Groenen, P.: Modern Multidimensional Scaling: Theory and Applications. Springer (2005)
124. Claveau, V.: IRISA Participation in JRS 2012 Data-Mining Challenge: Lazy-Learning with Vectorization. In: Yao, J., Yang, Y., Słowiński, R., Greco, S., Li, H., Mitra, S., Polkowski, L. (eds.) RSCTC 2012. LNCS, vol. 7413, pp. 447–454. Springer, Heidelberg (2012)
125. Vempala, S.: The Random Projection Method. DIMACS Series in Discrete Mathematics and Theoretical Computer Science. American Mathematical Society (2004)
126. Greco, S., Matarazzo, B., Słowiński, R.: Dominance-based rough set approach to case-based reasoning. In: Torra, V., Narukawa, Y., Valls, A., Domingo-Ferrer, J. (eds.) MDAI 2006. LNCS (LNAI), vol. 3885, pp. 7–18. Springer, Heidelberg (2006)
127. Kaufman, L., Rousseeuw, P.: Finding Groups in Data: An Introduction to Cluster Analysis. Wiley Interscience, New York (1990)
128. Böhm, C., Faloutsos, C., Plant, C.: Outlier-robust clustering using independent components. In: SIGMOD Conference, pp. 185–198 (2008)
129. Gabrilovich, E., Markovitch, S.: Computing semantic relatedness using wikipedia-based explicit semantic analysis. In: Proceedings of The Twentieth International Joint Conference for Artificial Intelligence, Hyderabad, India, pp. 1606–1611 (2007)

130. Ślęzak, D.: Rough sets and few-objects-many-attributes problem: The case study of analysis of gene expression data sets. Frontiers in the Convergence of Bioscience and Information Technologies, 437–442 (2007)
131. Deutsch, J.M.: Evolutionary algorithms for finding optimal gene sets in microarray prediction. BMC Bioinformatics 19(1), 45–52 (2003)
132. Jirapech-Umpai, T., Aitken, S.: Feature selection and classification for microarray data analysis: Evolutionary methods for identifying predictive genes. BMC Bioinformatics 6(148) (2005) (online)
133. Jolliffe, I.T.: Principal Component Analysis, 2nd edn. Springer (October 2002)
134. John, G.H., Kohavi, R., Pfleger, K.: Irrelevant Features and the Subset Selection Problem. In: Proceeding of 11th International Conference on Machine Learning, pp. 121–129. Morgan Kaufmann (1994)
135. Hall, M.: Correlation-based Feature Selection for Machine Learning. PhD thesis, University of Waikato (1999)
136. Liao, C., Li, S., Luo, Z.: Gene selection using wilcoxon rank sum test and support vector machine for cancer classification. In: Wang, Y., Cheung, Y.-m., Liu, H. (eds.) CIS 2006. LNCS (LNAI), vol. 4456, pp. 57–66. Springer, Heidelberg (2007)
137. Peng, H., Long, F., Ding, C.: Feature selection based on mutual information: Criteria of max-dependency, max-relevance, and min-redundancy. IEEE Transactions on Pattern Analysis and Machine Intelligence 27, 1226–1238 (2005)
138. Kira, K., Rendell, L.A.: A practical approach to feature selection. In: Proceedings of the Ninth International Workshop on Machine Learning, ML 1992, pp. 249–256. Morgan Kaufmann Publishers Inc., San Francisco (1992)
139. Ding, C., Peng, H.: Minimum redundancy feature selection from microarray gene expression data. In: Proceedings of the 2003 IEEE Bioinformatics Conference, pp. 523–528 (2003)
140. Breiman, L.: Random forests. Machine Learning 45(1), 5–32 (2001)
141. Dramiński, M., Kierczak, M., Koronacki, J., Komorowski, J.: Monte Carlo Feature Selection and Interdependency Discovery in Supervised Classification. In: Koronacki, J., Raś, Z.W., Wierzchoń, S.T., Kacprzyk, J. (eds.) Advances in Machine Learning II. SCI, vol. 263, pp. 371–385. Springer, Heidelberg (2010)
142. Marill, T., Green, D.: On the effectiveness of receptors in recognition systems. IEEE Transactions on Information Theory 9(1), 11–17 (1963)
143. Whitney, A.W.: A Direct Method of Nonparametric Measurement Selection. IEEE Transactions on Computers 20, 1100–1103 (1971)
144. Siedlecki, W., Sklansky, J.: Handbook of pattern recognition & computer vision, pp. 63–87. World Scientific Publishing Co., Inc., River Edge (1993)
145. Furey, T.S., Duffy, N., David, W., Haussler, D.: Support vector machine classification and validation of cancer tissue samples using microarray expression data (2000)
146. Vapnik, V.N.: The nature of statistical learning theory. Springer-Verlag New York, Inc., New York (1995)
147. Schölkopf, B.: The kernel trick for distances. In: Leen, T.K., Dietterich, T.G., Tresp, V. (eds.) Advances in Neural Information Processing Systems 13, Papers from Neural Information Processing Systems (NIPS) 2000, Denver, CO, USA, pp. 301–307. MIT Press (2000)
148. Graupe, D.: Principles of Artificial Neural Networks, 2nd edn. World Scientific Publishing Co., Inc., River Edge (2007)
149. Wojnarski, M.: LTF-C: Architecture, training algorithm and applications of new neural classifier. Fundamenta Informaticae 54(1), 89–105 (2003)

150. Michalewicz, Z.: Genetic Algorithms + Data Structures = Evolution Programs. Springer (1996)
151. Skowron, A., Stepaniuk, J., Peters, J.F., Swiniarski, R.W.: Calculi of approximation spaces. Fundamenta Informaticae 72(1-3), 363–378 (2006)
152. Deerwester, S., Dumais, S.T., Furnas, G.W., Landauer, T.K., Harshman, R.: Indexing by latent semantic analysis. Journal of the American Society for Information Science 41(6), 391–407 (1990)
153. United States National Library of Medicine: Introduction to MeSH - 2011 (2011), http://www.nlm.nih.gov/mesh/introduction.html
154. Nguyen, H.S.: On efficient handling of continuous attributes in large data bases. Fundamenta Informaticae 48(1), 61–81 (2001)
155. Jensen, R., Shen, Q.: New approaches to fuzzy-rough feature selection. IEEE Transactions on Fuzzy Systems 17(4), 824–838 (2009)
156. Ganter, B., Stumme, G., Wille, R. (eds.): Formal Concept Analysis. LNCS (LNAI), vol. 3626. Springer, Heidelberg (2005)
157. Ganter, B., Wille, R.: Formal Concept Analysis: Mathematical Foundations. Springer (1998)
158. R Development Core Team: R: A Language and Environment for Statistical Computing. R Foundation for Statistical Computing, Vienna, Austria (2008)
159. Frank, A., Asuncion, A.: UCI machine learning repository (2010)
160. Parkinson, H.E., et al.: ArrayExpress update - from an archive of functional genomics experiments to the atlas of gene expression. Nucleic Acids Research 37(Database-Issue), 868–872 (2009)
161. Ben-Dor, A., Bruhn, L., Friedman, N., Nachman, I., Schummer, M., Yakhini, Z.: Tissue classification with gene expression profiles. Journal of Computational Biology 7(3-4), 559–583 (2000)
162. Bouckaert, R.R.: Choosing between two learning algorithms based on calibrated tests. In: Fawcett, T., Mishra, N. (eds.) Machine Learning, Proceedings of the Twentieth International Conference, ICML 2003, August 21-24, pp. 51–58. AAAI Press, Washington, DC (2003)
163. Kohavi, R.: A study of cross-validation and bootstrap for accuracy estimation and model selection. In: IJCAI, pp. 1137–1145 (1995)
164. Demšar, J.: Statistical Comparisons of Classifiers over Multiple Data Sets. Journal of Machine Learning Research 7, 1–30 (2006)
165. Baldi, P., Hatfield, G.W.: DNA Microarrays and Gene Expression: From Experiments to Data Analysis and Modeling. Cambridge University Press (2002)
166. Brazma, A., Hingamp, P., Quackenbush, J., Sherlock, G., Spellman, P., Stoeckert, C., Aach, J., Ansorge, W., Ball, C.A., Causton, H.C., Gaasterland, T., Glenisson, P., Holstege, F.C., Kim, I.F., Markowitz, V., Matese, J.C., Parkinson, H., Robinson, A., Sarkans, U., Schulze-Kremer, S., Stewart, J., Taylor, R., Vilo, J., Vingron, M.: Minimum Information About a Microarray Experiment (MIAME) - Toward Standards for Microarray Data. Nature Genetics 29(4), 365–371 (2001)
167. Diaz-Uriarte, R., Alvarez de Andres, S.: Gene selection and classification of microarray data using random forest. BMC Bioinformatics 7(3) (2006) (online)
168. Roberts, R.J.: PubMed Central: The GenBank of the published literature. Proceedings of the National Academy of Sciences of the United States of America 98(2), 381–382 (2001)
169. Spearman, C.: The proof and measurement of association between two things. By C. Spearman, 1904. The American Journal of Psychology 100(3-4), 441–471 (1987)

170. Stawicki, S., Widz, S.: Decision bireducts and approximate decision reducts: Comparison of two approaches to attribute subset ensemble construction. In: Ganzha, M., Maciaszek, L.A., Paprzycki, M. (eds.) Proceedings of Federated Conference on Computer Science and Information Systems - FedCSIS 2012, Wrocław, Poland, September 9-12, pp. 331–338 (2012)
171. Bazan, J., Nguyen, S.H., Nguyen, H.S., Skowron, A.: Rough set methods in approximation of hierarchical concepts. In: Tsumoto, S., Słowiński, R., Komorowski, J., Grzymała-Busse, J.W. (eds.) RSCTC 2004. LNCS (LNAI), vol. 3066, pp. 346–355. Springer, Heidelberg (2004)
172. Sarawagi, S., Thomas, S., Agrawal, R.: Integrating association rule mining with relational database systems: Alternatives and implications. Data Mining and Knowledge Discovery 4(2/3), 89–125 (2000)
173. Ślęzak, D., Synak, P., Borkowski, J., Wroblewski, J., Toppin, G.: A rough-columnar rdbms engine – a case study of correlated subqueries. IEEE Data Engineering Bulletin 35(1), 34–39 (2012)
174. Bazan, J., Szczuka, M.S.: RSES and rSESlib - A collection of tools for rough set computations. In: Ziarko, W.P., Yao, Y. (eds.) RSCTC 2000. LNCS (LNAI), vol. 2005, pp. 106–113. Springer, Heidelberg (2001)
175. Ahrn, A., Komorowski, J.: ROSETTA – a rough set toolkit for analysis of data. In: Proceedings Third International Joint Conference on Information Sciences, pp. 403–407 (1997)

Author Index